R. J. Ehrig
Plastics Recycling

SPE Books from Hanser Publishers

Ulrich, *Introduction to Industrial Polymers*

Saechtling, *International Plastics Handbook for the Technologist, Engineer and User*

Stoeckhert, *Mold-Making Handbook for the Plastics Engineer*

Bernhardt, *Computer Aided Engineering for Injection Molding*

Michaeli, *Extrusion Dies—Design and Engineering Computations* [out of print]

Rauwendaal, *Polymer Extrusion*

Brostow/Corneliussen, *Failure of Plastics*

Menges/Mohren, *How to Make Injection Molds*

Throne, *Thermoforming*

Manzione, *Applications of Computer Aided Engineering in Injection Molding*

Macosko, *Fundamentals of Reaction Injection Molding*

Tucker, *Fundamentals of Computer Modeling for Polymer Processing*

Charrier, *Polymeric Materials and Processing—Plastics, Elastomers and Composites*

Wright, *Molded Thermosets: A Handbook for Plastics Engineers, Molders and Designers*

Michaeli, *Extrusion Dies for Plastic and Rubber—Design and Engineering Computations*

Ehrig, *Plastics Recycling*

R. J. Ehrig, Editor

Plastics Recycling

Products and Processes

With 77 Figures and 98 Tables

Hanser Publishers, Munich Vienna New York Barcelona

Distributed in the United States of America and Canada
by Oxford University Press, New York

Distributed in USA and in Canada by
Oxford University Press
200 Madison Avenue, New York, N. Y. 10016

Distributed in all other countries by
Carl Hanser Verlag
Kolbergerstr. 22
D-8000 München 80

Die Deutsche Bibliothek – CIP-Einheitsaufnahme

Plastics Recycling : Products and Processes ; with 98 Tables /
R. J. Ehrig, ed. [Contributors : G. A. Baum...] . – Munich ; Vienna ;
New York ; Barcelona : Hanser, 1992
    ISBN 3-446-15882-0
NE: Ehrig, Raymond J. [Hrsg.] ; Baum, Gerald A.

    ISBN 3-446-15882-0 Hanser
    ISBN 0-19-520934-6 Oxford University Press

Printed on recycled paper.

Front cover illustration:

Mierle Laderman Ukeles, DETAIL, *Recycle Works,* 1989,
courtesy Ronald Feldman Fine Arts, New York

Copyright © Carl Hanser Verlag, Munich Vienna New York Barcelona 1992
Printed in Germany by Grafische Kunstanstalt Josef C. Huber KG, Dießen

Contributors

G. A. Baum
Engineering Plastics Division, Hoechst Celanese Corporation, 86 Morris Avenue, Summit, NJ 07901

R. H. Burnett
The Vinyl Institute, Wayne Interchange Plaza II, 155 Route 46 West, Wayne, NJ 07470 (Retired from Hoechst Celanese Corporation)

W. F. Carroll, Jr.
Occidental Chemical Corporation, 5005 LBJ Freeway, P.O. Box 809050, Dallas, TX 75244

R. Coughanour
Polystyrene Research and Development, Arco Chemical Company, 3801 West Chester Pike, Newtown, PA 19073

M. J. Curry
Plastics Institute of America, Inc., Suite 100, 277 Fairfield Road, Fairfield, NJ 07004

R. J. Ehrig
Aristech Chemical Corporation, 1000 Tech Center Drive, Monroeville, PA 15146

R. G. Elcik
Occidental Chemical Corporation, 5005 LBJ Freeway, P.O. Box 809050, Dallas, TX 75244

W. J. Farrissey
Dow Chemical U.S.A., B-1610, Freeport, TX 77541

D. Goodman
Occidental Chemical Corporation, Armand Hammer Boulevard, P.O. Box 699, Pottstown, PA 19464

G. A. Mackey
Dow Chemical U.S.A., Granville Research and Development Center, P.O. Box 515, Granville, OH 43023

J. Milgrom
Walden Research, Inc., P.O. Box 69, Concord, MA 01742

W. P. Moore
Waste Management of North America, 3003 Butterfield Road, Oak Brook, IL 60521

T. J. Nosker
Center for Plastics Recycling Research, Rutgers, The State University of New Jersey, P.O. Box 1179, Piscataway, NJ 08855

B. D. Perlson
Quantum Chemical Corporation, 11500 Northlake Drive, P.O. Box 429550, Cincinnati, OH 45249

C. C. Schababerle
Quantum Chemical Corporation, 11500 Northlake Drive, P.O. Box 429550, Cincinnati, OH 45249

J. B. Schneider
154 Hidden Valley Drive, Pittsburgh, PA 15237
(Formerly with Aristech Chemical Corporation)

K. E. Van Ness
Department of Physics and Engineering, Washington and Lee University, Lexington, VA 24450

R. C. Westphal
Hunstman Chemical Corporation, 5100 Bainbridge Boulevard, Chesapeake, VA 23320

Preface

The plastics industry has experienced phenomenal growth during the last decade. In 1976, plastics became the most widely used material in the United States, surpassing even steel. Today plastics is a $140-billion-per-year industry—an industry greater than steel and aluminum combined. Plastics are now used in every segment of American business and in the daily life of every one of us. They are found either in the form of the entire product as containers, and fast food packaging or in combination with other materials as parts of transportation vehicles, computers, tools, recreational equipment, etc.

The significant increase in plastics production and product generation has effected a similar significant increase in plastics disposal. Estimated at more than 7% by weight, but possibly twice that by volume, the amount of plastics in the nation's solid waste stream has more than doubled in the last 15 years; a greater percentage rise is expected as more plastic packaging and durable plastic components of automobiles and appliances come to the end of their service life.

Since 73% of all solid waste generated in the U.S. goes to landfill, the amount of plastic in this stream is considered a contributing factor to the nation's increasing solid waste burden and decreasing availability of landfill sites. The Environmental Protection Agency's strategy is to reduce all solid waste by 25% by 1992 and to foster markets for recycled materials. Lawmakers in nine states have passed mandatory plastic beverage bottle recycling bills; eighteen other state legislatures are in various stages of planning or formulating recycling bills; it is expected that all fifty states will have some form of plastics recycling legislation in place by the year 2000.

The plastics industry is responding, but it has a long way to go. Although a high percentage of scrap plastic has been recycled within the industry for many years, the recycling of consumer plastic waste is in an embryonic stage. About 1% of plastics is recovered from the solid-waste stream. Technology for the reclamation of polyethylene terephthalate (PET) bottles has been developed; similar technology exists for reclaiming other plastic bottles; film and molded fabricated parts reclaim technology also exists. Other technology has been developed for commingled plastics. PET is currently the major recyclable plastic material, followed by high-density polyethylene (HDPE), polypropylene (PP), and polyvinyl chloride (PVC). Some styrenics, acrylics, polycarbonates, and polyurethanes are reclaimed and recycled. New products and end-use markets are developing rapidly for these materials. Fibers, moldings, construction materials, chemical intermediates, and blending components are just a few.

Organizations, sponsored by many industrial corporations, have been formed to foster plastics recycling through education and research development funding. The same corporations, and others, have become part of the plastic recycling industry by formation of recycling entities within their own organizations and/or through joint ventures. The recycling industry, however, remains primarily an entrepreneurial one.

This book is intended for the plastic or materials engineer, specialist, or entrepreneur in the plastics or recycling industry. It is hoped also that the book will help many environmentalists, organizations, local, state, and federal governmental people who seek information on plastics and plastics recycling. Finally, the editor and authors wish it to be used by researchers, faculty members, and students as a starting point for future technological developments in this very young industry.

I wish to pause here for a moment to express a sincere appreciation and thank you to my "co-editor," Marge Ehrig, for her literary guidance and criticism, for many hours of word processing, proof reading, letter typing, etc., and especially for her patience and understanding.

Many thanks are extended also to the authors and their secretarial staffs, to Peter Prescott, and Marcia Sanders.

R. J. Ehrig

Foreword

Despite the large number of journal articles, technical meetings and general conferences on plastics waste and recycling since 1985, a sound understanding of this extensive and complex subject has been difficult to achieve. *Plastics Recycling* presents the first comprehensive survey of the technical, business and environmental components involved with recovery for reuse of discarded plastics.

The editor, Raymond Ehrig, and his collaborators have undertaken and completed a very demanding task. Their work will provide an important guide to those in the plastics industry, in government, and in other organizations, faced with decisions on the best choices for action.

The publication of *Plastics Recycling* comes at a time of rapid change for this field. Most large plastics producers and users are now heavily involved compared to the smaller companies just five years ago. Progress in the technology for recovery and for new applications has been remarkable.

Recycling of plastics from waste automobiles, appliances and construction is now receiving major attention. The effect of recycling on types and volume of virgin plastics which will be needed in the years ahead is under intensive study.

The Plastics Institute of America is pleased to introduce this volume as an important contribution to this challenging and exciting area for new developments.

William Sacks
Executive Director
Plastics Institute of America, Inc.
Fairfield, New Jersey

Foreword

The Society of Plastics Engineers (SPE) and its recycling division are particularly pleased to sponsor *Plastics Recycling*. It is the first definitive SPE technical volume detailing recycling technology currently available as a means of reducing the burgeoning volume of mankind's solid waste. The editor, Dr. Raymond Ehrig, is a distinguished scientist, ultimately qualified in the techniques of recycling and product applications of reclaimed plastics.

Dr. Ehrig's credentials include service as the co-founder and Chairperson of the Plastics Recycling Division and Chairman of the Plastics Institute of America. In the latter role, he has been involved both as an author and educator in publications, conferences and tutorial seminars on the subject of recycling technology.

SPE, through its Technical Volumes Committee, has long sponsored books on various aspects of plastics. Its involvement has ranged from identification of needed volumes and recruitment of authors to peer review and approval and publication of new books.

Technical competence pervades all SPE activities, not only in the publication of books, but also in other areas such as sponsorship of technical conferences and educational programs. In addition, the Society publishes periodicals, including *Plastics Engineering, Polymer Engineering Science, Polymer Processing and Rheology, Journal of Vinyl Technology,* and *Polymer Composites,* as well as conference proceedings and other publications, all of which are subject to rigorous technical review procedures.

The resource of some 37,000 practicing plastics engineers has made SPE the largest organization of its type worldwide. Further information is available from the Society at 14 Fairfield Drive, Brookfield, Connecticut 06804, U.S.A.

Robert D. Forger
Executive Director
Society of Plastics Engineers

Robert E. Biskowski
Chairperson, 1991–92
Plastics Recycling Division

Contents

Contributors v
Preface vii
Foreword: Plastics Institute of America, Inc. ix
Foreword: Society of Plastics Engineers, Inc. xi

1 Introduction and History 1
   R. J. Ehrig, M. J. Curry

2 Collection and Separation 17
   W. P. Moore

3 Polyethylene Terephthalate (PET) 45
   J. Milgrom

4 Polyolefins 73
   B. D. Perlson, C. C. Schababerle

5 Polystyrene (PS) 109
   G. A. Mackey, R. C. Westphal, R. Coughanour

6 Polyvinyl Chloride (PVC) 131
   W. F. Carroll, Jr., R. G. Elcik, D. Goodman

7 Engineering Thermoplastics 151
   R. H. Burnett, G. A. Baum

8 Acrylics 169
   J. B. Schneider

9 Commingled Plastics 187
   K. E. Van Ness, T. J. Nosker

10 Thermosets 231
   W. J. Farrissey

Abbreviations 263

Appendices 267
   Appendix 1 Standard Guide for the Development of Standards Relating to the
               Proper Use of Recycled Plastics 267
   Appendix 2 Coding and Labeling Plastics for Recycling 268
   Appendix 3 Organizations Involved in Plastics Recycling 271
   Appendix 4 University Programs 275
   References 278

Index 281

1 Introduction and History

R. J. Ehrig, M. J. Curry

Contents

1.1   Introduction 3

1.2   Thermoplastics, Thermosets 5

1.3   Terminology 5

1.4   Coding and Labeling...6

1.5   Organizations Involved in Plastics Recycling 7

1.6   University Programs 8

1.7   History of Plastics Recycling 8

        1.7.1 Recycling into the 1980s 8

References 14

This chapter begins with an overview of plastics in municipal solid waste and presents the current status and future role of plastics recycling. Thermo-plastics and thermosets are then defined; the terminology, coding and labeling of plastics related to recycling are discussed; organizations in-volved and current university research programs are introduced. The chapter ends with a brief history of plastics recycling. Some interesting, innovative, entrepreneurial ventures of the past two decades are presented. Research and development efforts to obtain alternate sources of fuel and raw materials for plastics and to use plastics waste as a feedstock source are reviewed. The influence of past work on present developments is noted.

1.1   Introduction

Many parts of the industrial world face serious problems of managing the generation and disposal of municipal solid waste (MSW). In 1989, the European community generated approximately 110 million tons (99.8×10^6 t) of MSW, and Japan's industrial and municipal waste totaled about 330 million tons (299×10^6 t) [1]. The problem is most serious in the United States where the populace generated 180 million tons (163×10^6 t) of MSW in 1989. This amounts to approximately 1400 pounds of discarded solid waste per person for the year. Since 1977, the amount of discards entering the waste stream increased by 35%. It is estimated that 73% of the MSW went to landfill or was otherwise disposed of, 13% was recycled, and 14% incinerated. The disproportionate amount of waste going to landfills, the closing of many of these sites, and the lack of equal numbers of replacement sites is the disposal aspect of the solid waste problem. This has received much attention in the media and is generally known as the "landfill crisis." In the past two decades, the number of active operating landfills has decreased by 75%. Within the next five years, one-third of the landfills currently in operation will be forced to close because of stricter governmental regulations on ground and water contamination and on landfill capacity. Few new sites will be opened since new site selection, preparation, maintenance, and operation also must meet more rigid standards and regulations. This could lead to an approximate shortfall of space for 50 million tons (45×10^6 t) within the next few years. Shortages of landfill space have already occurred in many parts of the northeastern United States. In 1970, New Jersey had 370 operating landfills; by 1989, fewer than eight were accepting more than 90% of the state's MSW [2].

The latest study on the characterization of MSW in the United States estimates that 40% by weight of discarded material is paper, 18% is yard waste, and a 7–9% weight range each for metal, glass, plastics, and food [3]. On a volume basis, the composition of the MSW stream is somewhat different. Paper is still the largest contributor at 34%, while plastic goes from 7% by weight to estimated values of 10% by volume [4] to 20% by volume [5]. Metals are estimated at 12%, yard waste at 10%, glass and food at 2–3%, and all other at 18%. The volume basis of solid waste is an important factor in landfilling. Although the cost of landfilling is done on a weight basis (i.e., tipping fees are determined by weight), the capacity of a landfill is determined by its volume. The relatively large volume that plastics occupy in the MSW stream has caused the plastics industry to come under severe attack in the last few years by environmentalists and solid-waste professionals.

The composition of the MSW stream has also been characterized by product type [6]. Discarded packaging products constitute about 30% by volume; nondurable goods such as newspaper, clothing, single-serve cups, etc., make up another 34%; durable goods such as appliances, furniture, tires, and consumer electronics, 22%; and food, yard, and inorganic wastes make up the remaining 14%. Breaking down the 30% that is packaging products in the MSW, 8% is plastics, 14% is paper, and 8% is metals, glass, etc.

Current recovery rates [7] for all materials in the solid-waste stream with the exception of food wastes exceed that of plastics. Paper is estimated at a 26% recovery rate, aluminum at

32%, and glass at 12%. Plastic is just now reaching an approximate recovery rate of 1%. This low rate has been a major concern to the plastics industry since it reinforces the popular belief that plastics are not recyclable. During the last few years, the plastics industry has been addressing these concerns through the formation and working of various organizations as well as individual companies within the industry. The most notable of these has been The Council for Solid Waste Solutions (CSWS). A discussion of the Council and other organizations concerned with plastics recycling is given in Appendix 3.

Plastics recycling in the United States during the last few years and up to the present time has focused mainly on plastics packaging and primarily on plastic bottles and containers. There are a number of reasons for this. As mentioned above, plastics packaging is considered a significant component of the solid-waste stream at 8% by volume. Packaging has an estimated life cycle of less than one year. As a consequence, plastics packaging continuously enters the solid-waste stream on a short turnaround time. Plastics in packaging has had a sustained growth period and will continue to grow, increasing the volume going to landfill unless increased recycling, as well as source reduction or incineration, is put into effect. In 1990, about 14.8 billion pounds (6.7×10^6 t) of plastics went into packaging applications [8]. This is a 6.5% increase over 1989. Rigid containers consumed 7.5 billion pounds (3.4×10^6 t) of which 4.0 billion pounds (1.8×10^6 t) went into containers and 2.2 billion pounds (1.0×10^6 t) into beverage containers.

Plastics packaging is a very visible part of litter. This, together with its visibility and volume in landfills, has made plastics packaging an easy target for environmentalists, governing bodies, the citizenry, and the media to call for its reduction, removal, or elimination.

In the early eighties, a number of states enacted return bottle bills. This was followed in the late eighties by other states enacting recycling legislation which decreed that local governments establish and provide collection services for the citizens. A growth of curbside collection programs has resulted, and a similar growth in recycling of plastic containers and bottles has occurred when these items are included in the programs. Curbside collection systems allow for easy separation of plastic containers and bottles either by the consumer at curbside or through collection and subsequent separation at municipal recycling facilities (MRFs). This is discussed in the following chapter.

Based on the factors cited above, the primary focus of recycling packaging materials probably will continue for a number of years. This means a continued high level of activity in the technology of recycling of the associated plastics. There are six commodity plastics that constitute 97% of all plastics packaging. These are commonly found in residential household items in the form of rigid bottles and containers or as flexible packaging and wrapping films. A number of chapters are devoted to a discussion of these commodity plastics. Individual chapters review the products and processes of recycling polyethylene terephthalate (PET); the polyolefins, which include the polyethylenes (PE) and polypropylene (PP); polyvinyl chloride (PVC); and polystyrene (PS). Another chapter describes in detail the commingled fraction of these materials after one or more selected resins have been removed (i.e., the mixture of resins after PET and high-density polyethylene (HDPE) containers have been removed).

Packaging items are not the only products of the commodity plastics. Many are used in the fabrication of component parts of durable finished products, such as appliances, furniture, computer housings, and electrical and electronic equipment. Other polymers also used include acrylonitrile-butadiene-styrene (ABS), acrylics, nylon, polycarbonate (PC), phenolic, polyurethanes (PURs), filled and reinforced unsaturated polyesters, etc. As mentioned before, these durable products currently constitute approximately 22 volume % of the MSW stream. Building, construction, and automotive products are the other durables. Building and construction materials are disposed of at sites other than MSW landfills. Automotive waste, such as junked automobiles, are first dismantled, then shredded, ferrous metals removed, and the remainder or automotive shredder residue (ASR) sent to municipal landfills. The recovery of durables from the solid-waste stream has not yet received the focus of attention accorded the nondurables. However, some companies already looking to the future have focused on the recycling of durables. Many companies involved in the manufacture of these polymers and in part fabrication have active programs on recovery processes and new product generation. The remaining chapters of this book review many of these activities—Chapter 7 on engineering resins, Chapter 8 on acrylics, and Chapter 10 on secondary and tertiary recycling (defined in Section 1.3) of phenolics, PURs, and unsaturated polyesters.

1.2   Thermoplastics, Thermosets

The organization of this volume is based on different polymeric material types, the market applications, and current and future recycling of these materials. All the polymers discussed in the ensuing chapters, with the exception of Chapter 10, are thermoplastic.

Historically, polymer materials were classified as thermoplastics or thermosets long before there was a realization of the chemical nature of these materials. The terms are based on the physical changes that occur when the materials are subjected to heating and cooling. Thermoplastics are materials that (1) become soft or "plastic" when heated, (2) are molded or shaped with pressure when in the plastic state, and (3) solidify when cooled to retain the mold or shape. Since this is a physical change with no chemical change occurring, the process is reversible and can be repeated. Such repeating heating cycles will eventually cause decomposition of the polymer. On a molecular level, these thermoplastics are linear or slightly branched polymers.

Thermosets are materials that can be softened, molded, and then hardened or "set" when heated once. The thermosets are infusible solids that decompose on reheating. The hardening or curing process is an irreversible chemical reaction known as cross-linking. On a molecular level, the thermosets are branched polymers that form three-dimensional networks through the cross-linking reaction. Examples from Chapter 10 include the phenolics, epoxies, and PURs.

Thermoplastics have been further subdivided on a cost-property performance basis into commodity and engineering plastics. As new plastics have been developed, further divisions such as advanced polymers, high-heat resistant polymers, etc., have been recognized. Interest here is only on the commodity and engineering plastics. Commodity plastics are less costly and have much greater volume sales than the engineering plastics. The commodity plastics also have lower physical properties and are used in applications with less demanding performance requirements. The engineering plastics possess physical characteristics that allow these materials to be used for structural applications where high strength, impact, chemical, or heat resistance may be required. Some polymeric materials possess characteristics of both commodity and engineering plastics, so the distinction between the two is not always clear. For example, ABS and polymethyl methacrylate (PMMA) may be considered in either category. For this volume, ABS is discussed in the engineering resin chapter.

1.3   Terminology

Through continued use and reference in technical discussions, papers, reports, and at technical conferences, various terms have become a part of the language of plastics recycling. A number of authors [9–11] have defined and refined some of these terms. There are descriptive terms such as reuse, recover, reconstitute; terms that differentiate plastic streams as waste, scrap, nuisance, industrial, and postconsumer; and terms that classify plastics recycling technology. The American Society for Testing and Materials (ASTM) published in September 1990 a standard guide under Designation D-5033-90 [12] that defines many terms and provides information on important factors relevant to the development of standards for the proper use of recycled plastics. Many of the terms defined in the guide are given in Appendix 1. A few terms not included in the guide are redefined here based on earlier definitions of Milgrom [13] and Leidner [14].

1. Waste plastics—that fraction of plastics that enters the solid-waste stream and must be recovered, recycled, incinerated or otherwise disposed of.
2. Scrap plastics—that fraction of plastics generated by various plastic operations that can be recycled into viable commercial products using standard plastic-processing techniques.
3. Nuisance plastics—a byproduct fraction of any plastic operation that cannot be reprocessed into a viable commercial product under existing technical-economic conditions.

Terms also have been developed for classifying plastics recycling into four types of technologies: primary, secondary, tertiary, and quaternary. Milgrom [15] was probably the first to define these terms. He simply defined primary, secondary, and tertiary recycling as scrap plastics reformed into their original form, reprocessed into another form, and converted into energy or another nonplastic form, respectively. Versions of these terms given in this book are based on Leidner's more detailed definitions [16]. ASTM D-5033-90 also defines these terms.

1. Primary recycling—the conversion of scrap plastics by standard processing methods into products having performance characteristics equivalent to the original products made of virgin plastics.
2. Secondary recycling—the conversion of scrap or waste plastics by one or a combination of process operations into products having less demanding performance requirements than the original material.
3. Tertiary recycling—the process technologies of producing chemicals and fuels from scrap or waste plastics.
4. Quaternary recycling—the process technologies of recovering energy from scrap or waste plastics by incineration.

Although this present work is not structured according to the above classification, the process technologies cited are representative of all four types and are easily recognized throughout. Primary recycling is not emphasized since it is not normally a postconsumer plastics recycling technology. The technology is practiced mainly in the manufacturing sector for recycling in-house scrap. Historically, the recycling of in-plant scrap has not been considered part of plastics recycling since such scrap does not normally enter the solid-waste stream. Primary recycling is cited in the historical section of this chapter and when the technology has served as a basis for a secondary recycling process as is given in Chapter 4. Secondary recycling is emphasized throughout the chapters on commodity plastics and also the chapter on commingled plastics. The processes of tertiary recycling are discussed in the acrylic and thermoset chapters and in the history of plastics recycling (Section 1.7). Since quaternary recycling is considered a resource recovery rather than a recycling technology, it has received only minor attention. It is discussed, however, in the PVC chapter. Here, incineration is an important factor.

1.4   Coding and Labeling

In 1988, the Society of the Plastics Industry Inc. (SPI) recognized the need for a simple and easy method to identify the type of plastic used in individual bottles and containers. Recycling of PET beverage bottles had already reached a 20% rate; HDPE milk and water jugs recycling was underway, and many other different plastic bottles and containers were entering the recycling stream. Acting on this need, the SPI developed a simple code system known as the SPI's Voluntary Plastic Container Coding System [17]. Since 1988, the system has gained wide acceptance, and the code also is now imprinted on other forms of packaging, such as supermarket plastic bags, which are intended to be recycled. As of June 1991, 32 states [18] had enacted legislation either accepting or specifying that the SPI plastic identification code system be adopted on bottles of 16 ounces (~0.5 liter) or more in capacity and rigid containers of 8 ounces (~0.25 liter) or more in capacity. Details of the coding system are given in Appendix 2A.

The coding system was developed to help recyclers sort plastic bottles and containers by specific resin and was intended as an interim solution until automatic identification and sorting technology developed. Some of this technology is now in existence. The SPI system should remain, however, since it serves as an excellent means for the consumer to identify the specific plastic items to be separated from other plastics if required as part of a municipal curbside collection program.

Another system for marking plastic products for resin identification was issued as a standard by the ASTM in 1991 under the designation D-1972-91 [19]. The markings are used to assist in product handling, disposal, or recovery. The system is similar to that of the SPI's but broader in scope. Acronyms for 92 plastic family resins and another 37 for commercial plastic blends are listed. Details of the system are given in Appendix 2B.

The Society of Automotive Engineers, Inc., (SAE) also has a system for marking plastic parts to identify the type of material used in the fabrication of the parts [20]. The purpose of the marking is to provide information for recycling as well as repairing or repainting. Standard acronyms for plastics published by the International Organization for Standardization (ISO 1043) and those shown in ASTM D-300[01600] are used. Details of the SAE system are given in Appendix 2C.

Before leaving this subject, some mention should be made about the labeling of products and packages, including plastics, as recyclable, recycled, reusable, etc. It is now accepted by packaging manufacturers that consumers are more environmentally conscious than ever before and will become more so in the years ahead. Manufacturers are reacting by making products and packages more environmentally acceptable and labeling these as recyclable, reusable, recycled, etc. State governments are reacting by introducing or enacting labeling legislation that defines terms and sets requirements for legal use of official labels or emblems. The New York State Department of Environmental Conservation filed for the establishment and use of official emblems in the state. This became effective in December 1990. To use the "recycled" emblem, plastics packaging must contain 30 wt % minimum secondary material and 15 wt % minimum postconsumer material; for plastic products, a 50 wt % secondary material content and a 15 wt % postconsumer material content is required. "Secondary material" or "recovered material" is defined similar to a recovered material as defined in ASTM D-5033-90 (see Appendix 1).

Rhode Island has recycling emblem regulations, and the Indiana State House has proposed legislation providing definitions for environmental labeling terms. Wisconsin and Oregon have bills pending with labeling definitions; Illinois, New Jersey, and California are studying such legislation.

The Northeast Recycling Council, representing ten northeastern state governments, adopted standards to be used by these states as guidelines for the labeling of packages and products in regard to being reusable, recyclable, or having a certain recycled content [21].

Independent certification organizations and associations representing manufacturers also are involved in labeling and submitting guidelines and recommendations.

These diverse labeling measures and recommendations have caused confusion and conflict, prompting industry and consumer groups to call for federal government action to establish uniform labeling standards and guidelines. Further action by state and federal lawmakers on labeling is to be expected.

1.5 Organizations Involved in Plastics Recycling

There are several organizations within the plastics industry involved in plastics recycling education, research, and development. Some of these organizations have been formed in recent years by companies and trade associations with various business interests in the industry (e.g., The Council for Solid Waste Solutions and the Plastics Recycling Foundation). The Polystyrene

Packaging Council was formed by a member group with interest in a specific resin, and another group with a particular product interest formed the National Association for Plastic Container Recovery. A few organizations are newly formed entities or previously established special interest groups placing a new emphasis on recycling (e.g., The Polyurethanes Recycle and Recovery Council and The Vinyl Institute of the Society of the Plastics Industry Inc.). There are also two well-established independents serving the needs and interests of the industry—both The Plastics Institute of America Inc. and The Society of Plastics Engineers Inc. have strong plastics recycling educational programs. Information on these organizations is given in Appendix 3.

1.6   University Programs

There are many excellent university and college research programs underway, and many of these are cited by the various authors throughout this book. However, because of time constraints and the rapid pace of research and development in this field, some of the most recent research and latest findings could not be cited. We have, therefore, prepared in Appendix 4 an overview with references of some of the most recently published and current research being carried out at a number of academic institutions.

1.7   History of Plastics Recycling

There is no documented evidence as to when plastics recycling first began, but a form of it probably started within the industry sometime during the early development stages of fabricating synthetic thermoplastics. It was quickly learned that rejected parts, trim, and flash from the fabrication operations were valuable materials. It also was determined that certain percentages of these materials could be ground, blended with virgin material, and molded into acceptable parts. This process could be repeated a number of times, provided the percentage of regrinds remained low. As the plastics industry grew through the 1900s, greater quantities of scrap plastics came into existence from new fabrication processes as well as from other sources within the industry. Resin producers began to generate larger amounts of waste material from their manufacturing operations. Compounders, converters, packagers, and distributors also began to generate waste plastics. Leidner [22] gives tabulated data of plastic wastes generated by the various segments of the plastics industry from 1978, 1980, and projected to the year 2000.

Up until the end of the Second World War, much of the plastics scrap generated was recycled within the industry. At that time, the plastics industry was still rather small and just emerging as a significant part of industrial America. In 1946, total U. S. plastics production was 734 million pounds (333×10^3 t), of which only 235 million pounds (107×10^3 t) was synthetic thermoplastics [23]. As long as the plastic scrap generated by the industry was clean and uncontaminated with other plastics, reprocessing it within the industry continued to grow. Independent reprocessors or scrap dealers came on the scene. When primary fabricators did not wish to increase the percentage of scrap in their products, the material was sold directly to other independent processors or indirectly to third parties. Usually, the intended uses for the scrap did not conflict with the markets of the original scrap generator. Sometimes, however, the users were competitors, and this often led to landfilling the scrap rather than selling it. Further, in the 1960s, virgin plastic prices decreased, profit margins for scrap reprocessors decreased, and the growth of industrial recycling diminished. Greater amounts of scrap plastics were disposed of in landfills or by incineration.

1.7.1   Recycling into the 1980s

The first intense effort directed to plastics recycling by the industry, other than the internal consumption of plastic scrap, came about during the latter part of the 1960s and into the 1970s.

A number of factors caused this to occur. By 1969, total plastics production had reached 18.5 billion pounds (3.3×10^6 t), of which 14.6 billion pounds (6.6×10^6 t) or 78% of production was in thermoplastics [24]. Plastics packaging began to show dramatic gains, and, by 1970, an estimated 4 billion pounds (1.8×10^6 t) per year of such waste was being generated [25]. Then, during the 1970s, plastics prices began to rise, followed by a supply shortage as a result of the oil embargo and OPEC price increases on petroleum feedstocks. This combination of factors resulted in an increased effort to recycle and to obtain material for plastic products at reasonable costs. These approaches are well documented in the literature of the 1970s. The recycling of plastics carried out during that time showed a wide range of approaches and sophistication. Much of the research, development, and technological innovations of the time period formed the basis of the new plastics recycling industry of today. Many projects never emerged from the laboratory, others reached only the pilot-plant stage, while a few went on to commercialization.

Entrepreneurs as well as large industrial companies were involved. Representative of the recycling ventures and typical of an enterprising business at the time was Gary Plastics Corporation in inner New York City, a manufacturer of polystyrene drinking cups. Gary brought back used cups after each delivery of new cups to major centers, such as Lincoln Center. These returned cups were washed, shredded, and then molded with virgin resin using one part of recycled to two or three parts of virgin resin. The molded items were curios which were sold in Gary's retail store. Because the business could not use all of the recycled polystyrene, the rest was sold to other molders of polystyrene. This simple recycling operation was feasible because Gary was a low-labor-paying company in the inner city working with local large users of polystyrene cups [26].

At the other extreme of recycling activity was the study conducted by the Union Carbide Corporation in New Jersey on the pyrolysis of waste plastics to obtain chemicals and fuels. The laboratory-scale work emphasized the pyrolytic decomposition of PE, PP, and PVC wastes to low molecular weight polymers suitable as waxes, greases, oils, etc. Among the products isolated were styrene monomer and hydrogen chloride. It was noted that the styrene isolated from the pyrolysate of polystyrene could be used as a monomer source for the manufacture of the resin. Mixed plastics wastes were pyrolyzed also [27, 28].

The recycling of ABS telephone sets was undertaken by Western Electric and Bell Laboratories. The recovery system for these ABS housings and handsets consisted of a sorting conveyor, hammer milling, cycloning, and screening, followed by a magnetic separation, a flotation purifier, granulation, and a final screening before pelletizing. The recycled materials surpassed the Bell System's virgin ABS minimum specifications for the critical properties required. It is not known if the process is still being used in view of the breakup of the Western Electric Company [29, 30].

AT&T also studied the reuse of PVC and copper from coated wires. One approach was to dissolve the PVC in solvents, recover the wire, and isolate the PVC from solution. A second and preferred approach was to chop the coated wire to "pop out" the wire. However, a small amount of wire remained in the PVC scrap [31]. The U. S. Bureau of Mines at Rolla, Missouri, experimented with a pulsating liquid-solid bed to reduce the traces of metal remaining in waste wire insulation tailings [32]. Work has continued to this day in many laboratories outside of AT&T to develop a better separation system. (See Chapter 6.)

Other interesting and somewhat futuristic work applicable to today's recycling efforts was carried out by the Bureau in the early 1970s. For example, an electrostatic drum-type separator was used in an experimental separation of mixtures of polyethylene film/styrofoam and paper/plastic. The Bureau also developed a separation technique for mixed-plastics waste. A sink-float system was devised to separate polyolefins, PS, and PVC. The separation was done using an elutriation column and water as the sole parting medium. A more complicated theoretical sink-float system was devised to separate a five-component mixture of low-density polyethylene (LDPE), HDPE, PP, PS, and PVC. Laboratory experiments were conducted on hand-sorted

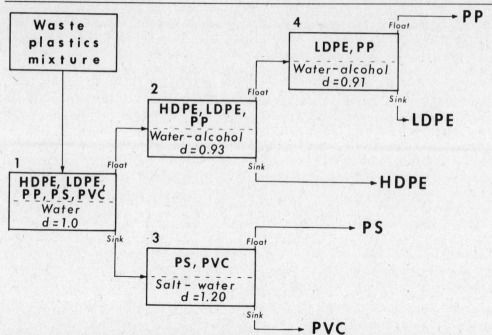

Figure 1.1   Theoretical four-stage sink-float scheme for separating waste plastics mixture [33]
Courtesy: Bureau of Mines, U.S. Department of the Interior

plastics waste in unagitated tanks using a water, salt-water solution, and two water-alcohol mixtures. A schematic of the sink-float system is given in Figure 1.1 [33].

General Electric, through a joint venture with Polymer Machinery Corp., was involved in a program to reuse thermoset scrap [34]. Urea-formaldehyde scrap was recovered from the flash and runners of production moldings of fluorescent light sockets. The scrap was ground and pulverized to a fine powder by a special Polymer Machinery cyclone mill. Particle size of the regrind was critical to the remolding operation. At 10% filler loading, no detrimental effects were noted on resin processibility or product quality. The use of various thermoset regrinds has received renewed interest from the industry during the past two years.

During this time period also, a considerable amount of research and development was expended in an effort to obtain alternative sources of fuel and raw materials for plastics and to evaluate plastics wastes as a feedstock source. This becomes apparent through the literature survey of the 1970s and into the 1980s. Articles and patents teach methods and techniques used to obtain valuable chemical raw materials and fuel through chemical and thermal decomposition of plastics waste. The pyrolysis study by Union Carbide mentioned above is an example. Another typical example is the pyrolysis process developed by Procedyne in 1975 to convert waste polymers to fuel oils. The Department of Energy partially funded a pilot-plant scale-up in 1977 and in 1980 provided start-up and operating analysis of a commercial unit at a PP facility in LaPorte, Texas, operated by U. S. S. Chemicals Division of U. S. Steel. The process utilized a tubular coil embedded in a gas-heated fluidized bed of sand. A flow diagram of the process is shown in Figure 1.2. In the process, molten byproduct atactic PP from an isotactic PP unit was heated to 425–510 °C at 50–250 psi (0.34–1.7 MPa) and a residence time of 30 minutes. About 94% of the feed polymer was converted to energy, with 90% to liquid similar to No. 2 and No. 6 type fuel oils, 4% gas, and another 6% gas, which is used to heat the reactor. The commercial plant had a design capacity to convert 17 million pounds (7.7×10^3 t) per year of atactic PP. Although the plant operated for a few years, it was not trouble free and eventually shut down because of a lack of sufficient byproduct feed from the main PP facility [35]. The U. S. Bureau of Mines at Rolla, Missouri, conducted a laboratory study on the pyrolysis of PVC

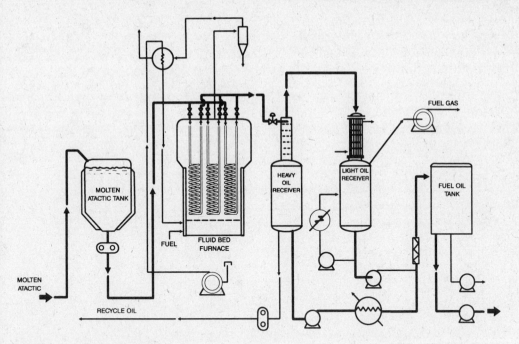

Figure 1.2   Tubular coil, fluidized bed pyrolysis process for conversion of waste polymers to fuel oils
Courtesy U.S. Department of Energy [35]

at various temperatures to determine the removal of hydrochloric acid. At temperatures up to 220 °C, only small amounts of acid were released; total liberation of acid occurred at 400 °C [36].

In Japan, a number of companies were involved, including Mitsui, Mitsubishi, and the Sekisui Chemical Company, which pyrolyzed PS and PE, and the Japan Steel Works Ltd., which operated a pyrolysis unit similar to Union Carbide's system [37].

In Germany, detailed studies were conducted at the Hamburg University on the pyrolysis of plastics waste and tires using first a molten salt bath and then a fluidized bed reactor [38, 39]. These fluid-bed studies, which were carried out in a three-step program from laboratory-scale through pilot plant and semi-commercial units, spanned more than a decade and culminated in the construction of a 10–20-million-pound (4.5–9.0 x 10^3 t)-per-year industrial-scale plant in Ebenhausen in 1984. Pilot-plant pyrolysis of the plastic fraction of household waste has been reported [40]. The waste, containing 67.6% polyolefins, 1.7% PVC, 1.0% PS, and the remainder a mixture of paper, metals, sand, etc., pyrolyzed at 717 °C, yielded methane, ethylene, benzene, toluene, and carbon black as main products.

Roy, Rollin, and Schreiber of the Ecolé Polytechnique (Montreal, Quebec) [41] reported on the furnace pyrolysis of mixtures of virgin LDPE and PS. In the presence of nitrogen at a pressure $\geq$ 200 torr (26.7 kPa) and with a gradual increase to the temperature of pyrolysis, the polymers pyrolyzed as single entities with no interactions; at lower pressures, some polymer or pyrolytic product interactions occurred; with a rapid increase of temperature to 600–800 °C, large quantities of carbon residue resulted from the pyrolysis.

In 1974, Swedlow, Inc., a cast-acrylic sheet manufacturer, started up a thermal cracking unit using molten metal as a liquid heat transfer medium to thermally depolymerize PMMA to the monomer. The yield of monomer was better than 95% and 99% purity, after distillation. Polymer scrap from Swedlow's own manufacturing operation as well as from outside sources was used to provide monomer for repolymerization [42]. A further discussion of acrylics recycling is given in Chapter 8.

Also in 1974, Liquid Nitrogen Processing Corp. developed a process for the pyrolysis of tetrafluorethylene and tetrafluorethylene composites using super-heated steam to decompose the polymer. In the process, the steam also acts as the condensable carrier gas to remove the monomer and other low molecular weight byproducts [43].

The decomposition of PUR foams by hydrolysis and glycolysis to form polyols suitable as partial replacement of virgin polyols in the manufacture of these polymers received considerable attention during this period. In a series of patents and papers, The Ford Motor Company and General Motors described processes for the hydrolysis of flexible foam to polyols and diamines using super-heated steam temperatures of 250–350 °C [44, 45]. A continuous hydrolysis process using a specially designed extruder was patented by Bayer AG in 1977 [46].

The Upjohn company patented a glycolysis process to recover polyols from rigid and flexible PUR-based foams [47]. In the latter process, no separation of products is required, and the recovered polyols could be used in the preparation of rigid foams. This process was licensed to Nippon Soflan, a subsidiary of Toyo Rubber Co. in Japan, which was operating a commercial plant with an annual capacity of 1.3 million pounds (585 t) in 1978. Further details on these processes and later developments are given in Chapter 10.

Similarly, a number of processes were developed on the chemical decomposition of PET to give monomer species and polyols for repolymerization to PET, unsaturated polyesters, and polyurethanes. For example, a patent issued to E.I. DuPont in 1970 describes the use of aqueous sodium hydroxide in ethylene glycol at 90–150 °C and at atmospheric pressure to decompose in-plant scrap to the disodium salt and then to terephthalate acid (TPA) [48]. The yield of disodium terephthalate is 95.7%. The purified acid when recycled back into a PET polymerization reaction at a 20% level showed no detrimental effects on the product. An earlier patent, also to DuPont, reacts waste fiber PET with ethylene glycol (EG) and a benzenesulfonate catalyst to give bis-hydroxyethyl terephthalate (BHET). A final polymer made with 45% of the recycled monomer gave fiber properties equivalent to PET produced with virgin monomer [49]. Less formation of diethylene glycol (DEG) and a faster depolymerization of the waste PET fiber (in excess EG) was obtained upon the addition of BHET and sodium acetate trihydrate to the reaction [50].

Barber Coleman Co. was depolymerizing waste PET photographic and X-ray film to TPA while reclaiming the silver. Michigan Technological University held the patents on the technology and adapted it for PET bottles. In this process, the PET is reacted with aqueous ammonium hydroxide at an elevated pressure and temperature. The diammonium salt is acidified with sulphuric acid to isolate TPA and EG. A purity of 99% for the TPA is claimed [51].

Farbwerke Hoechst disclosed a process for depolymerizing PET waste in any form (fiber, chips, ribbon, film) by reacting the molten PET with methanol in a two-stage process to obtain dimethyl terephthalate in 99% yield [52]. High-purity, colorless, depolymerized PET products were obtained by Eastman Kodak [53] by reacting scrap polyesters with various alcohols and then catalytically hydrogenating the products.

The glycolysis of reclaimed PET with propylene glycol (PG) to polyols suitable for reaction with maleic anhydride to produce unsaturated polyesters was described by Eastman Chemical Products [54]. The cast and laminated physical properties of the PET modified unsaturated polyester resin compared favorably with controlled resins containing phthalic anhydride and isophthalic acid. A shorter reaction time was observed; some haze was noted in the styrene resin solution. During this study, it was noted that reclaimed PET for unsaturated polyester preparation should be clean and free of HDPE, labels, etc. The economics based on 1980 raw material costs looked favorable. Scrap or waste PET may also be reacted with polyols to form prepolymers. In a 1977 patent [55] to Freeman Chemical Corp., PET reclaimed from desilvered photographic film is digested with organic diols and triols, and the digested product is further reacted with an organic polyisocyanate to produce a polyisocyanate-terminated prepolymer. Flexible PUR foams prepared with these prepolymers resulted in unexpected strength properties. The teachings of this patent and subsequent patents issued to Freeman on polyols prepared from reclaimed PET remain in practice today [56, 57].

A two-stage process for depolymerizing e-caprolactam derived polyamides (nylon-6), including scrap polymer, was patented by Allied Chemical Corp. in 1965 [58]. The ground scrap nylon is first solubilized with high-pressure steam at 125–130 psig (963–997 kPa) at 175–180 °C for 0.5 hours in a batch mode and then continuously hydrolyzed with super-heated steam at 350 °C and 100 psig (790 kPa) to e-caprolactam monomer at an overall recovery efficiency of 98%. The recovered monomer was of good quality and was repolymerized without further purification. A later patent teaches the hydrolysis of mixed PET and nylon-6 fiber scrap [59].

Much of the original work in the development of the processes and process equipment to fabricate products from commingled plastics was done during this time. Although the original concepts may have been intended to develop working systems for relatively uncontaminated and nonheterogeneous plastic scrap, the systems ultimately were used on contaminated and heterogeneous commingled plastics. Reclamat's Tufboard, Japan Steel Works Ltd., Reclamation Line, and the Regal Converter processes are typical; the Reverzer, the Remaker, and the Klobbie are extruder-type machines of the time. Many of these form the basis of the machines and systems in existence today (e.g., the ET-1, Superwood, Hammer's and Polymerix's systems). It is unnecessary to describe the early machines and systems since this has been done adequately by Leider [60]; later developments are discussed in Chapter 9.

Other recycling activities of the 1970s were smaller in size but not in importance to the development of today's technology. For example, the Golden Arrow Dairy Company (San Diego, CA) was probably the first company with a recycling program in HDPE milk bottles. The company started to collect the bottles in 1970 while delivering milk to individual homes. The bottles were returned to the plant, granulated, washed, and dried. The ground resin was sold to fabricators for the production of agricultural drainage tile. Eventually the operation was discontinued because the use of recycled resin was not permitted according to the material specification [61].

Another example of a postconsumer recycling effort was a one-time program sponsored by Owens-Illinois Corp. whereby Boy Scouts collected and cleaned PE bottles. The bottles were sold to a manufacture of drainage pipe.

In another approach to today's recycling activity, Mobil Chemical, a major producer of PE bags, purchased clean off-specification bags for reuse in the making of new bags. While this is a reuse of clean manufacturing scrap, it was unique in that the purchaser worked with the generator of the bags to segregate the scrap and keep it clean. Mobil Chemical also reprocessed PS foam egg cartons returned by consumers to supermarkets. At the Mobil plant, the cartons were washed, melted, extruded, and pelletized. The pellets were then blended with virgin PS and remolded [62].

The postconsumer recycling programs mentioned above took place in the early 1970s and were not sustained for one reason or another. By 1980, however, a number of companies that had been involved in industrial plastics waste/scrap in some manner via collection, separation, reprocessing, or fabrication were becoming involved in postconsumer plastics recycling. A few were already involved, and others could see the possibilities depending upon the cleanliness of the raw material available at the time. Primary interest was in postconsumer PET beverage bottles. A probable continuous source of this raw material had become a reality through the deposit laws passed by a number of state legislatures. By the end of the 1980 legislative sessions, seven states had passed such laws: Connecticut, Delaware, Iowa, Maine, Michigan, Oregon, and Vermont. Oregon's law had already existed for nine years. Postconsumer PET, at approximately half the cost of industrial scrap, was becoming more attractive as an inexpensive raw material. St. Jude Polymer (Frackville, PA), a small company founded in 1977, began to use postconsumer PET together with bottle scrap early in the operation. Wellman (Johnsonville, SC) and M. K. S. (Elizabethton, TN), two PET and nylon fiber producers, began to purchase postconsumer PET bottles in 1979 for the production of unbranded polyester fiber manufacture. DuPont, through the Material Reclamation Systems group, also became actively involved in postconsumer PET reclamation and reuse in 1979. The end-product application for the recycled PET by DuPont was proprietary and could have been used for fiber, film, or molding grade. Building Components, Inc., Pure Tech Industries, and Plastics Recycling, Inc., were a few other companies reprocessing, selling or fabricating postconsumer PET beverage bottles. It was estimated that in 1980 about 11 million pounds (4.9×10^3 t) of postconsumer PET was being recycled, of which 10 million pounds (4.5×10^3 t) went into fiberfill and 1 million pounds (0.45×10^3 t) into strapping [50].

Postconsumer HDPE recycling was still negligible in 1980, and no estimate of the amount being recycled is available. A few of the companies involved at the time obtained all the bottles through voluntary collection groups. For example, N. E. W. was experimenting with extruding reclaimed HDPE into boards for marine and agricultural applications. E. Z. Recycling was selling the material to manufacturers of plastic drain tiles.

The early 1980s passed as a period of very little new developmental activity in plastics recycling. One study carried on by The Plastics Institute of America (PIA) during that time has some relevance to the recycling of durable plastics today. The study [64] was funded by the U. S. Department of Energy, and the work was conducted at a number of graduate schools. The program's objective was to attempt to recover the plastics fraction from nonmetallic ASR and to use the mixed recovered material for low-grade plastics applications.

The study showed that plastics constituted less than one-half of the shredder residue, the remainder being metal, glass, wood, contaminating oil, etc. Experimental results gave indications that the shredder residue or "fluff" held the materials in a tight network that was difficult to separate. Products molded from the residue after the nonmetals were removed were found to be of poor quality compared to commercial resin-wood chip boards. Shredder waste was also used in polymer concrete and as a polyester filler. Products were usable but were of lower quality.

The work ended in the late 1980s with no definitive recovery approach for separating the contaminating waste from the shredder residue and for reusing the mixed plastic to produce a commercially acceptable material *per se* or in combination with other plastics components.

As the 1980s continued, plastics recycling gained momentum. The impetus this time came from the environmentalists, solid waste professionals, and local and state governments. The growth of plastics recycling that has occurred from the mid-1980s to today is in large part due to the work done in the two previous decades. Plastics recycling also has become more economically feasible due to the continued plastics technological growth and the increased plastics unit values that have occurred over those two decades.

The recycling of today, therefore, is the continuation of much of the recycling that started in the 1970s, together with more and better available technology and more concerns on the part of the government and the enlightened citizenry. The ensuing chapters review the recycling of specific plastics with particular emphasis on past important work and latest developments.

References

[1]   *Chem. Eng.* (Nov. 1990): p. 43.
[2]   The Council for Solid Waste Solutions. *The Solid Waste Management Problem.* 1989, p. 10.
[3]   U. S. Environmental Protection Agency. *Characterization of Municipal Solid Waste 1990 Update.* EPA/530-SW-90-042. June 1990, p. ES-4.
[4]   Rathje, W. L. " Once and Future Landfills." *National Geographic* (May 1991): p. 122.
[5]   Reference 3, p. ES-10.
[6]   Reference 3, p. ES-5.
[7]   Reference 3, p. ES-7.

[8]   *Modern Plastics, 68*(1) (January 1991): p. 120.

[9]   Milgrom, J. *Incentives for Recycling and Reuse of Plastics.* Report No. EPA-SW-41C-72. Arthur D. Little, Inc., Cambridge, MA, 1972, pp. xv, xvi.

[10] Leidner, J. *Plastic Waste.* New York: Marcel Dekker, Inc., 1981, pp. 64, 65.

[11] Curlee, T. R. *The Economic Feasibility of Recycling.* New York: Praeger, pp. 19, 20, 24, 27.

[12] *Standard Guide—The Development of Standards Relating to the Proper Use of Recycled Plastics.* Designation D 5033-90. American Society for Testing and Materials, Philadelphia.

[13] Reference 9.

[14] Reference 10.

[15] Reference 9.

[16] Reference 10.

[17] *SPI's Voluntary Plastic Container Coding System.* Brochure. The Society of the Plastics Industry, Inc., Washington, DC, 1988.

[18] *The SPI Resin Identification Code.* Brochure. The Plastic Bottle Industry Division. The Society of the Plastics Industry, Washington, DC.

[19] *Standard Practice for Generic Marking of Plastic Products.* Designation D 1972-91. American Society for Testing and Materials, Philadelphia.

[20] *Marking of Plastic Parts.* The Society of Automotive Engineers, Inc., Warrendale, PA, Revised March 1988.

[21] *Regional Labeling Standards and Labeling Resolutions.* The Northeast Recycling Council, Brattleboro, VT, 27 November 1990.

[22] Reference 10, pp. 69–72.

[23] *Modern Plastics 24* (1) (Jan. 1947): p. 62.

[24] *Modern Plastics 47* (1) (Jan. 1970): p. 70.

[25] Seymour, R. B., Sosa, J. M. *Chemtech* (Aug. 1977): p. 508.

[26] Lawrence, J. R. Special Technical Publication 533. American Society for Testing and Materials, 1973, p. 56.

[27] " Recycling Plastics." 1972 Studies for The Society of Plastics Industry Inc., by International Res. & Tech. Corp., June 1973, p. 31.

[28] Reference 10, pp. 248–249.

[29] Hancock, H. G., Hubbauer, P. Bell Laboratory Record. Dec. 1975, pp. 427, 428.

[30] Wehronber, R. H. *Materials Eng.* (Feb. 1982): p. 44–50.

[31] Donavan, S. C., Pompeo, A. J., and Scales, E. Bell Laboratory Record. September 1977, pp. 215–218.

[32] Holman, J. L., Stephenson, J. B., and Adam, M. J. "Recycling Plastics from Urban and Industrial Refuse." RI 7955 (1974). Rolla Metallurgy Research Center, Bureau of Mines, Rolla, MO.

[33] Holman, J. L., Stephenson, J. B., and Jensen, J. W. " Processing the Plastics Waste From Urban Refuse." TPR 50 (1972). Rolla Metallurgy Research Center, Bureau of Mines, Rolla, MO.

[34] *Modern Plastics 52* (9) (Sept. 1975): p. 64, 65.

[35] Amato, A. "Start-up and Monitoring of First Commerical Waste Atactic Polypropylene to Fuel Oil Conversion Process." Final Report to U. S. Department of Energy by USS Chemicals. DOE/CS/402T3-T3, 1983, p. 12.

[36] Reference 27, p. 32.

[37] Reference 10, pp. 252–256.

[38] Menzel, J., Perkow, H., and Sinn, H. *Chem. Ind.* (16 June 1973).

[39] Kaminsky, W., Menzel, J., and Sinn, H. *Cons. Res., 1* (1976).

[40] Kaminsky, W. *J. Anal. & Appl. Pyrol, 1* (1985): p. 439–448.

[41]   Roy, M., Rollin, A. L., and Schreiber, H. P. *Polym. Eng. & Sci., 18*(9) (July 1978): p. 721–727.

[42]   *Modern Plastics 51* (11) (Nov. 1974): p. 80.

[43]   Arkles, B. C., Bonnett, R. N. U. S. Patent 3 832 411, 1974.

[44]   Salloum, R. SPE Antec, 4 May 1981.

[45]   *Chem. Eng.* (16 Feb. 1976): p. 54, 55.

[46]   Geigat, E., Hetzel, H. U. S. Patent 4 051 212, 1977.

[47]   Ulrich, H., Odinak, A., Tucker, B., and Sayighaar. *Polym. Soc. & Eng., 18*(11) (1978): p. 844–848.

[48]   England, R. J. U. S. Patent 3 544 622, 1970.

[49]   MacDowell, J. T. U. S. Patent 3 222 299, 1965.

[50]   Malik, A., Most, E. E. U. S. Patent 4 078 143, 1978.

[51]   Lamparter, R. A., Barna, B. A., Johnsrud, D. R. U. S. Patent 4 542 239, 1985.

[52]   Grushke, H., Hammerschick, W., Medem, H. U. S. Patent 3 403 115, 1968.

[53]   Stevenson, G. M. U. S. Patent 3 501 420, 1970.

[54]   Calendine, R., Palmer, M., Bramer, P. V. *Modern Plastics 57* (5) (May 1980).

[55]   Svoboda, G. R., Suh, J. T., Carlstrom, W. L., Maechtle, G. L. U. S. Patent 4 048 104, 1977.

[56]   Carlstrom, W. L., Reineck, R. W., Svoboda, G. R. U. S. Patent 4 223 068, 1980.

[57]   Svoboda, G. R., Carlstrom, W. L., Stoehr, R. T. U. S. Patent 4 417 001, 1983.

[58]   Bonfield, J. H., Hecker, R. C., Snider, O. E., Apostle, B. G. U. S. Patent 3 182 055, 1965.

[59]   Lazarus, S. D., Twilley, I. C., Snider, O. E. U. S. Patent 3 317 514, 1967.

[60]   Reference 10, pp. 169–186.

[61]   Reference 9, pp. iv-17–18.

[62]   Kaufman, F. S. Proceedings of the National Materials Conservation Symposium, STP 592. ASTM, 1975.

[63]   "Plastics Recovery Trends and Opportunities in Post Consumer Segments." Report for The Society of the Plastics Industry—Technomic Consultants, 17 July 1981.

[64]   "Secondary Reclamation of Plastic Waste." ORNL Subcontract no. 19X-091000. Phase I, Phase II. Plastics Institute of America, 1987.

2 Collection and Separation

W. P. Moore

Contents

2.1	Introduction	19
2.2	Residential Recycling	19
	2.2.1 Collection Containers	20
	2.2.2 Education and Promotion	21
2.3	Which Resins to Pursue	22
2.4	Collection Vehicle Issues	24
	2.4.1 Vehicle Types	24
2.5	Material Recovery Facilities	29
	2.5.1 Role of the MRF	29
	2.5.2 Plastics Separation	29
	2.5.3 A Working MRF	29
2.6	Economics	32
2.7	Separation at the Reclamation Facility	36
2.8	Case Studies	36
	2.8.1 Case Study 1: San Jose, California	36
	2.8.2 Case Study 2: Camden County, New Jersey	37
2.9	Special Collection and Separation Systems	41
	2.9.1 Agricultural and Other Films and Agricultural Chemical Containers	41
	2.9.2 Plastic Automotive Parts	42
	2.9.3 Polycarbonate Water Jugs	42
	2.9.4 Polystyrene Foam Packaging	42
2.10	Global Activities	42
References		43

This chapter discusses the various types of programs and approaches for the collection and separation of postconsumer plastics waste. Starting with residential recycling, the effect of the program approach and equipment used for collection will be reviewed. The role that education and promotion plays in the cleanliness and amount of materials collected is presented. Critical collection vehicle issues and separation at the processing facilities are covered in regard to preparing the materials for the reclamation marketplace. The economics of the system and case studies are presented. The concluding section of the chapter discusses various specialized plastics collection programs in the commercial and industrial sectors.

2.1 Introduction

Consumer plastics recycling is in a very dynamic stage of development and implementation. In order to preserve our natural resources and reduce the disposal of solid waste, the combined efforts of individuals, businesses, and the government are moving forward to put in place a network of residential plastics recycling programs. Various approaches to collecting plastics are taking place in the United States today. The dynamics of plastics recycling however require more than just hardware and desire; education and various types of promotion play an important role in residential plastics recycling area. Getting the right materials in the right form for reclamation is one of the most difficult hurdles to overcome. This and the economics involved are discussed below.

2.2 Residential Recycling

In the early 1990s, the United States stands at the threshold of putting into place an infrastructure to introduce residential recycling on a large scale. Through the early to mid-1980s, recycling programs consisted largely of drop-off and buy-back centers, where people had to take their materials for recycling [1]. They were either motivated by the intrinsic value of the material (e.g., aluminum buy-back centers) or by the environmental or civic ethic (e.g., newspaper recycling drop-off facilities or Boy Scout drives). While these approaches were noteworthy as early methods to recycle, only a small percentage of the population, about 5%, would go to the trouble of recycling in this way [2]. In addition, plastics, because of their light weight and lack of end markets, did not have enough inherent value to justify the drop-off or buy-back system.

By the middle 1980s, nine states had a deposit system for beverage containers (commonly called "bottle bill" states) Connecticut, Delaware, Iowa, Maine, Massachusetts, Michigan, New York, Oregon, and Vermont. Most programs had a $0.05 return value ($0.10 in Michigan) for the containers. In these states, a network of buy-back and grocery store centers was established with a redemption value on all types of beverage containers: aluminum, glass, steel, and plastic. They were the mainstay of recycling programs for beverage containers throughout the 1980s. This did lead to large-scale collection of polyethylene terephthalate (PET) beverage bottles for recycling in those states. As of 1990, this supply stood at about 120 million pounds $(54 \times 10^6 \text{ kg})$ per year of PET [3].

Beginning in the middle 1980s, the country began to put in place a system called residential curbside recycling. This approach to recycling features periodic collection of recyclables from individual homes, obviating the need for people to take their materials to a center. By early 1990, 40 million people (15%) in the United States had this service [4]. The results were dramatic—an average of 70–80% of the people in many communities participate in these types of programs [5]. In addition, because of economy of scale, materials of lower value per unit weight, such as newspapers and plastics, are routinely included in the programs. The list below provides an idea of the amount of growth of curbside recycling programs [6]. Curbside recycling programs are expected to continue to grow at a 15–20% annual rate.

Number of Curbside Recycling Communities in the United States
1988 1,052
1989 1,518
1990 2,711

Most curbside programs started between 1985 and 1988 did not include any postconsumer plastics material. The lack of good end markets for the materials and, generally, an antiplastic sentiment by recyclers contributed to this. In the latter 1980s, the plastic, chemical, and packaging industries spent quite an effort in promoting and developing methods for the routine inclusion of plastics into these programs. The effort was highly successful, and, as of mid-1990, almost 75% of new curbside recycling programs included some types of plastic collection [7].

2.2.1 Collection Containers

It is ironic that a plastic, box-like household recycling container has become one of the key features in making curbside recycling programs successful. For the homeowner, having an attractive container of anywhere between 12 and 20 gallons (45 and 76 liters) is a key reminder to recycle and provides a place for the materials to be stored. Figures 2.1 and 2.2 show curbside recycling containers frequently utilized. On pick-up day, the container forms obvious on-the-street peer pressure support for recycling. When residents go out and see that all of their neighbors have put their containers out, they quickly fill theirs and place it at the curb, too.

It appears as if the household container, due to its appearance and ability to allow the homeowner to store materials, is a key part of the program. Without a container supplied by the program, participation rates drop to as little as 30–40% [8].

Figure 2.1   A single-bin residential curbside container
Source: Fremont/Newark, CA program

Figure 2.2   A three-stack residential curbside container
Source: Waste Management Inc.

As mentioned earlier, curbside recycling did not include large-scale plastics collection until after 1988. This raised a new group of issues, a key one being the need to go to larger collection containers. With the plastics containers most often used in the residential programs today, a container size of at least 17 gallons (64 liters) is necessary for an average family recycling on at least an every other week basis. Without plastics in a program, the typical container was 12–14 gallons (45–53 liters). The types of resins collected in the program will be discussed in Section 2.3, but as the number of materials that are collected is expanded, certainly even larger containers will be needed.

A group of programs beginning in 1990 using plastic bags instead of containers was begun. They are still under evaluation, and the outcome of this type of program is not yet conclusive.

2.2.2 Education and Promotion

Education and promotion are important factors in getting residential recycling programs to work. Early communication of the style and type of program, day of pick up, and acceptable materials are of course critical, but, once in place, residential recycling quickly becomes a routine habit. A continuing high level of education and promotion is not necessarily required. Most people rank education and promotion first in importance when regarding what would most likely make a program succeed. A study in 1987 by Lane County, Oregon, showed that education and promotion was indeed second in importance [9]. The most important factor was routine, frequent, and convenient service. Table 2.1 shows the effect of frequency of collection on participation. In addition, the study also pointed out that a household container was a vital element of the program (Table 2.2) [10].

With regard to education and promotion specific to plastics, there is an additional issue. It is called the "step-on-it" factor. In order to reduce the volume occupied by plastic bottles, residents are asked to step on plastic bottles twice to reduce the space they take up not only in the homeowner's collection container but also in the recycling collection vehicle. Stepping on

Table 2.1   Effect of frequency of collection on participation

Program Type	Collection Frequency	
	Weekly	Monthly
Voluntary Programs		
Number surveyed	17	14
Participation range, %	10–80	4–65
Average participation, %	46	29
Mandatory Programs		
Number surveyed	9	6
Participation range, %	40–98	25–85
Average participation, %	73	48

Sources: Resource Conservation Consultants; *Waste Age;* Ferrand-Scheinberg Associates
Note: Self-reported program participation data

Table 2.2   Relationship of household container use on participation

Test Community	Program Participation in %	
	With Container	Without Container
Champaign, IL	83	11
Kitchener, ON	75	65
San Jose, CA	75	48
Santa Rosa, CA	70	35
Toronto, ON	66	42

Source: Resource Conservation Consultants
Note: Data for studies performed in five communities; programs are voluntary

a container twice is necessary because of the container "memory." It has been shown to be effective—at least half of the people in the program will crush the bottles. An example of a brochure for educational and promotional purposes is shown in Figure 2.3 [11].

2.3   Which Resins to Pursue

In the deposit system of recycling, only PET beverage containers were involved. When residential curbside recycling came along, recycling of residential plastics containers became more viable. At first, only translucent high-density polyethylene (HDPE) milk jugs and clear PET beverage containers were included. By 1990, it was considered routine to include all PET beverage bottles, clear and colored, as well as HDPE milk jugs, spring water jugs, and other pigmented HDPE household cleaner jugs, food containers, etc.

It looks as if the trend in the early 1990s will be to continue to include more materials. The next most likely material that exists in only a few programs at this point, but which is spreading, is polyvinyl chloride (PVC) containers. It is hard not to accept PVC containers when collecting PET, since both are clear, rigid plastics. Separation of PVC containers after collection is discussed in Section 2.7 and further in Chapter 6.

Montgomery recycles plastics

...it's a smashing idea!

Comments anyone?
A primary objective of this pilot project is to determine the most convenient and efficient methods for households like yours to recycle. We will be conducting a survey during the project to get your feedback on the program. If you are experiencing a problem with the collection of your recyclables or have ideas or general comments about the project, use the following phone numbers/addresses to contact us:

A Collection Problem?
City of Rockville, Department of Public Works: 424-2224.

General Comments/Ideas?
Multi-Materials Pilot Program
Montgomery County Department
of Environmental Protection
101 Monroe Street, 6th Floor
Rockville, Maryland 20850

You may call and leave a message on our Recycling Line at 217-6990. Your call will be returned.

*A Cooperative Recycling Project
of Montgomery County Government
and City of Rockville*

*This flyer sponsored by
The Council for Solid Waste Solutions*
Printed on recycled paper

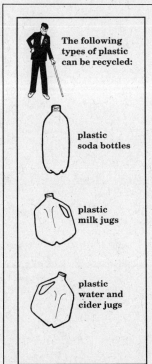

The following types of plastic can be recycled:

plastic
soda bottles

plastic
milk jugs

plastic
water and
cider jugs

Prepare them for set-out by:

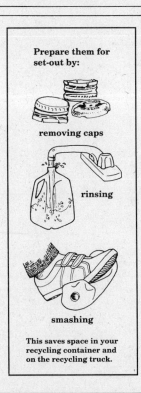

removing caps

rinsing

smashing

This saves space in your recycling container and on the recycling truck.

Reasons for recycling plastics:

▶ **ENERGY SAVINGS**
Recycling plastics saves 88% of the energy required to produce plastics from raw materials.

▶ **ENHANCES WASTE MANAGEMENT OPTIONS**
Our nation is facing a solid waste crisis. Recycling plastics saves space in the landfill and in the Resource Recovery Facility.

▶ **REDUCED DISPOSAL COSTS**
The cost for using the County's landfill is $53/ton and is increasing every year. Recycling helps keep down the cost of garbage disposal.

▶ **RESOURCE CONSERVATION**
Recycling plastics is a means to re-use petroleum products.

Figure 2.3   Flyer distributed to Montgomery County residents to promote plastics recycling and "step-on-it" program

It is the objective of the people working in the recycling collection business and in the plastics reclamation industry to expand the programs to collect all rigid containers. The cost and technology for all rigid container sorting are still developing issues. There will be some formidable end market challenges dealing with multilayer and composite containers, and the economical recyclability of those materials is yet to be proven. A few programs, mostly on the eastern coast of the United States, now also collect expandable polystyrene (EPS) foam containers and sheet. (See Chapter 5.) At this point, the widespread recycling of these materials has still not occurred [12]. More interest in these other types of materials is expected. Table 2.3 summarizes average estimated recovery levels of various resins from typical programs [13].

Table 2.3   Estimated recovery levels of postconsumer resins in a curbside program

Category Told to the Public	Resins Collected	Estimated Recovery Level lb/HH yr (kg/HH yr)		Estimated Density with Step-On-It Program lb/yd^3 (kg/m^3)	
Soft Drink Bottles	PET	5–11	(2.3–5.0)	30–45	(18–27)
Milk Jugs	HDPE Natural	5–10	(2.3–4.5)	25–30	(15–18)
Soft Drink Bottles & Milk Jugs	PET, HDPE Natural	8–17	(3.6–7.7)	30–40	(18–24)
Beverage Bottles	PET, HDPE Natural, PVC Clear	8–18	(3.6–8.2)	30–40	(18–24)
Detergent & Bleach Bottles	HDPE Colored	1–2	(0.5–0.9)	35–50	(21–30)
All Clean Bottles	PET, HDPE Natural, HDPE Colored, PVC, PP, Multilayer	15–24	(6.8–10.9)	40–50	(24–30)
All Rigid Containers	PET, HDPE, PS, PVC, PP, Multilayer	20–30	(9.1–13.6)	40–50	(24–30)
All Clean Plastics	PET, HDPE, LDPE, PS, PVC, PP, Multilayer	25–40	(11.3–18.1)	NA	NA

2.4   Collection Vehicle Issues

The low weight-to-volume ratio of plastics makes the choice of which type of collection vehicle to use a challenge. As large-scale recycling is still relatively young, a variety of evolutionary vehicles exist in the programs. With plastics collection becoming routine only within the last one to two years, the solution to issues surrounding the collection vehicles handling the extra volume are still in their infancy. A continued evolution towards more practical vehicles that can handle the extra volume of plastics is expected.

2.4.1   Vehicle Types

A summary of recycling truck body types typically utilized for residential curbside collection is presented in Table 2.4. As can be seen from Figures 2.4, 2.5, and 2.6, the styles of vehicles used in residential recycling vary quite widely in their appearances. One common thread in all

Table 2.4   Recycling vehicle body configurations

Vehicle Type	Manufacturer	Body Capacity yd³ (m³)	Advantage	Disadvantage
Open top, rear dump with movable dividers	Eager Beaver LODAL 3000 Rudco Kahn Timsco	15–30 (11–23)	Low capital cost; Flexible bin sizes; Low maintenance; Tarping required on open tops	Progressively higher loading heights; Generally lower capacity
Side dumping bodies	Koenig LODAL 3030 KANN Eager Beaver Marathon Timsco	15–30 (11–23)	Side dump allows for emptying compartments into a roll-off or other container; Plastics compactor easily installed	Limited flexibility to adjust bin sizes; High loading height
Semi-automated Top-loading	Dempster EZ Pack Jaeger Labrie Ltd. Rogers	25–40 (19–31)	Uniform (low) loading height; Capacity; Flexible bin sizes	Higher capital cost; Higher maintenance cost; Glass breakage
Rear load compactor	Dempster Heil Leach Loadmaster EZ Pack	20–32 (15–25) (compacted)	Low loading height	More glass breakage; Higher capital and maintenance costs; Loads from rear only; Typically limited to one material; Two-person crew required

Note: This listing is representative of some vehicle types and manufacturers.

Figure 2.4   Side-unloading residential recycling vehicle

Figure 2.5   Solid waste collection vehicle modified for recycling service
Courtesy: Delta, British Columbia

Figure 2.6   Low profile, front-wheel drive recycling vehicle

of these vehicle types is that they feature a low loading height, multiple compartments, and either side- or rear-unloading capabilities.

As with much of the equipment in this emerging industry, collection vehicle bodies are in an evolutionary stage. This is strikingly similar to the evolution of traditional solid-waste collection vehicles, with recyclables currently being collected in many communities in open self-dumping trucks, as was household waste prior to the development of the rear-loading compactor truck. In general, the low weight-to-volume ratio of commingled containers, especially in cases where programs include plastics, typically results in full loads with weights of 5 tons (4.5 t) or less. Thus, it is most practical to mount these bodies on lighter chassis than used for solid-waste collection, with single-mean axles and a gross vehicle weight of less than 30,000 pounds (13,600 kg).

The use of divided truck bodies and the need to minimize glass breakage in the collection of mixed recyclables has impeded the widespread use of compaction vehicles for curbside collections. Moreover, materials such as newspaper are not significantly reduced in volume through compaction.

The increased inclusion of plastics in curbside recycling programs is an important issue in collection vehicle design. It has been shown that adding plastics to a program can almost double the volume of the commingled container fraction. The density of commingled glass and metal cans falls in the range of 250–300 lb/yd^3 (148–178 kg/m^3), while commingled containers including plastic typically range in density from 130–200 lb/yd^3 (77–118 kg/m^3). One result of this has been a trend towards larger capacity truck bodies in the 30–40 yd^3 (23–31 m^3) range [14].

Another result has been the development of on-board compactors for reduction of the plastics volume. The effectiveness of commercially available units of this type has been a matter for debate. In the worst cases, the space taken up by the compactor unit is greater than the resultant truck capacity gained by compressing the plastics.

Figure 2.7 [15] shows a picture of a curbside recycling truck fitted with an on-board compactor. Table 2.5 provides insight into the various types of on-board compactors available and their effectiveness.

Figure 2.7   On-board plastic bottle compactor on a curbside recycling vehicle
Courtesy: Kohlman Hill Co., Inc.

Table 2.5   Comparison of on-board compaction units[a]

Manufacturer	Type [b]	Container Capacity yd³	(m³)	Compactor Unit Size yd³	(m³)	Current[c] Status	Units Operating	Cost Thousands $
Inger-Teco: LC60	C	0.60	(0.46)	1.70	(1.30)	A	60	5.90
Inger-Teco: LC100	C	1.00	(0.76)	2.70	(2.06)	A	5	6.80
Kohlman-Hill 3600	C	1.00	(0.76)	3.50	(2.68)	A	3	6.50
Kohlman-Hill 4200	C	1.50	(1.15)	4.10	(3.13)	A	1	6.50
Kohlman-Hill 4800	C	2.00	(1.53)	4.70	(3.59)	A	2	6.50
Reliable: SAC	C	0.62	(0.47)	1.48	(1.31)	A	1	45.00 [d]
SP Industries	C	2.00	(1.53)	5.00	(3.82)	A	2	7.80
Lummus PVC-100	FP	N/A	(N/A)	0.85	(0.65)	A	2	9.80
Labrie Ltd.: PC3	F	2.60	(1.99)	3.50	(2.68)	A	6	9.00
Tafco	F	4.40	(3.36)	3.70	(2.83)	T	4	N/A
Multitek	F	N/A	(N/A)	N/A	(N/A)	A	0	3.00
P/M Rudco	F	8.50	(6.50)	N/A	(N/A)	T	1	9.00-10.00
Midwest P-Cage	U	11.00	(8.41)	11.00	(8.41)	A	20	6.40
Midwest P-Cage	U	15.00	(11.47)	15.00	(11.47)	A	—	7.10
Buff Grove: Cage	U	15.00	(11.47)	14.00	(10.70)	A	16	1.80

[a] Manufacturers' data
[b] C = Compactor; FP = Flattener/Perforator; F = Flattener; U = Uncompacted
[c] A = Commercially available; T = Testing phase
[d] Cost is for collection truck with compactor

The "memory" or tendency of plastic bottles to expand to their original shape after being crushed also has been problematic. In any case, on-board compaction requires separate handling of plastics, thus penalizing it economically. With plastics having a substantial "cushioning" effect that reduces glass breakage, there is increased experimentation being conducted on compacting the entire commingled metal, glass, and plastic fraction. The results to date look promising [16]. Shredding on the vehicle is not practiced because of time, cost, and resin-mixing issues. The most common approach in use right now to accommodate the extra volume that plastics take up is to use a larger vehicle and encourage the homeowner to "step on it."

2.5 Material Recovery Facilities

Material recovery facilities (MRFs) have become the focal point of large-scale recycling programs in North America [17]. At the beginning of 1991, there were about 100 MRFs in existence in the United States. By the end of the 1990s, it is conceivable that almost every county with a population greater than 100,000 people will have some form of MRF.

2.5.1 Role of the MRF

The MRF performs two functions—one separating the various materials and the other preparing the materials for market. In the case of plastics, this means either separating the plastic fraction from the glass and metals in a commingled-style pick up or, where plastics are handled separately, simply routing it toward its final market preparation. Most MRFs are equipped to bale plastic materials; a few shred various bottles. Shredding must be approached with great caution because the mixing of resins will make for a low-value product.

2.5.2 Plastics Separation

Depending on the form of collection and the needs of the end market, plastics separation can, in its simplest form, involve removing the plastic bottles and containers and baling these together for later separation at the reclamation facilities. The other end of the spectrum is to separate by both resin and color via manual sorting techniques. In between, all variations exist, such as separating only PET and baling together green and clear materials, or separating out only milk jugs, etc. The separation at the MRF is largely done manually, with high-volume facilities increasingly utilizing air or mechanical devices to separate the light plastic and aluminum fraction from the heavier materials and, in some cases, utilizing eddy current magnets to further separate aluminum from plastic. Some identification techniques presently being looked at are summarized in Table 2.6. These may bring some mechanization to the separation of materials by resin and color, but the expenses involved may make the techniques applicable only at the reclamation facility, not the MRF.

2.5.3 A Working MRF

A typical commingled facility is presented in Figure 2.8, with indications of the process flow. Recyclable materials are delivered to this facility by specially designed two-compartment collection vehicles. This facility also is capable of processing materials that are collected in the form of separated glass, plastics, and cans.

When designing a MRF, careful consideration must be given to equipment layout so as to allow flexibility and expansion capability in the future. The characteristics of the recyclable materials that are processed generally dictate two separate processing lines. As shown in Figure 2.8, one line (A) is configured to process the light fraction of the recyclables, including

Table 2.6   Material identification techniques

Physical Identification Techniques	Electromagnetic Identification Techniques

Density/Specific Gravity/Molecular Weight
Thermal-Mechanical Properties:
    Service Temperature
    Glass Transition Point
    Depolymerization/Decomposition Point
	Volume Resistivity (DC)
	Dielectric Strength (DC through RF)
	Magnetic Properties (DC through RF)
	Far Infra-Red Transmission Spectra
	Near Infra-Red Transmission Spectra
	Visible Spectra
	    Narrow Band
	    Broad Band
	Ultraviolet
	    Transmission
	    Fluorescence
	X-ray
	    Soft X-ray Transmission
	Gamma Ray
	    Fluorescence

newspaper, cardboard, mixed paper, and plastic film. The second line (B) is designed for the separation of various types and sizes of glass, aluminum, steel, and plastic containers.

Newspaper is tipped directly onto the receiving floor. The material is then pushed by a front-end loader onto the feeder conveyor (A1), where it is conveyed up to the elevated sorting platform and picking stations (A2). Manual sorting pulls out containers and drops them down into storage bunkers beneath sorting stations. It is critical that these material storage bunkers (A3) be sized with adequate capacity to store enough material to make a full bale. The concept of separating sorting from baling operations maximizes the availability and efficiency of the baler. The best approach is to implement a negative-sort production concept, that is, to remove contaminants and smaller quantity recoverable materials and let the newspaper accumulate in the final bunker of the sorting line. Physically handling less material reduces labor costs and increases throughput capability of the processing line.

Commingled containers are tipped into a continuous feed floor conveyor, which is fitted with a rubber-lined skirt (B1) on the receiving edge. This skirt acts as a buffer to minimize glass breakage. Materials are conveyed up to the pre-sort station. Depending on the level of contamination of the incoming materials, it may be necessary to provide a pre-sort station before the overhead magnetic separation stage. This would allow for the removal of contaminants such as steel paint cans and plastic bags.

Steel cans are the first material to be separated from the commingled stream. This is accomplished through the use of a suspended magnetic separator (B2). The separator will remove 90–95% of the steel present in the commingled stream. The remaining commingled materials are passed on to a density separator (B3), which splits the materials into a light and heavy fraction. This separator could be an air separation system or a rotating brush.

The heavy fraction as it leaves the density separator is primarily glass containers. This glass is collected on a rotating sorting conveyor (B4) for subsequent manual color separation. The advantage of using the sorting ring is that temporary storage is gained. In terms of operations, it allows sorters to be allocated more effectively since they are not required on the sorting ring until it is at capacity.

The light fraction is passed through another screen, which separates according to size. The resultant materials on the divided sorting conveyor (B5) are large plastic containers on one side and aluminum and small plastics on the other. The divided conveyor allows for negative sorting for the aluminum cans, which can be easily transferred to the large or oversize side of the sorting belt. Manual separation is required for all of the materials except for the aluminum, which is left on the sorting conveyor. The aluminum undergoes a second magnetic sort to assure product

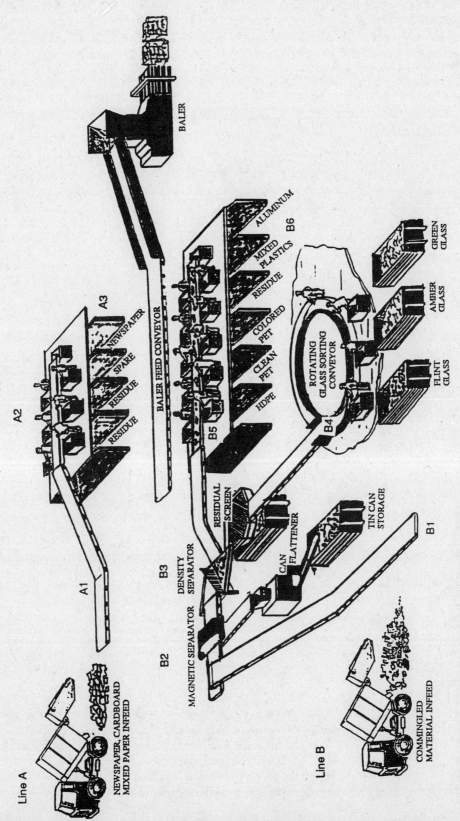

Figure 2.8   100 ton/day commingled materials recovery facility

purity and to remove any remaining steel cans not removed by the suspended magnetic separator.

All materials except for glass containers are stored in the bunkers (B6) located beneath the picking stations on the elevated sorting platforms. When sufficient quantities of material have been accumulated, the bunker is emptied by front-end loader and the material pushed onto the baler feed conveyor. Glass containers are sorted into bunkers beneath the sorting ring or, alternatively, can be dropped directly into roll-off containers for shipping.

2.6 Economics

Table 2.7 provides an estimate of the costs for processing collected curbside plastics at an operating MRF as described above.

Table 2.7   Estimated costs of processing plastics at a material recovery facility

Assumptions:
1.   Bailing/handling costs: Average $21.00/ton ($23.00/t)
2.   Labor costs: 1.5 sorters @ $10.00/h wages and benefits
3.   Productivity: Sorting of one trailer @ 2600 lb (1179 kg) in 5.5 h
4.   Plastics: @ 50% of commingled volume   @ 7% of commingled weight
5.   Fixed processing costs: $7.00/ton ($7.72/t)

Calculations:
$$\frac{5.5\,h}{2{,}600\,lb} = \frac{x}{2{,}000\,lb}, \quad x = 4.2\ h/ton\ (4.7\ h/t)\ \text{sorting time}$$
$15.00/hr total labor $\times$ 4.2 h = $63.00/ton ($69.00/t) labor costs
Facility fixed costs: $7.00/ton $\times$ 50% = $3.50/ton ($3.90/t) $7.00/ton $\times$ 7% = $0.50/ton ($0.55/t)
$21.00/ton($23.00/t) Bailing/handling $63.00/ton($69.00/t) Sorting labor $3.50/ton($3.90/t)   Apportionment of facility fixed costs
$87.50/ton($95.90/t) Plastics processing costs

While much time has been spent on the discussion of the incremental cost to add plastics to curbside recycling programs, this debate existed mainly on programs started and designed without plastics. Since most new programs include plastics from the start, this issue has calmed somewhat. Since the direction is maximum diversion of the waste stream, plastics recycling is a necessity.

Table 2.8 presents a detailed summary of the economics of curbside recycling and especially that associated with plastics. It can be seen from the table that the average cost of collecting curbside recyclables is $75/ton ($83/t) versus the average cost of collecting the plastics fraction being $500/ton ($551/t). Similar large differences are noted in the costs for processing plastics. These high costs are off-set by the higher-than-average material sale price for plastics and, when the landfill cost avoidance of the bulky materials is taken into account, the cost to collect plastics is not out of line with other curbside recyclable materials.

Table 2.8   Curbside recycling economics survey   (based on 50,000 household (HH) program collecting newspaper and commingled containers)

Material	Collection Costs $/ton ($/t)	Processing Costs $/ton ($/t)	Material Sale Price $/ton ($/t)	Net Costs $/ton ($/t)	Gross Collection/ Processing Costs ($/hh/mo)	Collection/ Processing Cost Net of Material Sale Revenue ($/hh/mo)	Landfill Cost Avoidance ($/hh/mo) By vol.	Landfill Cost Avoidance ($/hh/mo) By wt.	Collection/ Processing Costs Net of Material Sale Revenue and Landfill Cost Avoidance ($/hh/mo)
Newspaper	45 (50)	35 (39)	20 (22)	60 (66)	0.96	0.72	(0.33)	(0.36)	0.39
Aluminum	450 (496)	100 (110)	900 (992)	(350) [386]	0.16	(0.11)	(0.02)	(0.01)	(0.13)
Glass	34 (37)	56 (62)	30 (33)	60 (66)	0.50	0.34	(0.06)	(0.17)	0.28
Steel Cans	156 (172)	50 (55)	50 (55)	156 (172)	0.25	0.19	(0.04)	(0.04)	0.15
PET/HDPE Plastic	500 (551)	106 (117)	180 (198)	426 (470)	0.55	0.38	(0.06)	(0.03)	0.32
TOTAL	75 (83) (Average weighted)	46 (51) (Average weighted)	45 (50) (Average market value)	76 (84) (Average weighted)	2.42	1.52	(0.51)	(0.61)	1.01

Data Required for Calculations for Table 2.8

Material	Weight (%)	Volume (%)	Collection lb/HH mo	Collection (kg/HH mo)	Compacted Density lb/yd³	Compacted Density (kg/m³)
Newspaper	60.0	36.0	24.0	(10.8)	725	(486)
Aluminum	1.5	9.0	0.6	(0.3)	250	(148)
Glass	28.0	12.5	11.2	(5.1)	2000	(1186)
Steel Cans	6.0	12.5	2.4	(1.1)	560	(332)
PET/HDPE Plastic	4.5	30.0	1.8	(0.8)	300	(178)
TOTAL	100	100	40.0	(18.1)		

Assumptions and Calculations for Plastics

1. Collection Costs

Assumption: One collection per week of commingled containers and newspapers
Cost (including profit margin) per HH mo: $1.50
Cost per ton: $75.00; Cost per metric tonne (t): $83.00

Calculation:

a. Total collected/month $= \dfrac{\text{no. of HH} \times \text{lb/HH mo}}{\text{lb/ton}}$

$$= \frac{50,000 \times 40}{2000} = 1000 \text{ ton/mo; (905 t/mo)}$$

b. Total cost of collection = total collection/mo × cost/ton
   = 1000 × $75.00 = $75.00 × 10³/mo

c. Plastics collected/month = total collection/mo × wt.%
   = 1000 × 4.5 = 45 ton/mo; (40.8 t/mo)

d. Plastics collection cost $= \dfrac{\text{total cost of collection} \times \text{vol.\%}}{\text{plastics collected/mo}}$

$$= \frac{175.00 \times 10^3 \times 30}{45} = \$500.00/\text{ton; (\$551.00/t)}$$

2. Processing Costs

Assumption: Sorting, handling, volume reduction, baling of newspapers, metals and
           plastic
    Variable costs (i.e., labor, trucking, utilities) $37.50/ton   ($41.33/t)
    Fixed costs (i.e., building and equipment depreciation) $12.50/ton   ($13.77/t)
                            Total cost = $50.00/ton   ($55.10/t)

Operating costs are extremely variable depending upon size and throughput of facility, specific material, and material quantities being processed. Data presented in the table are based on a large-scale processing facility. The costs for processing individual materials are proprietary to Waste Management.

3. Material Sales Price ($/ton)

Price received for separated plastics per ton = $180.00; ($198.00/t)

(continued)

4. Net Costs ($/ton)

Calculation: collection cost/ton + processing cost/ton − plastics sales price/ton

$500 + $106 − $180 = $426/ton; ($470/t)

5. Gross Collection/Processing Costs ($/HH mo)

Calculation: $\dfrac{\text{collection cost/ton + processing cost/ton} \times \text{lb plastics collected/HH mo}}{\text{lb/ton}}$

$\dfrac{\$500 + \$106 \times 1.8}{2000}$ = $0.55/HH mo

6. Collection/Processing Cost Net of Material Sale Revenue ($/HH mo)

Calculation: $\dfrac{\text{Net cost/ton} \times \text{lb plastic collected/HH mo}}{\text{lb/ton}}$

$\dfrac{\$426 \times 1.8}{2000}$ = $0.38/HH mo

7a. Landfill Cost Avoidance/HH mo (Volume Basis)

Assumption: Density of compacted plastic = 300 lb/yd^3; (178 kg/m^3)
Amount of plastic collected/mo = 1.8 lb/HH mo; (0.8 kg/HH mo)
Landfill costs = $10/yd^3; ($13/m^3)

Calculation:
  a. Compacted Plastics $= \dfrac{\text{lb plastic collected/HH mo}}{\text{lb/yd}^3}$

  $= \dfrac{1.8}{300}$ = .006 yd^3/HH mo; (.0045 m^3/HH mo)

  b. Cost Avoidance   = compacted yd^3/HH mo × cost/yd^3
       = .006 × $10 = $0.06/HH mo

7b. Landfill Cost Avoidance/HH mo (Weight Basis)

Assumption: Compacted plastic = .006 yd^3/HH mo; (.0045 m^3/HH mo)
Landfill costs = $30/ton; ($33/t)

Calculation:
Cost Avoidance $= \dfrac{\text{compacted yd}^3\text{/HH mo} \times \text{cost/ton}}{\text{yd}^3\text{/ton}}$

  $= \dfrac{.006 \times \$30}{6.66}$ = $0.027/HH mo

8. Collection/Processing Costs Net of Materials Sale Revenue and Landfill Cost Avoidance ($/HH mo)

Calculation:
    collection/processing costs net plastic sale revenue + landfill cost avoidance

  = $0.38 + ($0.06) = $0.32/HH mo

2.7   Separation at the Reclamation Facility

As of 1991, it appears that most of the color and resin separation will take place at the reclamation facility. This will be the case because, as mentioned in Section 2.5.2, it will probably be too expensive to install mechanical sorting at each individual MRF. Also, to maintain a work force and to further job creation, it is expected that a high percentage of manual sorting will continue to be done at the MRF. Currently, many reclamation facilities accepting multiple bottles and containers also use manual sorting. However, some mechanization has already occurred, and further research and development work on automatic detection and mechanical separation is being conducted by the private sector and at the Center for Plastics Recycling Research (CPRR) [18] at Rutgers University. (See Chapter 4, Section 4.2.1.1.)

The type of systems being looked at use a combination of photo cells and detectors to determine opacity by light transmittance (e.g., milk jugs being translucent and PET beverage bottles being clear). In addition, specific detectors based on the chlorine atom are being developed for use in spotting PVC bottles. Chapter 6 discusses this further. Once detected, air pulses or mechanical gates are used to move the various materials into the appropriate conveyor or hopper for later processing. It is evident that one of the key areas of competitive technology in the reclamation business will be the development of a less labor-intensive separation system.

2.8   Case Studies

Since much has been written on various case studies, to paraphrase them here is not useful. Included in the following pages are two case studies on San Jose, California, and Camden County, New Jersey, excerpted from the Plastic Recycling Foundation's work. Also, Tables 2.9, 2.10, and 2.11 summarize data from different curbside recycling programs and their plastic collection [19].

2.8.1   Case Study 1: San Jose, California

Background: In 1984, San Jose's city council decided to do something about their solid-waste crisis before they were "up against the wall." Bids went out for the city's residential garbage collection and disposal. Waste Management, Inc., won a five-year contract for residential garbage collection. BFI Industries won a 30-year contract for city use of its landfill. The council determined that the bid process and subsequent awards meant substantial savings for the taxpayers: it would have cost the city $60 million above the price of the contract awards to collect and landfill San Jose's waste. Because the collection of recyclables was not part of the original contract with Waste Management, Inc., an additional agreement was made with that company upon its acquiring the Recycle America® subsidiary.

In 1985, the city (pop. 732,000) started a pilot curbside collection program for 20,000 households. Glass, newspaper, and cans were the only recyclables collected at that time. Two neighborhoods were chosen: One was given containers to separate out their recyclables; the other was not given specific containers. While participation rates in both neighborhoods were higher than expected, the neighborhood provided with containers participated more than twice as much—72% versus 35%—than the neighborhood that had to supply its own. Based on these findings, the city decided to provide all households with containers, a move that increased all participation significantly.

San Jose's voluntary curbside program was expanded to 60,000 households in May 1986 and finally went citywide in August 1987. In December 1987, plastic soft drink bottles were added to the curbside collection program.

Procedure: Overall, 58% of the San Jose population participates in the city's curbside recycling program. Each week on "trash day," glass, plastic bottles without their caps, newspapers, and cans are set out at the curb in three stackable 10-gallon containers owned and provided by the city. The plastic bottles and cans are put in one container, glass in a second, and newspapers in the third. The recyclables are collected by a Recycle America® open-body truck with three self-dumping bins. Regular trash pick-up is provided by a different Waste Management truck on the same day.

The recyclables are brought to an intermediate processing facility where plastic bottles and cans are separated and baled. Recycle America® is responsible for seeking the best markets and sales prices for all of the materials collected. Baled plastic bottles are sold to Plastics Recycling Corporation of California.

Although San Jose's curbside recycling program is subsidized by the city, revenues have exceeded their projections. Participation rates continued to grow; between June and July of 1988, the plastic bottle collections increased from 2.6 to 5.65 tons (2.4 to 5.13 t)—the biggest increase since the beginning of the program.

San Jose officials attributed the new participation rates to an increased awareness of the recyclables program. In addition to media coverage, both the city and Waste Management advertised on television promoting the recycling effort. The city's Solid Waste Department also distributed door hangars shaped like plastic bottles to provide individual residents with additional information about the program.

2.8.2 Case Study 2: Camden County, New Jersey

Background: Camden County's curbside recycling pilot program was the first of its kind in the nation. It was developed as a cooperative effort among Camden County's Solid Waste Department, its municipalities, The Center for Plastics Recycling Research (CPRR) at Rutgers University in Piscataway, NJ, and private recycling companies.

This pilot program, initiated and planned by CPRR, was actually a six-month field test to include plastics in a current curbside recycling process. Camden County was chosen for the test because of its previous success in collecting glass, aluminum, steel, and newspaper. The county also was one of the first to have a recycling center. Two county municipalities participating in the program are included in the case study: Berlin Township and Clementon Borough.

Procedure: The collected recyclables were taken to Camden County's MRF, an intermediate processing center, by the municipality and a private hauler. The facility hand-sorted and mechanically processed 60 tons (54 t) of bottles and cans per day for recycling. The plastic bottles, a new addition during the pilot test, were placed in a 40-foot (12-meter) trailer and hauled to the CPRR. The MRF was privately operated by Resource Recovery Systems and Giordana Waste Materials. The processing equipment was owned by Camden County.

Berlin Township: Although Berlin Township is small (pop. 5,700), it was chosen for the program because of its high recycling participation rates. Before the program began, Berlin recycled oil, newspapers, cardboard boxes, glass, and cans. Plastic bottles were added in February 1988 and continued to be recycled for the duration of the six-month pilot program.

Success rates were overwhelming. According to Berlin Township Mayor Paul F. Farms, participation rates were almost 100%. Farms credited the new 20-gallon (76-liter) recycling containers provided by CPRR as the reason for the program's success. Collection rates for all recyclables tripled.

Before the development of Berlin Township's pilot program, each material had to be separated by the resident into different containers. Since then, however, the town has switched to a single-container system in which all recyclables are placed in one bin. Each household now places all of its recyclables (except newspapers) into the special yellow, plastic recycling container on trash day. Newspapers are bundled and placed on top of the bin. Garbage pick-up is followed within an hour or two by the township's Eager Beaver truck, a compartmental-

Table 2.9   Curbside recycling programs—plastics recovery rates

Program	Lb/HH yr					Kg/HH yr				
	PET	HDPE Milk	All HDPE	Other Plastics	Total	PET	HDPE Milk	All HDPE	Other Plastics	Total
Arlington Heights, IL	7.3		16.2		23.5	3.3		7.4		10.7
Champaign, IL	1.3				1.3	0.6				0.6
Charlotte, NC	7.9		0.9		8.8	3.6		0.4		4.0
Cincinnati, OH	3.8	10.1			13.9	1.7	4.6			6.3
Edmonton (N), Alberta	√	√	√	√	5.8	√	√	√	√	2.6
Edmonton (S), Alberta	√	√	√	√	6.5	√	√	√	√	2.9
Gainesville, FL	√	√			12.3	√	√			5.6
Palm Beach, FL	√	√			10.9	√	√			4.9
Palo Alto, CA	0.5				0.5	0.2				0.2
Philadelphia, PA	√	√			11.1	√	√			5.0
Phoenix, AZ	5.5	10.8		17.2	33.5	2.5	4.9		7.8	15.2
Portland, OR		3.3		5.3	8.6		1.5		2.4	3.9
San Francisco, CA	1.3				1.3	0.6				0.6
Seattle (N), WA	1.6				1.6	0.7				0.7
Seattle (S), WA	3				3	1.4				1.4
Somerset Co., NJ	5.1	5.1	5.1	1.6	16.9	2.3	2.3	2.3	0.7	7.6
Toronto, Ontario	0.5		5.8		6.4	0.2		2.6		2.8
West Warwick, RI †	6.6	7.6			14.2	3.0	3.5			6.5
AVERAGES →	3.7	7.4	7.0	8.0		1.7	3.4	3.2	3.6	

* may include materials from noncurbside activity

√ splits unavailable

† West Warwick figures represent the breakdown in quantities at the MRF, which services many other communities; actual data from the original pilot showed similar PET recovery but a higher HDPE recovery of 10.6 lb/HH yr (4.8 kg/HH yr)

Source: Report on Collection Method Profiles and Material Recovery Facilities Assessment, July 1990, DRAFT, Resource Integration Systems Ltd.

Table 2.10   General characteristics of various plastics recycling programs

Program	Households Served (thousands)	Voluntary or Mandatory	Municipal or Private Operator	Collection Frequency	Same day as refuse?	Container Type and Number	Container Size Gal. (Liter)
Arlington Heights, IL	10.6	voluntary	private	weekly	yes	plastic bin	18 (68)
Champaign, IL	15.0	voluntary	private	weekly	no	plastic bin	14 (53)
Charlotte, NC	101.0	voluntary	municipal	weekly	yes	plastic bin	12 (45)
Cincinnati, OH	100.0	voluntary	private	weekly	yes	plastic bin	14 (53)
Edmonton (N), Alberta	61.1	voluntary	nonprofit	weekly	yes	plastic bin & bag	12 (45)
Edmonton (S), Alberta	65.7	voluntary	private	weekly	yes	plastic bin & bag	12 (45)
Gainesville, FL	22.0	voluntary	private	weekly	yes	plastic bin	14 (53)
Palm Beach, FL	181.0	voluntary	municipal	weekly	yes	plastic bin	14 (53)
Palo Alto, CA	20.3	voluntary	private	weekly	yes	2 burlap bags	—
Philadelphia, PA	169.0	mandatory	municipal	weekly	no	bucket	6.5 (25)
Phoenix, AZ	11.0	voluntary	municipal	weekly	no	roll-out cart	90 (340)
Portland, OR	10.0	voluntary	private	monthly	no	two plastic bags	—
San Francisco, CA	70.0	voluntary	private	weekly	yes	plastic bin	12 (45)
Seattle (N), WA	69.0	voluntary	private	weekly	no	3 stacked bins	—
Seattle (S), WA	79.5	voluntary	private	monthly	no	cart	60–90 (227–340)
Somerset Co., NJ	85.0	mandatory	nonprofit	biweekly	yes	household buys bags	—
Toronto, Ontario	143.0	voluntary	municipal	weekly	no	plastic bin	12 (45)
West Warwick, RI	8.9	mandatory	private	weekly	yes	plastic bin	12 (45)
AVERAGE →	67.9						

Source: Report on Collection Method Profiles and Material Recovery Facilities Assessment, July 1990, DRAFT, Resource Integration Systems Ltd.

Table 2.11   Characteristics of plastics collected according to type of collection program

Collection Program	Data Source	Lb material Collected/HH yr (kg/HH yr)					
		Soft Drink Bottles	Milk Jugs	All Beverage Bottles	All Bottles	All Rigid	All Clean Plastics
Mandatory, Drop-off	CPRR Glen Rock, NJ			5.5 (2.5)	8.8 (4.0)		
Voluntary, Curbside, Source separation, Monthly	Seattle, WA* Portland, OR*	3 (1.4)	3.3 (1.5)				
Voluntary, Curbside, Source separation, Biweekly	Minneapolis, MN				18 (8.2)	22–23 (10–10.4)	
Voluntary, Curbside, Source separation, Weekly	Arlington Heights, IL*	7.3 (3.3)					
	Cincinnati, OH*		10.1 (4.6)				
	Gainesville, FL*			12.3 (5.6)			
	CPRR						
	Phoenix, AR*	5.5 (2.5)					
	Palm Beach, FL*		10.8 (4.9)	10.9 (4.9)	22 (10.0)		
	Charlotte, NC	7.9 (3.6)					
Mandatory, Curbside, Source separation, Weekly	CPRR Pitman, NJ			10–13 (4.5–5.9)	26 (11.8)		
	Center Co., PA			10–13 (4.5–5.9)			
Mandatory, Curbside, Commingled, Biweekly	Milwaukee, WI			15 (6.8)	15.5 (7.0)		
	Somerset Co., NJ CPRR						
Mandatory, Curbside, Commingled, Weekly	RI DEM	11 (5.0)	7 (3.2)				
	CPRR						
	W. Warwick, RI*	6.6 (3.0)	7.6 (3.5)		30 (13.6)		36 (16.3)

Note: Figures obtained from municipal programs with high participation rates. Programs providing data include: CPRR, Rutgers University; RI DEM; Minneapolis, MN; Pitman, NJ; Somerset Co., NJ; Glen Rock, NJ; Milwaukee, WI; and Center Co., PA.
* Data from RIS study.
Source: Report on Collection Method Profiles and Material Recovery Facilities Assessment, July 1990, DRAFT, Resource Integration Systems Ltd.

ized flat-bed pulled by a pick-up truck. Owned and operated by the township, the Eager Beaver hauls the recyclables to the Camden County MRF to be sorted and reclaimed.

To promote the curbside recycling program, a letter from Camden County officials was sent to explain the program and identify recyclables. Letters also were sent from Berlin's mayor and the CPRR urging residents to participate and supplying them with more information on the program.

Clementon Borough: Clementon Borough (pop. 5,800) had been recycling for several years. It was one of the first municipalities to collect and recycle mixed bottles and cans. The borough had used a single small container system for mixed recyclables since it began its recycling program. Newspapers were placed in normal garbage bags and placed with the recycling container. Regular trash was picked up once a week, while the recyclables were picked up every other week. Pick-up was on the same day for everything to make it easier for the resident.

Clementon had no investment in equipment; it paid a monthly fee to O'Connor Corp. Division of Waste Management, Inc., to haul both trash and recyclables. The Clementon program, according to General Manager Rob Allen, also was an experiment for the hauler, which collected the mixed plastic, bottles, and cans for the first time.

According to Mayor Richard Wooster, Clementon Borough estimated a savings of 25–40% at the end of the six-month pilot program. The borough had the potential to save $40,000 in landfill costs. The goal was to have a "cost-avoidance" program versus a profit-oriented one.

To promote the pilot test, ads were placed in the local newspaper. Letters from the CPRR and Camden County officials also were sent to residents, explaining the program and what should be recycled.

2.9   Special Collection and Separation Systems

This section examines some highly specialized commercial postconsumer applications of plastics recycling collection. The traditional practice of collecting industrial plant scrap is not included in this discussion.

2.9.1   Agricultural and Other Films and Agricultural Chemical Containers

The market for secondary plastic films in North America began to emerge in the late 1980s. On the West Coast, some MRFs began to separate out films and sort them by resin type for export to the Far East. This sorting was done exclusively by hand. Coincident with the political uprising in China in 1989, this market turned down, and the collection of these films largely stopped. After a hiatus of about a year, interest again seems to be growing in recovering film. Information from various sources indicates that most of these films are being sorted manually by resin type.

Several commercial MRFs under construction in the United States and Canada are intended to separate out both shrink wrap and other types of plastic films from the commercial waste stream. Beyond corrugated paper and steel, plastic films probably make up the single largest identifiable material in the commercial/industrial waste stream (HDPE and PVC are the common resins). Collection of these films is primarily from what are called dump-and-pick operations—hand picking select commercial/industrial solid waste off conveyors.

At this point, no appreciable reclamation capacity for film plastics exists in continental North America. In 1990, Union Carbide announced that a plant in northern New Jersey that would reclaim films was to be on-line by the end of 1991.

Agricultural films made of polyethylene are widely used in North America. If a market emerges for this material, it could easily be collected, baled, and shipped. Most agricultural films are either burned or dumped in pits on existing farm sites. In some states, film is being disposed of in sanitary landfills. In areas where it is being collected for appropriate landfill, the infrastructure already exists to get the material into the recycling stream. With a good end market, agricultural film could be handled like any other commercial or industrial material, either in compactor trucks or in roll-off boxes. The materials would then be baled for shipment to reclamation markets.

A new material that is just beginning to be recycled is agricultural chemical containers. The pressures to recycle these HDPE containers come from the need to appropriately dispose of them. At the present time, the practice is to triple rinse and dispose of the containers in municipal sanitary landfills or on farm sites. Open burning also is widely practiced. The agricultural chemical industry has been sponsoring a project to get these materials back into the recycling stream. Most likely, the collection scenario for agricultural chemical containers will consist of the seller, wholesale warehouser, etc., where the materials are purchased becoming the collection point. The collected containers will be brought to a centralized place for shredding and then shipped for reclamation.

The recycling of polyethylene drums used for chemical raw materials has been ongoing for a number of years. In some cases, they are washed out with surfactant solutions for direct reuse; in other cases, they are shredded and recovered in HDPE cleaning systems.

2.9.2 Plastic Automotive Parts

Several pilot programs are under way by resin producers to recycle plastic automotive parts. The early thinking has addressed very large parts such as plastic bumpers. At this point, the programs have not entered commercial scale, and their future is unclear.

2.9.3 Polycarbonate Water Jugs

General Electric Plastics has experimented with programs to buy back LEXAN® PC resin used in returnable milk and five-gallon water bottle systems. The lightweight, shatter-resistant bottles reduce solid waste in two ways. First, the clear bottles are sterilized and refilled up to 100 times, translating to an eight- to ten-year life cycle during which they do not enter the solid-waste stream. Second, the milk and water bottles are eventually reground and reprocessed into other applications such as automobile and building components, or even the cases in which new bottles are transported. GE is evaluating whether to move forward with this program.

2.9.4 Polystyrene Foam Packaging

The polystyrene foam packaging industry has quite a challenge: to recycle a lightweight, potentially food-contaminated product widely dispersed at a number of end locations. Chapter 5 goes into further detail on the reclamation of these materials.

The prime activities to date include the collection of polystyrene trays and food containers at large institutions where the items are being used. The separation of these items is performed at the point of use, and they are collected in a separate bin for recycling. Strong educational information is targeted at the users in order to acquire a high rate of compliance. Polystyrene containers are then bagged, and moved to a conventional solid-waste compactor at the institution. The handling from this point is similar to normal solid-waste disposal. Polystyrene containers in bags are compacted separately and delivered to the reclamation facility in conventional solid waste equipment.

2.10   Global Activities

Curbside recycling programs in Canada, especially the province of Ontario, are well advanced. The "blue box system" in Ontario was really the forerunner in the United States for residential curbside recycling. At this point, almost 60% of the residents in the province of Ontario have residential curbside recycling, with plastics being introduced in only the last two years and reaching now about one-third of those residents. A further discussion of curbside collection in Canada is given in the next chapter in Section 3.13.2.

Although residential recycling of plastics does not exist on any large scale overseas, there are however, a group of programs in Europe, most notably in northern Italy and in Germany. The programs that exist in northern Italy mostly consist of drop-off operations, which receive few materials and show poor economics [20]. The programs in Germany range from curbside recycling, similar to the United States, to a more advanced system called "wet-dry recycling." In this form of recycling, the household trash is separated into two fractions. The "wet" fraction contains yard waste, food waste, and any wet paper or other wet items, including diapers. The "dry" fraction contains all of the other materials, including not only plastic bottles but all rigid containers, films, paper, and other plastics. This dry fraction is then processed at a technologically sophisticated MRF to separate the various materials. One key here is the difference from the United States in that the paper fraction is mixed with all of the other materials, requiring a more sophisticated separation system. Once the plastics are separated, either by hand or by air classification, they are handled as in the United States. (See the earlier discussion under material recycling facilities.) The main difference in the German programs is that they do go well beyond plastic bottles, since the approach is to try to recycle the entire dry fraction. At this point in time, work on the wet-dry recycling approach in the United States is experimental.

An unique system of collecting and separating PET soft drink bottles for recycling in Australia is described in Chapter 3, Section 3.13.3.

References

[1]   Glenn, J. "Junior—Take Out The Recyclables." *Biocycle* (May/June 1988): pp. 26–31.
[2]   Glenn, J. "New Age Drop-off Programs." *Biocycle* (February 1989): pp. 42–45.
[3]   Bennett, R., University of Toledo. Telephone interview. January 1991.
[4]   Glenn, J. "The State of Garbage in America–Part I" *Biocycle* (March 1990): pp. 48–53.
[5]   Barbero, C., Waste Management of North America. Personal interview. 15 February 1991.
[6]   Reference 4.
[7]   Witheridge, J. G., Waste Management of North America. Personal interview. 15 February 1991.
[8]   Glenn, J. "Containers at Curbside." *Biocycle* (March 1988).
[9]   Foran, B. "1986–87 Oregon/California Residential Curbside Recycling Survey Report." Eugene, OR: Lane County Waste Management Division, August 1987.
[10] Barney, C. R. "A Guide to Residential Curbside Recycling for Waste Management of North America, Inc." Oak Brook, IL: 1987.
[11] *Montgomery Recycles Plastics*. A Cooperative Recycling Project of Montgomery County Government and City of Rockville. The Council for Solid Waste Solutions, 1987.
[12] Pearson, W., Plastic Recycling Foundation, Inc. Telephone interview. 15 March 1991.
[13] Council for Solid Waste Solutions by Frontier Recycling Systems, Inc. & Moore Associates, July 1990.
[14] Reference 5.
[15] Reference 13.
[16] Reference 12.
[17] Glenn, J. "The State of Garbage in America–Part 2." *Biocycle* (April 1990): pp. 34–41.

[18]   Pearson, W., Plastics Recycling Foundation, Inc. Letter to author. 16 March 1991.

[19]   Resource Integration Systems Ltd. *Report on Collection Method Profiles and Material Recovery Facilities Assessment.* DRAFT. July 1990.

[20]   Good, I., Waste Management International, Inc. Telephone interview. 17 February 1991.

3 Polyethylene Terephthalate (PET)

J. Milgrom

Contents

3.1	Introduction	47
3.2	Chemistry of PET	47
	3.2.1 Morphology	48
3.3	The PET Soft Drink Bottle	48
	3.3.1 Characteristics	48
	3.3.2 Markets	49
3.4	PET Film	50
	3.4.1 Characteristics	50
	3.4.2 Markets	50
3.5	PET Recycling Rates	50
	3.5.1 Bottles	50
	3.5.2 Film	51
3.6	Effect of Contaminants on PET Reprocessing	51
3.7	Reprocessing Technologies	51
	3.7.1 Flotation or Hydrocyclone Process	51
	3.7.2 Water Bath/Hydrocyclone Process	54
	3.7.3 Solution/Washing Process	55
	3.7.4 Solvent/Flotation Process	56
	3.7.5 Other Reprocessing Systems	57
3.8	Film	58
3.9	Characteristics of Recycled PET	58
3.10	Depolymerization Technology	59
	3.10.1 Hydrolysis/Methanolysis	59
	3.10.2 Glycolysis	59
3.11	End-Use Markets and Applications	60
	3.11.1 Fiber	61
	3.11.2 Strapping	62
	3.11.3 Sheeting	62

         3.11.4 Alloys and Compounds 63

         3.11.5 Polyols 64

         3.11.6 Bottles and Containers 65

3.12   Economics 66

3.13   Global Activities 68

         3.13.1 Europe 68

         3.13.2 Canada 69

         3.13.3 Australia 69

         3.13.4 Far East 70

References 71

The recycling of polyethylene terephthalate (PET) soft drink bottles began soon after their introduction in 1977 because some states had laws requiring a deposit on all beverage containers. By 1989, the recycling rate had increased to 23%, up from only 10% in 1982. More than 90% of the bottles were collected from deposit states in 1989. As curbside collection proliferates in the 1990s, we expect more bottles to be collected from nondeposit states. Over the past decade, the technology for recycling PET soft drink bottles has been advancing. Though most commercial recycling systems depend on some flotation or hydrocyclone system to separate PET from the high-density polyethylene (HDPE) base-cup resin, alternative systems have been developed. One of the serious contaminants in PET recycling is the adhesive used to attach the base cup and the label to the PET bottle. Today, new technology has minimized this problem and has allowed the recycling industry to produce very pure recycled PET. With the recent introduction of a sophisticated metal detector, recycled PET can be made with insignificant metal contamination. Today, new markets for recycled PET have changed the recyclers' outlook. In the past, markets were a concern; today, supply is the concern. The use of recycled bottle-grade PET to make new bottles for nonfood applications, and now the use of recycled PET to make new soft drink bottles, has opened a huge market for recycled PET. These markets are especially attractive because they are closed-loop recycling—recycling that the public understands.

3.1   Introduction

PET, a polyester, has been known for many years simply as "polyester" and was only known as fiber. The first laboratory samples of this fiber were developed by a small English company in 1941. Polyester research only began in earnest in Europe and the United States after World War II. In the 1950s, this research was based almost entirely on textiles—DuPont's Dacron™ and ICI's Terylene™. In 1962, Goodyear introduced the first polyester tire fabric, and it was only in the late 1960s and early 1970s that polyesters were developed specifically for packaging—film, sheet, coatings, and bottles [1]—although oriented PET film was available in the 1950s.

3.2   Chemistry of PET

PET is a condensation polymer derived from terephthalic acid (TPA) or dimethyl terephthalate (DMT) and ethylene glycol (EG). Polymerization occurs by heating these systems, typically with an antimony catalyst, and removing either water or methanol. The reaction of the acid and glycol is shown in Equation 3.1.

Note that this is a reversible reaction. In the forward direction, polymerization occurs by esterification, and in the reverse direction, depolymerization occurs by hydrolysis. Copolyesters, which are produced commercially to reduce the crystallinity of PET, are made by replacing a portion of the TPA or EG with another dibasic acid or glycol or both.

$$HOOC-\!\!\left\langle\bigcirc\right\rangle\!\!-COO\boxed{-H \quad HO-}CH_2-CH_2-OH \qquad\qquad (3.1)$$

$$\rightleftharpoons\ HO-\!\!\left[\begin{smallmatrix}O\\\|\\C\end{smallmatrix}-\!\!\left\langle\bigcirc\right\rangle\!\!-\begin{smallmatrix}O\\\|\\C\end{smallmatrix}-O-CH_2-CH_2-O\right]_n\!\!-H + H_2O,$$

where n = 100–220.

The basic raw materials for PET production are crude oil and natural gas liquids (ethane, propane, and butane). Para-xylene is derived from crude oil and is oxidized to TPA, which is then purified or esterified to DMT. Ethylene from either crude oil or natural gas liquids is oxidized to ethylene oxide and then hydrated to EG.

The condensation reaction occurs in two steps. First, a low molecular weight precursor is formed, which is then transesterified to form a high molecular weight reactor grade resin. The molecular weight of this polymer, as measured by its intrinsic viscosity (IV), reaches 0.58–0.67 dl/g at this stage. Initially, this was the only grade that was made, and it was used to produce Dacron™ and Terylene™. Today, this grade also is used for oriented film and glass-fiber reinforced PET.

In the 1970s, the industry learned that they would need a higher molecular weight PET to serve the container market—polymers with an IV > 0.7 dl/g. This high molecular weight polymer was made by further polymerizing the reactor grade polyester outside the melt phase reaction by a process called "solid stating." Different end uses require different molecular weight PET. The higher the molecular weight needed, the longer the solid-stating time required. Solid-stated PET is used mostly for bottles, but it also is needed for dual ovenable trays, where it is coated onto ovenable board. Strapping and nonoriented film also require solid-stated PET [2].

3.2.1 Morphology

PET is a linear molecule that exists either in an amorphous or a crystalline state. In the crystalline state, the molecules are highly organized and form crystallites, which are crystalline regions that extend no more than a few hundred angstrom units. The maximum crystallinity level that can be achieved is probably no more than 55%. The crystallinity in the PET soft drink bottle is normally about 25%. PET produced by solid stating comes from the reactor in crystalline form. It is in this form when it is shipped to the fabricator. Molecules of either amorphous or crystalline polymers can be uniaxially or biaxially oriented. In either case, orientation greatly increases the strength of PET, because strain-induced orientation usually imparts some crystallinity [3].

Because the crystalline state is the normal state for PET, the preparation of amorphous PET is deliberate. Amorphous PET is prepared by rapidly cooling the molten resin from a melt temperature of 260 °C to a temperature below the transition temperature of 73 °C. On the other hand, slow cooling of the molten resin will produce a crystalline polymer. A recycler of PET who produces pellets by extrusion will normally produce crystalline polymer. It is important to do so because pellets of crystalline PET are preferred by the processor, who normally dries the recycled PET before using it. Amorphous resin tends to soften and stick at the elevated temperatures of drying, forming clumps and adhering to the walls of the drying unit.

The crystallization rate of PET is of great concern in processing. Crystallinity greatly affects product clarity and processibility. However, if the size of the crystallite is small enough to minimize light scattering, clarity can be achieved in spite of the crystallinity of the polymer. The resin industry has learned how to control the crystallization rate and, therefore, the processing of PET by using small amounts of co-monomers. Many bottle-grade PET resins used to make bottles today contain a small percentage of a second glycol monomer. For example, high-speed blow molding equipment users frequently prefer these polymers.

3.3 The PET Soft Drink Bottle

3.3.1 Characteristics

The PET soft drink bottle is available either as a two-piece container with an HDPE base cup or as a base-free one-piece container. In 1990, the split was about 60%:40%, respectively. Over the years, the industry has made steady progress in reducing the weight of the PET used in these bottles. Today, as shown in Table 3.1, the average two-piece, two-liter bottle weighs 68 g, and the one-piece version weighs 56 g. Because a stiffer PET resin is needed for the one-piece bottle, a higher IV resin (0.80–0.84 dl/g) is used to blow mold these bottles; the two-piece bottle uses a PET resin with an IV of 0.71–0.74 dl/g. Recyclers have to consider these differences in bottle specifications in their economics and in designing a recycling line.

Table 3.1   1990 PET soft drink bottle specifications

Plastic	2-Piece Bottle 2-liter 16-ounce (0.5 liter)		1-Piece Bottle 2-liter 16-ounce (0.5 liter)	
PET (g)	47–50	23–24	54–58	27–29
HDPE (g)	19.5	5.5	—	—
TOTAL	66.5–69.5	28.5–29.5	54–58	27–29
HDPE (%)	29	19		

Source: Industry contacts

Another characteristic of the PET soft drink bottle that has to be considered in designing a recycling line is the nature of the bottle closure. Historically, the PET bottle only came with an aluminum roll-on closure (1.5 g), but these closures are gradually being replaced by polypropylene (PP) ones. In 1990, the plastic closure had almost 45% of the beverage container market. However, the use of a plastic or a metal closure varies widely depending on the bottler and the bottler's location. Today, the liner in the closure is made from ethylene vinyl acetate (EVA) copolymer.

All PET soft drink bottles use labels. Historically, only paper labels were used. Today, a paper/plastic label is very popular, and some bottlers use all-plastic labels. The nature of the label material is not too significant for the recycler. Labels, whatever the material, are effectively removed by all recycling systems. However, paper labels can be troublesome if they are finely shredded because the shreds are not easily removed and can contaminate the recycled PET.

3.3.2 Markets

The market for solid-stated PET has been growing at a healthy rate since the introduction of the PET plastic beverage container in 1977. Initially, there were the one- and two-liter sizes; in 1981, a one-half-liter size became available, which is now a 16-oz size. In 1988, as much as 721 million pounds (324×10^3 t) of PET were used to make these containers [4]. As shown in Table 3.2, the two-liter bottle is dominant. In 1989, because of the rapid run-up in PET resin prices, 705 million pounds (320×10^3 t) of PET were used to make soft drink bottles. However, this volume was up again in 1990 to 754 million pounds (342×10^3 t).

PET is also used to make bottles for liquor, food, toiletries, cosmetics, and medicinal and health products. This market has been growing faster than the soft drink bottle market. In 1990, about 335 million pounds (152×10^3 t) of PET went into this market—up from 212 million pounds (96×10^3 t) in 1988 [5].

Table 3.2   1988 PET container market

Soft Drinks	Units billion	PET Demand million pounds	(thousand tonnes)
16-ounce	1.87	104	(47)
1-liter	0.71	62	(28)
2-liter	4.34	501	(227)
3-liter	0.32	54	(25)
Subtotal	7.24	721	(327)
Non-soft drink	2.32	212	(96)
TOTAL	9.56	933	(423)

Source: Eastman Chemical Products Inc./Plastics Division

3.4   PET Film

3.4.1   Characteristics

PET film does not require the use of solid-stated resin. Typically, the IV of PET used for this application varies from 0.6–0.65 dl/g. The film is available in a variety of grades and is used in packaging, such as photographic and X-ray film, and as magnetic recording film, including video and computer tapes. Recycling of photographic and X-ray film has prevailed for many years because of the silver content. This PET film is a composite structure containing gelatin, silver, and a coating of polyvinylidene chloride (PVDC).

3.4.2   Markets

The PET film market consumed about 600 million pounds of PET resin in the United States in 1989, but the U.S. market is larger than this. PET film is imported into the United States from Japan, South Korea, and Taiwan. This U.S. market is growing at 8% per year. Two of the largest U.S. producers are Eastman Kodak and 3M, but together they captively consume their 200-million-pound-per-year (90×10^3 t) output [6]. In 1990 about 85 million pounds (38.5×10^3 t) of PET were used in magnetic recording film [7].

3.5   PET Recycling Rates

3.5.1   Bottles

Recycling of PET soft drink bottles collected from the public began soon after their introduction because some states had laws requiring a deposit on all beverage containers. In 1984, when perhaps 98% of the PET bottles were collected from the nine states that had deposit laws—Oregon, Vermont, Maine, Michigan, Connecticut, Massachusetts, Iowa, Delaware, and New York—an estimated 100 million pounds (45.5×10^3 t) of postconsumer PET was collected and recycled. This corresponded to a recovery rate of 18%, up from 10% in 1982 and 5% in 1979 [8].

In 1989, when PET bottles began to be retrieved from some curbside collection systems, the recovery rate increased to 23%, up from 21% in 1988. More than 90% of these bottles were collected from the deposit states. Of the reported 190 million pounds (86×10^3 t) of PET soft drink bottles recycled in 1989, an estimated 162 million pounds (73×10^3 t) was PET, and some 28 million pounds (13×10^3 t) was HDPE derived from the base cups [9]. In 1989, some 12–15 million pounds (5.4–6.8×10^3 t) of PET soft drink bottles were also imported into the United

States from Canada. This is in addition to the 190 million pounds (86×10^3 t) recovered in the United States. Most of these bottles were one-piece bottles.

3.5.2   Film

Estimating the recycling rate of PET film is more difficult because of imports. According to our best estimates, some 70 million pounds per year of X-ray film is recovered for silver removal. This accounts for about 10% of the total PET film consumed in the United States.

3.6   Effect of Contaminants on PET Reprocessing

A major concern during the reprocessing of PET is to remove all contaminants that can catalyze the hydrolysis of PET (Equation 3.1 in the reverse direction). Needless to say, the reprocessor must also avoid adding such cleaning agents as caustic soda or an alkaline detergent-containing caustic soda in the wash step. These compounds are sometimes used to aid in the removal of labels. They help dissolve or disperse the adhesives used both with the labels and the HDPE base cups that are used on about 60% of the PET bottles currently. Though recyclers may subsequently wash the PET with hot water, the alkaline compounds are frequently entrapped in the PET granules. Consequently, the recycler may produce a clean PET with a minimum degradation of the polymer chain, but when next extruded, the dried PET may degrade substantially in molecular weight due to the presence of the catalyst.

Hydrolysis catalysts are acids or bases, which promote hydrolysis at elevated temperatures but below 205 °C. Once hydrolysis occurs, the reaction is autocatalytic. Degradation of the polymer chain leads to the formation of low molecular weight polymers with carboxylic acid end groups. These acids further catalyze hydrolysis.

Simply heating PET at elevated temperatures in the absence of moisture can lead to the formation of carboxylic acid end groups, as shown in Equation 3.2.

$$\sim\!\!\!\!\!\sim\!\!\bigcirc\!\!-\!\overset{\overset{\displaystyle O}{\|}}{C}\!-\!OCH_2CH_2OH \xrightarrow{\ \Delta\ } \sim\!\!\!\!\!\sim\!\!\bigcirc\!\!-\!\overset{\overset{\displaystyle O}{\|}}{C}\!-\!OH + CH_3CHO \qquad (3.2)$$

This reaction converts a benign hydroxyl end group into a carboxylic acid end group that can catalyze hydrolysis.

Taking this chemistry into account, the industry has recognized that extruding virgin PET that has been adequately dried (a moisture content of 0.005% or less) still will cause a modest reduction of the molecular weight of the PET, as measured by the IV—a drop of 0.02–0.03 units. Each time virgin PET is extruded, a similar drop in IV occurs.

The most troublesome contaminants present in PET bottles are the adhesives used for the labels and the base cup. Most commercial washing systems used to recover PET from bottles inevitably leave traces of adhesives in the PET. Often, adhesive residues are trapped in the PET granules and remain there after washing. These are typically hot-melt adhesives that normally contain rosin acids and esters, rosin, EVA, and some elastomer. Unfortunately, the rosin acids and the acetic acid from the hydrolysis of the EVA can also catalyze the hydrolysis of PET, as noted above. Furthermore, because these adhesives darken when treated at PET extrusion temperatures, the recycled PET becomes discolored and hazy.

3.7   Reprocessing Technologies

3.7.1   Flotation or Hydrocyclone Process

Today, most PET is reclaimed by a system that washes the PET and HDPE and separates them by their differences in density (see Figure 3.1). These systems are fed with crushed, baled bottles with and without caps. First, the bales are broken, normally with a good mechanical bale breaker. Then, if the bales consist of both green and colorless bottles, the bottles are color sorted by hand. Occasionally, the reclaimer can purchase color-sorted bales of PET bottles (usually at a higher cost) and avoid the trouble and cost of color sorting. Sometime in 1991, some recyclers may begin mechanical color sorting using photo cells. The industry has been cautious in moving to mechanical color sorting because it takes only one colored bottle to ruin a batch of colorless PET.

---

1. Sortation of dirty baled bottles—by color, by polymer type

2. Shredding

3. Granulation

4. Air classification   Labels↑

5. Washing/rinsing

6. Flotation or hydrycyclone   HDPE↑   PET + Aluminum↓

7. Dewatering/drying

8. Electrostatiic separator/metal detector   Aluminum↑

    Product: High purity PET flake and moderate purity HDPE
               (contaminated with PP, EVA, and adhesive)

Figure 3.1   PET reprocessing system; flotation or hydrocyclone process

As the PET bottle recycling industry begins to recover PET bottles from curbside collection, the industry is facing the greater challenge of hand sorting plastic bottles not only by color but by polymer type. Bales of unsorted plastic bottles from curbside collection contain not only bottles made from PET but bottles made from HDPE and other resins such as polyvinyl chloride (PVC), which can be especially troublesome. As little as 1 ppm of PVC in PET can discolor PET. Separation of PVC from PET bottles is possible with some commercial and near-commercial systems but, to date, none can reduce contamination to this limit.

The dirty, sorted bottles are first reduced to 0.125–0.375 in (3.2–9.5 mm) flake by being processed through a granulator. Some recyclers first shred and then granulate, which increases the production capacity of the granulator. In this size reduction process, most label material is freed from the plastic; it is subsequently removed by air classification. In most processes, this is simply carried out on a vibrating screen that is part of an air table. The flow of low pressure air removes the loose dirt and most label material. Some recyclers use granulators that also blow away most of the loose label material.

The contaminated flake is then metered into an agitated washing tank along with a hot nonfoaming detergent solution. Some recyclers may use more than one washing tank. Washing is a very critical step. All recyclers have their own detergent recipes, a preferred solids concentration in the slurry, and a preferred temperature and wash cycle. As noted in Section 3.6, the use of caustic soda in the wash solution is not recommended. Nevertheless, two of the largest PET bottle recyclers—Wellman (Johnsonville, SC) and Nicon (Long Island City, NY)—use a flotation process, add caustic soda to the wash solution to reduce adhesive contamination, and produce a clean recycled PET. However, because these recyclers primarily serve the PET fiber market, some drop in the IV of the PET due to hydrolysis can be tolerated. On the other hand, recyclers who use this process to produce recycled PET for markets that require higher IV (see below) avoid adding caustic soda to the wash. These recyclers include Day Products (Bridgeport, NJ), St. Jude Polymer (Frackville, PA), and wTe Recycling/Star (Albany, NY).

The washing step removes the last traces of label material and disperses and sometimes dissolves the adhesives. These solid contaminants are then typically removed from the wash solution by screening the coarser crude polymer flake from polymer fines, dirt, and the finer fibrous label wastes. The wash solution is filtered and reused, and the sludge remaining on the filter is discarded.

Next, the crude polymer flake (PET, HDPE, and aluminum) is thoroughly rinsed with fresh water to remove residual wash solution, label and other materials, and the rinse water is pumped through a filter to remove solids.

Now cleaned, the crude flake or chip moves either into a flotation tank or a hydrocyclone that separates the heavy PET and aluminum from light HDPE in a water medium. HDPE floats in water while PET and aluminum sink. EVA, if present from the cap liner, stays with the HDPE. Typically, the flotation tank is equipped with a paddle wheel to improve the separation. A low-foam detergent is added to the water in the tank to minimize the capillary action that carries some lights to the bottom of the tank and some heavies to the top. Hydrocyclones are used in the newer recycling systems; it is claimed that they improve the separation of HDPE from PET. The hydrocyclone simply accentuates the action of a sink-float tank. It is a centrifuge device with a greater gravity force. Its effectiveness depends on the concentration of the solids and the speed of the centrifuge. Day Products and Johnson Controls (Novi, MI), which operate new PET recycling facilities, and Polymer Resource Group of Baltimore, which is constructing a new plant in Maryland to recycle PET bottles, use hydrocyclone technology.

The "heavy" and "light" product streams from the tank or the hydrocyclone are typically flushed once more with fresh water and processed first through spin dryers and then through hot-air dryers. The polyethylene (PE) from these processes is frequently contaminated with adhesive and some label materials. Depending upon the HDPE market, some recyclers may simply sell this lower quality HDPE. Alternatively, it may be further washed or simply extruded through screen packs to remove solid contaminants and sold as pellets.

The dried heavy fraction containing PET and aluminum chips is finally fed into an electrostatic separator, which is a multistage device. Here, a thin layer of PET and aluminum chips are fed onto a series of rotating drums. While on the drums, the chips are exposed to high voltage electricity. The plastic holds the charge because of its low conductivity, while the aluminum chips rapidly discharge. Therefore, the PET adheres to the drum until the chips are knocked off by a rotating fiber brush and the aluminum chips are thrown by the rotating drum into a chute. This process is repeated when the plastic chips drop to the second rotating drum (a four-stage unit is preferred). In this way, recycled PET flake can be freed of aluminum. Typically, the aluminum content of the resulting flake is 25–100 ppm. Most of the remaining aluminum can be removed during extrusion through screen packs if the PET is sold as pellets. The aluminum, thus removed, is contaminated with PET and can be sold as such or further purified by passing it through a two-stage electrostatic separator or an air-classification device.

In 1990, Carpco Inc. (Jacksonville, FL), a manufacturer of equipment for the electrostatic separation of metals from plastics, introduced a new metal detector/separator that further reduces the aluminum content of PET flake from 25–100 ppm to less than 5 ppm with less than 2% weight loss. This equipment detects metal from a flowing "sheet" of PET flake on a conveyor. When a metallic particle is detected, a jet of air propels it out of the flowing sheet (see Figure 3.2). Some recyclers may have already purchased this equipment in 1991.

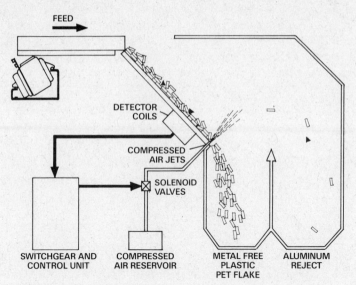

Figure 3.2   ELPAC metal detector/separator (Courtesy of Carpco Inc.)

An interesting variation on the flotation or hydrocyclone process is the addition of a step that granulates or grinds the PET bottles cryogenically. Because adhesive contaminants are embrittled at cryogenic temperatures, whereas PET is not, adhesive contaminants in a cryogenic process become a fine powder. The fines are then easily removed from the coarser PET flake by screening. This process can produce a very clean recycled PET free of adhesives and is used commercially by Western Environmental Plastics (Lewisville, TX) in a facility with an annual capacity of about 6 million pounds (2.7×10^3 t). Secondary Polymers Ltd. (Detroit, MI) uses a similar process, but in early 1991 it was still at the pilot-plant stage. Air Products and Chemicals Co. (Allentown, PA), a liquid nitrogen supplier, has been actively investigating the advantage of cryogenic grinding as a means of separating and purifying various materials over the past 20 years.

3.7.2   Water Bath/Hydrocyclone Process

The process shown in Figure 3.3 was developed by Reko, a division of DSM in Holland, and is being practiced by Johnson Controls under a license from Reko. This process operates either with PET bottles that have plastic caps or with cap-free bottles. Because Johnson Controls retrieves the PET bottles from its bottler customers in deposit states, they can insure that these bottles have no aluminum caps.

In this process, bottle components are substantially separated before granulation. Color-sorted crushed bottles from the bale move continuously through a hot water bath (1–1.5 min) that is at least 70 °C and close to 100 °C. At these temperatures, the PET bottles, which are blow-molded by a process that orients the PET, shrink. As a result, the HDPE base cups, labels, and many of the PP caps, which do not shrink, separate from the PET bottles.

From the immersion tank, the separated components are deposited on a vibrating screen that removes the detached labels. The remaining bottles and base cups then pass through another screen with openings of a diameter that separates cups from bottles. Next, the bottles and base cups are granulated separately and washed. After washing and rinsing, the PET flake in a water medium moves through a hydrocyclone that removes any residual PE and adhesives (i.e., "lights"). Finally, the clean and dried recycled PET passes through a metal detector to insure the absence of aluminum.

```
    1.   Sortation of dirty baled bottles without metal caps

    2.   Hot-water bath (1.0–1.5 min)

    3.   Vibrating screen        Base cups↑   Labels↑  Plastic closures↑

    4.   Granulation

    5.   Washing/rinsing

    6.   Hydrycyclone

    7.   Dewatering/drying

    8.   Metal detector

         Product:   High purity PET flake
                    HDPE purity depends upon operation
```

Figure 3.3   PET reprocessing system; water bath/hydrocyclone process

3.7.3 Solution/Washing Process

This process resembles the continuous water bath process (see Figure 3.4), but the solution process was developed first. In the early 1970s, this process was used to de-ink plastic bottles and was later modified to recycle PET bottles. It was developed by Pure Tech International and is a batch rather than a continuous process.

```
    1.   Sortation by color

    2.   Hot Solution (Proprietary)
         Wash (30 min)        Base cups↑   Caps↑      Labels↑

    3.   Granulation

         Product:   High purity PET flake (minimal adhesive)
                    HDPE purity HDPE (some adhesive)
```

Figure 3.4   PET reprocessing system; solution/washing process

Color-sorted crushed bottles are first thoroughly washed in a hot solution containing a proprietary noncaustic cleaning agent in what is essentially a commercial laundry washing machine. Bottles free of aluminum caps are preferred. If aluminum is present, a metal detector or an electrostatic separator can be installed at the end of the recycling line to remove the metal. The wash and rinse cycle takes 30 minutes and processes about 500 pounds (227 kg) of bottles per machine. In this step, labels and base cups come loose from the bottles. The base cups are removed mechanically, and the label materials are removed by filtration. In this process, adhesives are also separated from the bottles and subsequently removed when the cleaning solution is filtered. The Pure Tech process is unique in its ability to disperse not only adhesives but PVDC coatings.

The clean HDPE and PET thus prepared are granulated separately into flake. The wash solution can always be reused after filtration as long as the concentration of the cleaning agent is maintained and the washing solution is free of contaminants. A full-scale, 15-million-pound-per-year (6.8×10^3 t) plant using this process is in operation in Taiwan, and a U.S. plant is in the planning stage.

3.7.4 Solvent/Flotation Process

This system, shown in Figure 3.5, was developed by Dow Chemical. The process begins like the conventional flotation process shown in Figure 3.1, but it is followed by a series of float/sink steps using chlorinated solvents. After the water-flotation step that separates the PE and some labels from the PET flake, the "heavies" move first through a float/sink step with 1,1,1-trichloroethane as the solvent and then through another float/sink step using a mixture of perchloroethylene and trichloroethane. The trichloroethane dissolves the adhesive and floats any remaining HDPE and label materials, whereas the last float/sink step separates clean PET from aluminum. Finally, the solvents are removed and recovered in a closed distillation system and the adhesive-free PET is dried.

Dow, together with Domtar of Canada, had planned to use this process in a joint venture that would recycle both PET and HDPE bottles. Unfortunately, the economics of the Dow process are unfavorable compared to the systems described above, and the joint venture was abandoned in 1990.

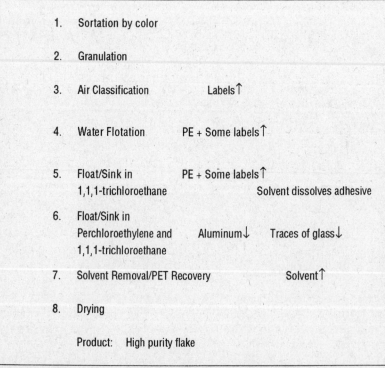

Figure 3.5 PET reprocessing system; solvent/flotation process

3.7.5 Other Reprocessing Systems

The recycling systems described above were designed to process PET bottles at the rate of 10–40 million pounds ($4.5–18 \times 10^3$ t) per year. However, some believe that there is a market for systems that can economically process as little as 3–5 million pounds ($1.4–2.3 \times 10^3$ t) per year of PET bottles. These systems process only unbaled, whole, and modestly crushed PET bottles. Such systems may be appropriate for bottlers who collect uncrushed PET bottles from the supermarkets in deposit states, and perhaps they may be of interest to some material recovery facilities (MRFs) that can collect uncrushed bottles from curbside systems.

Recycled Polymers (Madison Heights, MI) developed and patented an inexpensive PET recycling process in 1990 that relies on uncrushed PET bottles as the feed. The equipment is simply an old soft drink filler that washes the interior of used bottles with hot water and a nonfoaming detergent. Prior to washing, the noncrushed bottles are processed through a mechanical de-capper that removes the caps, if present. As a result of the hot-water wash, base cups and labels are loosened. These are completely removed in the next stage, when the bottles are subject to a water spray. Most of the adhesive remains with the base cups and labels, while the adhesive remaining on the PET bottle is removed during spraying. The cleaned PET and HDPE from the base cups are then granulated separately and dried. We would expect this process to produce a recycled PET that is somewhat contaminated with adhesive. This process is only at the pilot plant stage [10].

Still another system that relies on unbaled bottles is that developed by Automated Recycling Corp. (Sarasota, FL) (Figure 3.6). This system is fed by a line that supports the bottles at the neck ring. Few baled bottles, however, have developed necks, so baled bottles cannot be processed. In this system, caps and tamper-evident rings are removed first as the containers pass through a proprietary rotary table. Base cups, labels, and adhesive are removed in the next step, which is a combination of aeration and high velocity hot water jets that scrub the bottles inside and outside. The cleaned bottles, free of contaminants, are drained and dried before granulation. This system is to be installed by a Pepsi-Cola bottler in Long Island City, New York, in

Figure 3.6   The Automated Recycling Corp. ARC 7220 machine (Courtesy of Automated Recyling Corp.)

1991 to process 5–6 million pounds ($2.3-2.7 \times 10^3$ t) per year of PET bottles. The installation will be operated as a joint venture of Automatic Recycling Corp. and the Pepsi-Cola bottler [11].

3.8   Film

PET X-ray and lithographic film is the only significant PET film that is recovered and recycled today. The main interest in these films is in their silver. The most common desilvering process uses enzymes to destroy the gelation on the film and thus liberate the silver. The use of cyanides to complex the silver is an alternative process.

Several companies desilver film, the major ones being DuPont, Eastman Kodak, Gemark (Newburgh, NY), and Chicagoland (Elk Grove, IL). The granulated PET film that remains after desilvering is coated with PVDC. Some of this film scrap is sold to companies that glycolyze the PET to produce polyols (Section 3.10.2). Eastman uses a methanolysis process that regenerates the monomers from the PET scrap film. At this time, only Gemark has a proprietary process for removing PVDC from the PET film. The PVDC-free film is sold to others who can use PET. Some of the smaller companies that desilver simply discard the desilvered film into landfill.

3.9   Characteristics of Recycled PET

A purchaser of recycled PET has to be concerned with its IV, its aluminum content, and its color. Typical characteristics of recycled PET bottle flake appear in Table 3.3. Generally, contamination with paper fibers from labels and HDPE from base cups is minimal. Because HDPE is incompatible with PET, it can produce a hazy PET. Occasionally, depending on the ink used on the printed labels, inks will leach out of the labels in the hot-washing step and produce a faintly colored recycled PET resin.

Table 3.3   Typical characteristics of recycled PET bottle flake

Company	IV (dl/g)	Aluminum (ppm)	Other (ppm)
Pure Tech	0.74	Nil	<2 (glue undetectable)
Nicon	0.65–0.68	25	15–20
Day Products	0.72–0.74	25–125	225 (mostly PP)

Source: Industry contacts

The most troublesome contaminant in recycled PET is the adhesive, as discussed in Section 3.6. Its presence can be noted in a number of ways. An adhesive will produce a hazy PET— one that will discolor on heating, usually producing a tan color. When PET contaminated with adhesive is heated in an extruder, it will degrade and produce a lower IV resin as well as cause discoloration. Similarly, caustic soda, if used in the washing step, could contaminate the PET and cause it to degrade during extrusion.

Acetaldehyde is another degradation product in recycled or virgin PET, as illustrated in Equation 3.2. Because of its volatility, acetaldehyde is easily removed from PET. The presence of acetaldehyde is not a problem in recycled PET because, until recently, recycled PET has not been considered for direct contact with food. Acetaldehyde affects taste. If the resin producers make PET resin from depolymerized recycled PET, they should be able to control its acetaldehyde content in much the same way as they control it in virgin PET.

With the new advances in aluminum separation and with more plastic closures being used, most recyclers in the future will be able to produce not only pelletized recycled PET with negligible aluminum contamination but also recycled PET flake with 5 ppm or less aluminum contamination.

3.10 Depolymerization Technology

As noted in Equation 3.1, the condensation of EG with TPA is a reversible reaction. In terms of recycling, this equation is significant not only because PET recyclers have to insure that they do not hydrolyze the PET during recycling but because companies are now also interested in depolymerizing PET back to its monomers or oligomers in order to repolymerize them into a food-grade recycled PET. Companies which carry out the reversible reaction with glycols (glycolysis) instead of water can produce aromatic polyols that can be further treated with isocyanates or unsaturated dibasic acids to produce polyurethanes (PURs) or unsaturated polyesters, respectively. Goodyear apparently uses glycolysis to produce its "Repete" resin. The company is seeking FDA approval for Repete's use in food-contact applications.

3.10.1 Hydrolysis/Methanolysis

Treating PET with water in excess at an elevated temperature of 150–250 °C in the presence of sodium acetate as catalyst produces TPA and EG in four hours. Catalysts for hydrolysis are either acids (such as sulfuric) or bases (such as ammonium hydroxide) [12]. An acid catalyst will promote the hydrolysis in 10–30 minutes at 60–95 °C [13].

Alternately, PET can be treated with an excess of methanol, as shown in Equation 3.3, to produce DMT and EG. A typical PET/methanol ratio is 1:4.

$$\text{PET} + \text{CH}_3\text{OH} \xrightarrow[\Delta]{\text{CAT.}} \text{CH}_3\text{O}-\overset{\overset{\text{O}}{\|}}{\text{C}}-\underset{}{\bigcirc}-\overset{\overset{\text{O}}{\|}}{\text{C}}-\text{OCH}_3 + \text{HO}-\text{CH}_2-\text{CH}_2-\text{OH} \qquad (3.3)$$

In a typical methanolysis process, molten PET is mixed with methanol and, if desired, a catalyst, and heated at 160–240 °C for less than an hour under a pressure of 20–70 atm. A 99% yield of monomers is claimed [14]. Hoechst Celanese uses a methanolysis process for depolymerizing recycled PET to make its repolymerized PET for food-contact applications. Eastman Chemical and DuPont are now building methanolysis pilot plants to supply recycled PET pure enough for food-contact applications.

To prepare high-quality monomers from recycled PET, contaminant-free recycled PET is preferred in flake form. The methanolysis process is more tolerant of contaminants than is the glycolysis process described in 3.10.2. For example, the methanolysis process can tolerate green recycled PET to produce colorless food-grade recycled PET, whereas the glycolysis process can use only colorless recycled PET to produce colorless food-grade recycled PET. Recognizing that the use of clean recycled PET will increase costs compared to making PET from virgin monomers, Eastman Chemical has launched an extensive research effort to develop a methanolysis process that can use partially cleaned, postconsumer PET as its raw material for producing a food-grade recycled PET [15]. DuPont has claimed that its technology is more tolerant of potential contaminants in the PET feedstock than other methanolysis processes [16].

3.10.2 Glycolysis

If PET or recycled PET is treated not with an excess of water or methanol but with an excess of a glycol, a transesterification reaction takes place. The reduction of high molecular weight

PET to short-chain fragments is achieved by heating the PET with a glycol such as propylene glycol (PG) in the presence of a catalyst. Typical catalysts are amines, alkoxides, or metal salts of acetic acid. During this glycolysis reaction, some of the free PG replaces EG in the polyester chain by a process of chain scission and glycol exchange. Eventually, the polymer is reduced to hydroxy-terminated short-chain fragments. With PG as the free glycol for glycolysis, the resulting major products are bis-hydroxyethyl terephthalate, bis-hydroxypropyl terephthalate, and mixed EG/PG terephthalate diesters plus some free EG and PG. Typically, this glycolysis reaction takes place over an 8-hour period at 200 °C (reflux temperature) with a PG/PET ratio of 1.5:1. The reaction is carried out under a continuous nitrogen purge to inhibit degradation of the resulting polyols [17].

Under these reaction conditions, the resulting polyol has a number average molecular weight of 480 and a hydroxyl number of 480. These polyols can be subsequently treated with unsaturated dibasic acids or anhydrides to produce unsaturated polyesters (Section 3.11.5). If a higher molecular weight polyol is desired, the PG/PET ratio is lowered—less PG is used per mole of PET.

The glycolysis reaction also proceeds with glycerol, which produces a polyol with a higher hydroxyl number, or diethylene or dipropylene glycol. Polyols made with diethylene glycol (DEG) can produce PURs with improved properties as discussed below. Glycolysis with DEG to produce polyols for PURs typically use a DEG/PET ratio of 1:3. These polyols are made using manganese acetate catalyst at 205–220 °C under a nitrogen purge. Glycolysis is complete within 3.5 hours. The resulting polyol has a number average molecular weight of 556 and a hydroxyl number of 202 [18].

The major source of PET for glycolysis reactions to produce PURs is PET film wastes generated by the desilvering industry and film manufacturers. Other important sources of PET are postconsumer green soft drink PET bottles and industrial waste.

3.11 End-Use Markets and Applications

Considering that an estimated 215 million pounds (98×10^3 t) of PET soft drink bottles were recovered in 1990 in the United States and that about 183 million pounds (83×10^3 t) of PET derived from these bottles were recycled (Section 3.5.1), some 165 million pounds (75×10^3 t) of recycled PET flake and pellets were sold in the U.S. market that year. This lower figure is based on an average yield of 90% in the recycling process due to losses of material during processing. As shown in Table 3.4, almost 80% of the recycled PET produced in 1990 from

Table 3.4   Recycled PET markets[a] (1990)

Product	Million Pounds	(Thousand Tonnes)
Fiber	129	(58.5)
Strapping	12	(5.4)
Alloys and Compounds	11	(5.0)
Polyols	10	(4.5)
Sheet	2	(0.9)
Bottle	1	(0.5)
TOTAL	165	(74.8)

[a] Postconsumer PET bottles recovered and recycled in the United States; excludes Canadian imports; assumed a 90% yield of recycled PET based on input of 183 million pounds (83×10^3 t) of PET from soft drink bottles
Source: Walden Research, Inc.

bottles was used for fiber. The use of recycled bottle-grade PET in the sheeting, bottle, alloy, and compounds markets is in early development.

As noted earlier (Section 3.5) an estimated 10–15 million pounds (4.5–6.8×10^3 t) of PET soft drink bottles were imported into the United States from Canada in 1990. Most of these bottles were recycled into the fiber market.

3.11.1 Fiber

Though the polyester fiber market is a 3-billion-pound-per-year (1.36×10^6 t) market for PET, recycled PET is used in only selected fiber applications. For example, the use of recycled bottle-grade PET to produce coarse (> denier) staple fiber such as fiberfill is a well-developed market in the United States. Wellman pioneered the use of recycled bottle-grade PET for fiberfill. Wellman recycles PET bottles and makes fiberfill for pillows, sleeping bags, ski-jacket insulation, shoulder pads, etc. PET staple fiber also is used today for carpet backing, nonwoven blankets, and some geotextiles. Recycled bottle-grade PET competes in these markets with PET derived from inexpensive textile waste and off-grade PET resins, because fiberfill does not require high IV PET—an IV of 0.58–0.65 dl/g is adequate. Furthermore, fiberfill is a relatively low-value product; therefore, the economics dictate the use of low-cost raw materials. Wellman further reduces its raw material costs by recycling discarded PET soft drink bottles rather than purchasing recycled PET flake from others. Recently, Eastman Chemicals joined with Martin Color-Fi (Edgefield, SC) to make staple fiber from recycled PET [19].

The fiberfill application nevertheless requires a very clean recycled PET—free from adhesives, paper, or metal. This application has little tolerance for green recycled PET. Very few fiberfill markets can use green fiberfill.

On the other hand, some green recycled bottle-grade PET is used for nonwoven blankets and for some carpet backing. When given a choice, however, manufacturers of these products prefer colorless recycled PET. These are smaller markets than the fiberfill market.

Geotextiles, which are used in such applications as stabilizers for railroad beds and paved roads and for erosion control and retainer walls, are normally pigmented with carbon black and, therefore, can use green recycled PET. In this application, staple fiber, 1.5–6 in lengths (3.8–15 cm), is used to make a dry-laid nonwoven. The use of these geotextiles is more popular in Europe than in the United States, where geotextiles made from PP fiber dominate. A few are made from PET fiber today.

Looking to the future, the U.S. Department of Agriculture's Forest Products laboratory in Madison, WI, is investigating the use of lignocellulosic fibers (short-wood fibers) together with PET staple fiber as a nonwoven. The resulting composite is a flexible wooden mat that can be mold-pressed into any desired size or shape [20].

Only recently is recycled bottle-grade PET being used as a continuous filament or yarn in such applications as carpet face yarns, woven carpet backing, and some spun-bonded non-woven geotextiles. The carpet face yarn market is a significant new market for postconsumer PET—one that could perhaps consume as much as 100 million pounds (45×10^3 t) per year of recycled PET. However, this is a market for colorless PET. Image Carpets (Armuchee, GA) has opened up this new market.

In 1991, a new company planned to place a number of facilities strategically across the country to produce an insulation material from recycled PET that is spun into a fiber by a proprietary centrifugal process. The product is as energy efficient as fiberglass insulation material. Embrace Systems (Buffalo, NY) has started to sell its Puffiber™ sandwiched between two layers of plastic. Insulblank™ is used in winter construction to maintain the 18 °C temperature needed to cure poured concrete. The first commercial application of Puffiber™ was insulation for plant walls in refrigeration and cold storage units at food processing plants. These are very interesting markets that can utilize green recycled bottle-grade PET [21].

3.11.2 Strapping

Another market for PET that was developed in the early years of PET bottle recycling is plastic strapping, which replaces metal strapping used in pallet wrapping. PET strapping is made by an extrusion process and requires high IV, metal-free PET. The PET is used in pellet or compacted form. In contrast, PET sold to the fiber industry is normally sold in flake form. The strapping market is a good one for green recycled PET, because most strapping is black. Often, recycled PET is blended with virgin PET in this application, although strapping can be made from 100% recycled PET.

This is a significant market for recycled bottle-grade PET if the IV of the postconsumer PET is maintained. In some instances, the recycler may have to increase the IV of the recycled PET by the solid-stating process mentioned previously (Section 3.2). Strapping is a 32-million-pound-per-year (14.4×10^3 t) market that is growing at an average annual rate of 6–7%. An estimated 10 million pounds (4.5×10^3 t) per year of postconsumer PET are sold in this market. Signode Corp (Glenview, IL), a leader in this market, probably consumes almost 5 million pounds (2.25×10^3 t) of recycled PET annually, if it is available.

3.11.3 Sheeting

PET is becoming a popular extrusion resin for sheeting, where it often displaces PVC and high-impact styrene. It is used in blister packaging, clamshell packages, cups, food trays, etc. A high IV resin is needed for this application free of contaminants. This is a fast-growing market that consumed at least 80 million pounds (36×10^3 t) of virgin PET in 1990 [22]. But, today, the pressure to use recycled materials is forcing thermoforming companies with extrusion lines to use postconsumer PET. For example, Creative Forming, Inc. (Ripon, WI) recently installed an extrusion line specially designed to extrude recycled PET. Companies that thermoform and extrude are evaluating processing not only 100% postconsumer PET but blends containing industrial postconsumer and virgin PET.

One of the earliest applications for sheet made from postconsumer PET was its use as base cups for PET soft drink bottles. Desmacon in Holland thermoforms these base cups for the European market and has been doing so for the past four years. The recycled resin is made by Reko of Holland, and the base cups are made from impact-modified recycled PET on Gaber GmbH (Germany) thermoforming equipment. In Europe, these clear base cups are ultrasonically welded to the PET bottle, eliminating the need for adhesives. Consequently, with both bottle and base cup made from PET, and minimum usage of adhesive, recycling is made easy.

In 1990 Pepsi-Cola Canada Ltd. introduced the same approach in the Canadian market with a Montreal bottle maker (Constar International) [23]. A frosted PET base cup is used in Canada. To date, no one in the United States is interested in the thermoformed PET base cup. First, recyclers are concerned that some bottler would use a colored PET base cup that would then reduce the value of the clear PET bottle in recycling, and second, with the availability of relatively low-cost, injection-moldable recycled HDPE recovered by the PET bottle recyclers for recycling into new base cups, the economics would not make sense. Finally, marketing groups at soft drink companies prefer colored rather than clear base cups.

In 1989, Trio Products (Elyria, OH) introduced a clear egg carton weighing about 1.2 ounces (35 g) made from postconsumer PET purchased in pellet form [24]. This is another application that cannot tolerate green PET. The cartons carry eggs from Ise Farms, Holly Farms, Tyson, and Hillandale and are marketed in parts of such states as Georgia, New Jersey, New York, Ohio, South Carolina, and Tennessee. These cartons cost grocers about three cents more than regular cartons. Trio is also experimenting with postconsumer PET sheet for blister packaging cosmetics and carriers for beverage containers.

A number of thermoforming companies are experimenting with sheet made from postconsumer PET. For example, a northeastern company makes a clear clamshell container from this resin for packaging athletic tape. Solo Cup (Chicago, IL) has used industrial recycled PET to make sheet for thermoforming into cups. Others use recycled PET for audio cassette cases and corrugated awnings.

International Container Systems (ICS) (Tampa, FL) switched from high-impact polystyrene to postconsumer PET to produce reusable trays for its Traymate™ can packaging system. These shipping trays, thermoformed from recycled PET, accommodate 24 loose or multipacked soft drink cans. This application, still considered viable by ICS, was abandoned in 1990 because their equipment was not well-suited for manufacturing this product. However, ICS will license their technology [25].

Although all of these sheeting applications require primarily colorless recycled PET, the use of postconsumer PET in the form of foamed sheet to provide insulation can use green PET. The process for producing such products is available under license from two chemists in Switzerland. To foam PET, an additive that increases PET's extremely low melt viscosity index is required. A high-melt viscosity resin is needed to entrap the bubbles made by a foaming agent. A nucleating agent is added to the PET as a foam stabilizer. The inventors expect the foamed PET product will have the same insulation properties as foamed polystyrene but will be superior on a cost/performance basis. PET has a lower degree of flammability than polystyrene and, in contrast to polystyrene, does not emit a dark smoke when burned [26].

Though the sheeting market in 1989 consumed an estimated 2 million pounds (0.9×10^3 t) of postconsumer PET, we expect the sheeting market to become very important in the 1990s as more PET becomes available and as new applications are developed.

3.11.4 Alloys and Compounds

This is a high-value market for recycled PET. Once the recycled PET is compounded and fabricated, products sell for $0.90–2.50/lb ($1.90–5.50/kg). This is clearly a very desirable market for recycled PET, both colorless and green, if the recycled PET has an IV of at least 0.63 dl/g and is essentially free of contaminants [27].

Some of the major resin producers who are developing and selling high-performance engineered compounds based on postconsumer PET include DuPont, General Electric, and Allied-Signal. However, two smaller companies—MRC Polymers (Chicago, IL) and MA Industries (Peachtree City, GA)—also participate in this market, as well as Wellman.

PET, whether virgin or recycled, is not easily injection molded; it has a tendency to randomly crystallize during normal thick-wall molding. As a result, the molded part becomes brittle, warpage occurs, and the part fails. Adding an impact modifier to the PET can overcome these difficulties, and companies have developed proprietary compounds based on this approach.

Alloying polycarbonate with PET can produce an injection-moldable compound. This approach was taken by GE, which has used recycled PET in such alloys in automotive applications for items such as automobile bumpers for a number of years. It also has been tested as a body panel material. MRC sells these alloys for automobile wheel covers and uses both postconsumer PET and postconsumer polycarbonates in the alloys. Typical MRC alloys, called Stanuloy, combine 50–60% postconsumer PET with 50–40% recycled polycarbonate resins. These are 25–30% less expensive than alloys made from virgin resins and are injection molded into decorative body panels (simulated wood paneling), trim, and spoilers for automobiles. Other potential applications for Stanuloy are housings for office machines and cartridges for copy machines. These compounds are available in different colors with and without flame retardants. Other MRC compounds are claimed to be alternatives to acrylonitrile-butadiene-styrene (ABS) but cost considerably less.

Allied-Signal is the major user of postconsumer PET for glass-filled compounds, which it sells in the merchant market. Wellman, MRC, and MA Industries sell similar compounds based on postconsumer PET. The glass-filled PET compound can use both colorless and green PET.

This market for postconsumer PET in alloys and compounds is an exciting one. As with the sheeting market, this market will grow rapidly during the next few years. In 1989, this market accounted for an estimated 10 million pounds (4.5×10^3 t) of postconsumer PET. By 1993 this could be a 30-million-pound-per-year (5.9×10^3 t) market.

3.11.5 Polyols

Polyols, especially those made by glycolysis of postconsumer PET with DEG, are used very extensively for the production of rigid PUR/polyisocyanurate (PIR) foams. Currently, more than 50% of all rigid PIR insulation board uses polyols made from waste PET X-ray and lithographic film and PET soft drink bottles [28]. The Chardonol Division of Cook Composites and Polymers, formerly Freeman Chemical Corp. (Port Washington, WI), is the most important player in this market. This company alone converts 25–30 million pounds ($11.3–13.5 \times 10^3$ t) of postconsumer PET, mostly film wastes, into aromatic polyols for this market that can use colored PET because the insulation is not visible to the consumer [29]. The PIR foam is used as insulation in roofing and walls. All products in Chardonol's PET polyol line are derived from PET wastes.

Rigid PUR and PIR hybrid foam made from aromatic polyols derived from PET wastes have good cost/performance characteristics. Not only are costs reduced, but there is a marked improvement in physical properties of these foams—properties such as compressive strength and modulus. Flame spread and smoke generation are also reduced, and the foam is also less friable.

Whereas polyols derived from PET wastes are generally combined with other commercial polyols in the manufacture of rigid PUR foam, PET-based polyols can be used to a greater extent in urethane-modified isocyanurate foams because of their low functionality and low hydroxyl number. They can even be used as the sole polyol in this application.

PET-based polyols also can be used in flexible PUR foams, typically as modifiers combined with other commercial polyols. Again, adding PET-based polyols provides significant improvements in flexible PUR foams in tear strength, elongation, and indentation-load deflection. These polyols also have been used to produce integral-skin flexible foams and cast elastomers based on PUR chemistry.

If a PET-based polyol is treated not with a poly- or diisocyanate but with a dibasic unsaturated acid such as maleic anhydride, an unsaturated polyester is formed. However, in this case, PG rather than DEG is preferred for glycolysis. In the preparation of the unsaturated polyester, glycolysis and esterification can proceed in the same reactor by a two-stage reaction. Glycolysis usually occurs at a higher temperature than esterification—about 180 °C vs. about 150 °C. After cooling, the PET-based polyols and residual PG are treated with maleic anhydride to form a polyester. The esterification step also can include phthalic anhydride (PA) and/or isophthalic acid (IPA) [30]. Companies using postconsumer PET to make polyesters include Ashland Chemical (Columbus, OH), Alpha Corp. (Collierville, TN), Ruco Polymer Corp. (Hicksville, NY), and Plexmar Inc. (Houston, TX).

The reaction of maleic anhydride with PG and PET-based polyols is faster than the reaction of maleic anhydride with PG and either PA or IPA, which are normally used. The PET-based resins require about twelve hours to form unsaturated esters with a molecular weight of 2000–2500 and an acid value of 25–30. The conventional resins require about twenty hours.

To prepare conventional thermoset unsaturated polyesters, the esters are mixed with styrene. The viscosity of these solutions is important. A comparison of the viscosity of a conventional polyester with one made from a PET-based polyol in 45% styrene showed them to be comparable. The physical properties of castings made by curing the polyester/styrene mixtures made from PET-based polyols or from PG with maleic anhydride and either PA or IPA were also comparable.

Today, postconsumer PET derived from soft drink bottles is used infrequently for unsaturated polyesters. Postconsumer PET film wastes are preferred because they are adequate and are less costly than postconsumer PET derived from clear bottles. Glass-fiber-reinforced unsaturated polyesters are used in such products as bath tubs, shower stalls, and boat hulls, where a green-colored plastic is not desirable. On the other hand, green postconsumer PET derived from discarded bottles is being used for rigid PURs and PIR foams, as noted above. Nevertheless, polyols are a minor market (about 10 million pounds (4.5×10^3 t) per year) today for postconsumer bottle-grade PET. In the future, as the price for used PET bottles increases, we expect this market to be served for the most part by postconsumer PET film wastes, especially when higher value products are developed based on green postconsumer, bottle-grade PET.

3.11.6 Bottles and Containers

The earliest example of a PET container made from postconsumer bottle-grade PET is the stretch blow-molded tennis ball container made by Continental Can. It has been on the market since the early 1980s and is made from a blend of recycled PET and virgin PET. Continental Can never announced this activity. In the early years, they had problems due to the variability of the quality of the recycled resin. Wilson Sporting Goods Company, together with The National Association for Plastic Container Recovery (NAPCOR), began collecting these containers nationwide in 1991 at tennis clubs, specialty stores, and at major tennis tournaments.

The next example of a container made from recycled PET came in late 1988 when Procter & Gamble (P&G) announced a test program with a 15-ounce (425 g) bottle made from 100% postconsumer PET for its Spic and Span™. In the past, this bottle was made from virgin PET. The bottles are made by Plastipak Packaging, Inc. (Plymouth, MI). In late 1990, this bottle was essentially in national distribution. Plastipak hopes to become a PET bottle recycler, but in 1990 it either purchased postconsumer PET or it tolled used PET soft drink bottles through other recyclers to produce the high-quality recycled resin it needs for stretch blow molding. This application requires clean PET free of all contaminants with an IV of at least 0.68 dl/g.

Other companies followed P&G's lead in late 1990. Plastipak began supplying S.C. Johnson (Racine, WI) with 16- and 27-oz bottles made from 100% postconsumer PET for its Future™ acrylic floor polish [31]. Eventually, S.C. Johnson hopes to have its largest container for this product made from 100% postconsumer PET.

Amway Products also announced in late 1990 that it will use a plastic bottle made from 100% postconsumer PET for its laundry pre-wash product. These bottles will be made by Johnson Controls, which will use its own recycled, postconsumer, bottle-grade PET [32]. In October 1990, Johnson Controls (Novi, MI) started up its recycling facility based on the Dutch Reko BV Technology described in Section 3.7.2.

In the coming years, more companies that use clear plastic bottles for nonfood products will move to bottles made from 100% recycled PET. The availability of high-quality recycled PET will be a major constraint.

The idea of using postconsumer PET in closed-loop recycling for food and beverage containers has always been the dream of food and beverage packagers. They prefer plastic packaging but have not been able to satisfy the public that plastic containers, like metal and glass containers, can be recycled again and again back into plastic containers. One approach that the author has frequently proposed is to make a multilayer PET container with postconsumer PET sandwiched between two layers of virgin PET [33]. This approach is now used to make HDPE bottles for nonfood use. If it can be demonstrated that there is no migration of deleterious substances from such PET structures, the multilayer container should meet FDA standards.

Nissei ASB has made such multilayer PET bottles on its blow molding machines. Nissei claims that in the injection molding of preforms for these bottles, the layers of virgin and recycled PET do not mix, because PET has a high melt viscosity. However, further testing will be needed to insure that migration of contaminants from the middle layer of recycled PET, if present, does not in fact occur [34].

In late 1990, the major PET resin producers announced that their customers (the blow molders) and their customer's customers (i.e., Coca-Cola and Pepsi-Cola), would use PET soft drink bottles made from recycled PET. The recycled PET would be made by depolymerizing postconsumer PET to monomers or to an intermediate stage and then repolymerizing it [35]. Goodyear uses a glycolysis process, and Hoechst Celanese and Eastman Chemical announced they would use methanolysis (Section 3.10) depolymerization technology for the depolymerization reaction. In the Goodyear process, depolymerization leads to low molecular weight polyols; whereas, in methanolysis, depolymerization goes all the way to starting monomers. Since mid-1990, Goodyear has been selling a PET grade (Repete) consisting of both virgin and 10–20% repolymerized postconsumer PET for a premium price—10% above virgin PET.

Pepsi-Cola bottles will be made by Sewell Plastics, Inc. from Goodyear's Repete, containing about 25% postconsumer PET, while Coca-Cola uses a Hoechst Celanese repolymerized postconsumer PET in a blend with virgin PET. The soft drink companies hope to have their bottles made from postconsumer PET in test markets as early as the last quarter of 1991 [36].

Whether these new markets for food-grade recycled PET will in fact develop in 1991 hinges on a number of factors. The FDA plans to issue guidelines on plastics recycling processes and products by the end of 1991. On January 9, 1991, the FDA provided Hoechst Celanese with a letter of nonobjection that permits this company to sell its recycled PET made by methanolysis for food-contact applications. This is only the first hurdle [37]. There is also the question of liability if a consumer claims illness ingesting a soft drink from such bottles. And the biggest question is supply—all PET that is recycled today is being sold.

3.12 Economics

Postconsumer PET is a commodity whether it is purchased contaminated as a soft drink bottle from a bottler or a MRF, as discarded X-ray film from a desilverer, or as clean recycled PET in flake or pellet form from a recycler. However, selling prices of these materials do not necessarily follow the selling price of virgin PET. Recycled PET has its own markets. Generally, clean recycled PET sells for lower prices than virgin PET, but, if the demand is there, recycled PET can sell for prices equivalent to virgin PET.

Raw materials for the PET recycler are discarded bottles or film. Prices paid for these materials depend on demand and the selling prices of the recycler's PET. Historically, bales of used PET soft drink bottles were purchased from bottlers for 5–6¢/lb (11–13¢/kg) (F.O.B.) if not color-sorted, and 10–12¢/lb (22–26¢/kg) for bales of color-sorted clear PET bottles, especially without base cups. In early 1990, demand for used PET bottles jumped to 10–11¢/lb (22–24¢/kg) for mixed-color bales. Then, in late 1990, the fiber market softened, and bottle prices dropped to 7–8¢/lb (15–17¢/kg). If bales of unsorted plastic bottles are purchased from a MRF, they are available for 5–6¢/lb (11–13¢/kg) but consist of HDPE, PET, and other plastics that have to be sorted. In 1990, PET film from desilvering, still coated with PVDC, could be purchased for as low as 4¢/lb (9¢/kg) (F.O.B.)

Processing costs to convert used PET bottles have remained relatively constant over the past several years at 9–11¢/lb (20–24¢/kg). These costs include labor but do not include overhead. The recycler also must pay the cost of transporting the used PET bottles from the supplier to the recycling facility. On average, these costs amount to about 2¢/lb (4¢/kg). Other costs include packaging, usually gaylords (frequently used ones), to ship finished product to the customer. These cost about 0.7–0.8¢/lb (1.5–1.7¢/kg) of recycled PET.

The selling prices (F.O.B. recycling plant) for recycled PET have gradually increased over the past few years at a rate somewhat faster than virgin PET. In 1986, the average price for good quality, colorless, postconsumer granulate or flake was 27¢/lb (59¢/kg). In 1990, it was as high

as 40¢/lb (88¢/kg). Green postconsumer PET flake always sells for about 6–7¢/lb (13–15¢/kg),
less than colorless flake. Pelletized recycled colorless PET sells for as low as 43–48¢/lb (95¢–
106¢/kg). However, some recyclers are trying to get prices equivalent to virgin PET.

The selling price of the injection-moldable HDPE from base cups closely follows the selling
price of virgin resin but at a very substantial discount. In 1990, this HDPE in granulated form
sold for 15–18¢/lb (33–40¢/kg). Pelletized, the recycled HDPE sells for 28–30¢/lb (62–66¢/kg).

PET bottle recyclers can also recycle aluminum closures. Today, however, few PET bottle
recyclers recycle aluminum, because plastic closures are becoming more popular and perhaps
only 60% of the incoming used bottles come with closures. Nevertheless, if the throughput is
substantial—greater than 10 million pounds (4.5×10^3 t) of bottles per year—it pays to recover
and sell the aluminum derived from the caps. Some recyclers have even justified the purchase
of auxiliary equipment to further clean the aluminum recovered by the electrostatic separator.
The prices obtained for the aluminum vary according to its level of contamination and the price
of aluminum ingot. Prices can vary from 16 to 40¢/lb (35 to 88¢/kg).

An example of the economics encountered in operating a PET bottle recycling facility for
bottles with and without base cups is given in Tables 3.5 and 3.6. Typically, successful
recyclers will operate with a margin of 8–15%, once they have developed a process and

Table 3.5   Economics of a facility for recycling PET bottles with base cups

Assumptions			
Raw Material Feed:		2	liter bottles with base cups
Annual Input:		15	million pounds (6.8×10^3 t)
Bottle Composition (%):			
PET		91	
HDPE		29	
Colorless		70	
Green		30	
Operating Costs, ¢/lb (¢/kg):			
Raw Material		11	(24)
Freight		2	(4)
Processing		10	(22)
Packaging		0.725	(1.5)
Product Yield (%):		85	
Product (Flake) Selling Price, ¢/lb (¢/kg)			
Colorless PET		39	(86)
Green PET		30	(66)
HDPE		12	(26)

Calculations			
Operating Costs (thousands $):			
Raw Material:	1650		
Freight	300		
Processing	1500		
Packaging	92		
TOTAL	3542		
Product Output, thousands lb (10^3 t)			Sales (thousands $):
Colorless PET	6337	(2.8)	2471
Green PET	2715	(1.2)	869
HDPE	3698	(1.8)	629
TOTAL	12,750	(5.8)	3969
Gross Margin, including taxes and overhead: $427,000			

Table 3.6 Economics of a facility for recycling PET bottles without base cups

Assumptions		
Raw Material Feed:	2	liter bottles without base cups
Annual Input:	15	million pounds (6.8×10^3 t)
Bottle Composition (%):		
Colorless	70	
Green	30	
Operating Costs, ¢/lb (¢/kg)		
Raw Material	11	(24)
Freight	2	(4)
Processing	10	(22)
Packaging	0.725	(1.5)
Product Yield (%):	85	
Product (Flake) Selling Price, ¢/lb (¢/kg)		
Colorless PET	39	(86)
Green PET	30	(66)

Calculations			
Operating Costs (thousands $):			
Raw Material:	1650		
Freight	300		
Processing	1500		
Packaging	92		
TOTAL	3542		
Product Output, thousands lb (10^3 t)			**Sales (thousands $):**
Colorless PET	8925	(4.1)	3481
Green PET	3825	(1.7)	1224
TOTAL	12,750	(5.8)	4705

Gross Margin, excluding taxes and overhead: $1,163,000

established a position in the market place. As can be seen from these tables, recycling of base-cup-free PET bottles instead of two-component bottles can substantially improve the profitability of the recycler's operation. Generally, a recycler processes both types of bottles. Facilities to process 15 million pounds (6.8×10^3 t) of bottles per year will vary in cost from $1 to 2 million.

3.13 Global Activities

3.13.1 Europe

As noted in Section 3.7.2, Reko BV, a subsidiary of DSM in the Netherlands, developed the water bath/hydrocyclone process being practiced by Johnson Controls. Reko is a key recycler of PET bottles in Europe. Reko recycles other plastics as well and accounts for almost half of the plastics recycling in the Netherlands today. In 1985, Reko formed Retech to recycle PET bottles by the unique process described earlier [38]. Most interestingly, Reko was able to convince European PET bottle manufacturers such as Desmacon, beverage producers, and appropriate government agencies to adopt a PET bottle design that was eminently recyclable. These European bottles are made only from clear PET with a clear thermoformed PET base cup. The sheet used to make the base cup is made from recycled bottle-grade PET, and the base cups are ultrasonically welded to the bottle, thus eliminating adhesives. Paper labels are used and are attached by a water-soluble glue. Closures are made from PP with an EVA liner.

In 1986 Retech constructed a plant in Beek, South Limburg in the Netherlands, to recycle 8 million pounds (3.6×10^3 t) of PET bottles per year, collected mostly from Germany and Holland under a deposit system. In 1990, Retech announced it was constructing a new plant in Beek to double its PET recycling capability. Today, Retech collects PET bottles from other Western European countries, including the Benelux countries and Austria [39]. West European demand for PET increased from an estimated 430 million pounds (195×10^3 t) in 1987 to some 570 million pounds (259×10^3 t) in 1989, and about 55% of this demand was used for soft drink bottles. PET demand in Western Europe is expected to increase at about 12% per year in the 1987–1995 period [40].

More recently, in 1990, Wellman International Ltd. (the European arm of Wellman, Inc.) installed a PET recycling operation in Spijk in the Netherlands. Participation in this company, called Welstar, includes Constar with a 44% share and EniChem Fibre S.P.A. with a 12% share. This new facility, which cost some $5 million, has a capacity of 22 million pounds (10×10^3 t) of PET bottles per year. Presumably, the technology for PET cleaning is Wellman technology. It produces clean PET flake from used PET bottles, jars, trays, and film collected from across Europe. The flake is being shipped to the Mullagh plant outside of Dublin in the Republic of Ireland, where Wellman uses recycled nylon-6/6 and PET to make fiber [41].

3.13.2 Canada

In Canada, which has deposit laws in the provinces of British Columbia, Alberta, and Quebec, and, in addition, has very active curbside collection programs, PET bottle collection is extensive. Historically, many of these collected PET bottles were shipped across the border to recyclers in the United States. The Canadian PET soft drink bottle market is small—only 48 million pounds (22×10^3 t) of solid-state PET were consumed in this market in 1989 [42]—but the market is growing at a double-digit annual rate.

One of the earliest recyclers of PET bottles in Canada was Applied Polymer Research Canada Ltd (APR) (Edmonton, Alberta). With its deposit system, Alberta collects about 90% of the PET soft drink bottles consumed in that province. APR, with a capacity of some 8 million pounds (3.6×10^3 t) per year, uses a flotation recovery system and sells recycled PET flake. More recently, this operation was taken over by Agra Industries, Ltd (Saskatoon, Saskatchewan). Agra Industries may construct a PET bottle recycling plant in Ontario.

Meanwhile, late in 1990, Pelo Plastique, Inc. (Berthierville, Quebec) began recycling PET bottles collected in the province of Quebec. This company is a division of Robert Peloquin et Fils Ltd (Montreal, Quebec). In the past, this company had sold baled bottles to Wellman. The plant will be producing recycled PET flake and has a capacity of some 12 million pounds (5.4×10^3 t) per year. In the future, Pelo Plastique, Inc. plans to use some of its flake captively to make PET sheet and hopes to extrude about 4–5 million pounds (1.8–2.3×10^3 t) of sheet per year. Pelo Plastique may source PET bottles from the Maritime provinces and New England. Quebec collects some 11 million pounds (5×10^3 t) of PET bottles annually by its deposit system [43].

3.13.3 Australia

Australia is becoming an important player in PET bottle recycling. With a population of only 17 million, Australia now uses about 50 million pounds (23×10^3 t) of PET for soft drink bottles annually. PET soft drink bottles, mostly in the 1.25-liter size, dominate the soft drink market in large food stores, with almost 50% of the packaged soft drink market. The 2-liter size has another 22% share.

Though the state of South Australia has had a deposit law for the past five years, the collection of PET bottles in Australia is unique. In the two most populated states of Victoria and New South Wales, the collection system resembles curbside collection. Collection is handled not by a waste hauler but by a private entrepreneur called a "bottle merchant." Plastic bags are provided to the householder by the bottle merchant, who then collects the bags filled with containers from the household every two weeks. Bags are used to collect glass and PET bottles. More recently, on a test basis, the collection system has been expanded to include PVC bottles. The bottle merchant sorts containers by material and color, sells refillables to the stores, and sells sorted one-way containers to container manufacturers. A semigovernment association has set the fee schedule. The bottle merchant receives 70¢ Australian per kilogram of sorted PET containers (equivalent to about 1.5¢ U.S. or 1.9¢ Australian per bottle) from the container manufacturer. As an offset, the soft drink companies deposit 0.5¢ Australian into a special fund for each new PET bottle they purchase from the container manufacturer. In turn, the container manufacturer can use these funds to purchase used PET bottles from the bottle merchant [44]. In late 1990, about 9% of the PET bottles consumed in these two states were recovered. As new recycling facilities are installed, this rate is expected to double by early 1992.

Only two companies make PET soft drink bottles in Australia today—ACI (BTR Nylex) and Smorgon Consolidated Industries. Both are interested in recycling PET soft drink bottles. ACI will start up its PET bottle recycling plant some time in the first quarter of 1991 at a cost of $4 million. This plant is located in Wadonga between Sydney and Melbourne and is expected to recycle some 8 million pounds (3.6×10^3 t) of PET bottles per year by 1992. The technology is based on a flotation system and has been engineered by Plastic Forming Systems (Manchester, NH). This facility includes state-of-the-art instrumentation controls to insure consistent quality. Smorgon is not as far along in its construction of a PET recycling line, but it also hopes to have a line in place in 1991. Recycled PET in Australia will be used in sheet and bottles, as well as in engineering applications [45].

3.13.4 Far East

Japan is often cited for its efforts in recycling. However, most of this effort is focused on industrial wastes. Very little postconsumer plastic is recycled in Japan. The number of municipalities with recycling programs declined from 25 in 1982 to only three in 1987 [46]. An estimated 70% of all plastic waste is incinerated in Japan, which has about 1,900 incinerators [47].

Though the Japanese used 110 million pounds (50×10^3 t) of PET in 1989 [48], and PET soft drink containers account for about 25% of all soft drink containers sold, the Japanese have yet to establish a commercial-scale PET recycling plant.

Elsewhere in the Far East, PET recycling has started up. As mentioned above (Section 3.7.3), a new recycling facility with a capacity of 15 million pounds (6.8×10^3 t) per year began operation in Taiwan in 1990 to recycle PET bottles using the Solution/Washing Process developed by PTI. Taiwan uses some 26 million pounds (12×10^3 t) per year of PET for soft drink bottles.

In 1989, in response to the solid-waste issue, the Taiwan EPA issued a directive that requires makers and users of PET containers to collect them. As a result, the soft drink industry established a PET Management Fund to install a countrywide collection system. This system uses seven plastic scrap companies to collect used containers, each in an assigned territory. Most of the used containers are gathered directly from landfills; others are collected from schools, restaurants, scrap dealers, recreation centers, etc.

The collected PET bottles are baled and shipped to the Taiwan Recycling Corporation's (TRC) plant, which is located some 31 miles (50 km) from Taipei. During the first year of operation, over 7.7 million pounds (3.5×10^3 t) of used PET bottles were collected. TRC is a 50:50 joint venture between Far Eastern Textile and the Shin Kong Group—two major producers of PET soft drink bottles. The recycling plant can process used PET bottles at a rate of some 3200 pounds (1450 kg) per hour. The recycled PET is used to make staple fiber [49].

References

[1]   *Cleartuf Polyester Product Manual.* Goodyear Chemicals.
[2]   Eastman's Green Book on the PET Container Market, 1989.
[3]   Reference 1.
[4]   Eastman Chemical Products, Inc./ Plastics Division, Eastman Kodak Company. Private communication.
[5]   *Modern Plastics 68* (1) (Jan. 1991): p. 116.
[6]   *Chemical Week* (23 May 1990): p. 10.
[7]   Reference 5.
[8]   Milgrom, J. Ninth International Conference on Oriented Plastic Containers, 1985.
[9]   Milgrom, J. "The Impact of Curbside Collection on PET Beverage Bottle Recycling." Bev-Pak '91.
[10] *Reuse/Recycle* (July 1990): p. 4.
[11] *Packaging Digest* (Nov. 1990).
[12] Lamparter, R. A., Barna, B. A., and Johnsrud, D. R. U. S. Patent 4 542 239, 1985.
[13] Mandoki, J. W. U.S. Patent 4 605 762, 1986.
[14] Grushke, H., *et al.* U.S. Patent 3 403 115, 1968.
[15] *Reuse/Recycle* (Jan. 1991): p. 1.
[16] *Plastic News* (28 Jan. 1991): p. 3.
[17] Vaidya, U. R. and Nadkarni, V. M. *Ind. Eng. Chem. Res., 26* (1987): 194–198.
[18] Eastman Chemical Products, Inc. Publication No. N 262 A. March 1984.
[19] *Chem. Eng. News* (28 Jan. 1991): p. 10.
[20] *Chem. Eng. News* (10 Sept. 1990): p. 19.
[21] Lauzon, M. "Recycled PET Being Used for Insulation Applications." *Plastics News* (20 Nov. 1989): p. 6.
[22] Reference 5.
[23] Reall, J. "Recycled PET Base Cups Hold Promise." *Plastics News* (19 June 1989): p. 4.
[24] Reall, J. "Trio Products Finds Recycled PET Works in Egg Cartons." *Plastics News* (12 June 1989): p. 1.
[25] *Beverage World* (June 1990).
[26] *Chem. Eng. News* (12 Feb. 1991): p. 25.
[27] Milgrom, J. "Markets for Recovered PET Scrap." *Beverage World* (June 1985): p. 86.
[28] Polyisocyanurate Insulation Manufacturers Association. Private communication.
[29] *Modern Plastics 65* (12) (Dec. 1988): p. 32.
[30] *Modern Plastics 66* (5) (May 1989): p. 64.
[31] *Reuse/Recycle* (Jan. 1991): p. 2.
[32] Reference 31.
[33] Reference 27.
[34] Milgrom, J. "Plastic Bottles from Scrap—Problems, Solutions, and Future." Bev-Pak '90.
[35] *Reuse/Recycle* (Dec. 1990): p. 2.
[36] Reference 31.
[37] *Reuse/Recycle* (Feb. 1991): p. 1.
[38] Brewer, G. "Going Dutch: Transatlantic Teamwork in Plastics Recycling." *Resource Recycling* (Dec. 1990): p. 96.
[39] Short, H. "DSM Doubling PET Recycling Capacity." *Plastics News* (19 Feb. 1990): p. 2.

[40]   Bauman, R. J. "Global Analysis of PET and HDPE for Beverage Containers—Supply, Demand, and Pricing Trends." Bev-Pak '90.
[41]   *Plastics News* (5 Feb. 1990).
[42]   Reference 40.
[43]   Lauzon, M. "Quebec Bottle Collector Pelo Becomes Plastics Recycler." *Plastics News* (12 Nov. 1990): p. 2.
[44]   Milgrom, J., ICI Australia. Private communication.
[45]   McIlwaine, K. "Australian Focus on Post-Consumer Waste." *Plastics News* (17 Sept. 1990): p. 21.
[46]   *Waste Age* (Nov. 1990): p. 168.
[47]   Schreffler, R. "Japanese Slow to Adopt Plastics Recycling." *Plastics News* (17 Sept. 1990): p. 22.
[48]   Reference 40.
[49]   Pure Tech International. Private communication.

4 Polyolefins

B. D. Perlson and C. C. Schababerle

Contents

4.1　Introduction 75

　　4.1.1　Polyethylenes 75

　　4.1.2　Polypropylenes 76

4.2　Reprocessing Technologies 77

　　4.2.1　Rigid Containers 77

　　　　　4.2.1.1　Postconsumer Containers 78

　　　　　4.2.1.2　Postcommercial Containers 88

　　4.2.2　Films 89

　　　　　4.2.2.1　Polyolefin-based Films 89

　　　　　4.2.2.2　Polyolefin-coated Materials 89

　　4.2.3　Foams 90

　　4.2.4　Wire and Cable Jacketing 91

　　4.2.5　Polypropylene Battery Cases 92

4.3　Effects of Multiple Processing on Polyolefin Resins 93

　　4.3.1　Polyethylenes 93

　　4.3.2　Polypropylenes 96

4.4　End Uses 97

　　4.4.1　Polyethylenes 97

　　4.4.2　Polypropylenes 98

4.5　Global Activities 98

　　4.5.1　United Kingdom 98

　　4.5.2　Continental Western Europe 99

　　4.5.3　Canada 100

　　4.5.4　Japan 101

4.6　Future Outlook 101

References 103

This chapter discusses postconsumer polyolefin items that are being collected for recycling in the United States; the processes used to reclaim these items into saleable resin; the end-use markets for the resulting reclaimed materials; the effects of multiple processing on polymer properties; a global perspective of postconsumer polyolefins recycling; and a perspective on the domestic outlook for postconsumer polyolefins recycling.

4.1   Introduction

Polyolefins are produced by the polymerization of small molecules called α-olefins [1,2]. α-olefins are unsaturated, aliphatic hydrocarbons possessing a single double bond between the first and second carbon atoms and having the general chemical formula C_nH_{2n}. Examples of commonly used a-olefins include ethylene ($CH_2=CH_2$), propylene ($CH_3-CH=CH_2$), and butene-1 ($CH_3-CH_2-CH=CH_2$). The polyolefins that are most widely produced commercially are polyethylene (PE) and polypropylene (PP). PE and PP will be the only polyolefins discussed in this chapter.

Polyolefins are members of the class of materials called thermoplastics, meaning that these materials can be melted or softened at elevated temperatures and molded or formed into new shapes repeatedly [3]. Throughout this chapter, the term "processing" will be applied to manufacturing operations used to melt and shape polyolefins to produce useful items.

The mechanical and physical properties and processing characteristics of polyolefins are primarily dependent on the average molecular weight, the distribution of molecular weights, and the number and length of "branches" of the polymer molecules. Polymerization conditions, catalysts and initiators, and co-monomers all strongly influence these characteristics of the polymer. Herein, the discussion will be limited to co-monomers that are also α-olefins, unless otherwise noted.

During the processing of polyolefins into end-use items, manufacturing scrap is produced that may either be reclaimed by the manufacturer or sold to others to be reclaimed. Herein, the reclaiming of these materials will be termed "industrial polyolefins recycling."

Once a manufactured polyolefin item has been sold and discarded by the buyer, the item is defined herein as a "postconsumer" item. The process of reclaiming these items into a saleable resin is termed "postconsumer polyolefins recycling." Postconsumer recycling of polyolefins is the principal topic of the remainder of this chapter.

4.1.1   Polyethylenes

In the case of PE, a terminology has been adopted to classify materials in this "family" according to their densities, as shown below.

Polyethylene Class	Density Range (g/cm³)	Acronym
Low density	0.910–0.925	LDPE
Medium density	0.926–0.940	MDPE
High density	≥ 0.941	HDPE

PEs are partially crystalline materials. The amount of crystallinity depends on the branching of the polymer chains and determines the density of the material.

There are two fundamental polymerization techniques used to produce PE: a free radical-initiated, high-pressure polymerization process; and a low-pressure, transition metal-catalyzed process. The latter process may be conducted either in the gas phase or in an inert hydrocarbon liquid.

Materials produced by the high-pressure process fall primarily in the LDPE and the lower density end of MDPE classifications. These materials are generally termed "high-pressure, low-density polyethylenes." Film for packaging is the largest single outlet for LDPE. Copolymers of ethylene with polar monomers, such as vinyl acetate, acrylic acid, methyl and ethyl acrylate, are also made by the high-pressure process. Copolymers of ethylene-vinyl acetate (EVA) are used as adhesives (e.g., for PET soft drink bottle end caps and labels.)

Materials produced by the low-pressure process span all three of the above classifications and are generally termed "linear polyethylene." The HDPE homopolymers are used for bottle and container applications. Most of the other polymers produced by this process are co-polymers of ethylene with higher α-olefins, such as butene-1, hexene-1, octene-1, and 4-meth-ylpentene-1. These copolymers, known as linear low-density polyethylene (LLDPE), are used in a variety of marketing applications, such as film, moldings, pipe, and wire and cable.

The constant temperature melt viscosity of PE is directly related to the molecular weight characteristics of the material. Therefore, PE can, in part, be characterized by this physical property. Because of ease of measurement, PEs are characterized by melt index (MI), which is the amount of melted polymer (in grams) flowing through an orifice in ten minutes at a designated temperature and static load (see ASTM Method D-1238 for a detailed description).

In addition to density and MI, PEs also are characterized by their mechanical properties, processing characteristics, (photo)chemical resistance, and surface and opacity properties. Additive systems frequently are used in PE to alter one or more of these properties.

PE is the dominant thermoplastic material consumed in the world [4]. Table 4.1 shows the consumption of this material in Japan, Western Europe, the United States, and Canada in 1989. In the United States, the largest use of PE is packaging. In 1989, an estimated 56 and 47% of HDPE and LDPE/MDPE were used in this application, respectively.

Table 4.1 Polyethylene consumption–1990

Country	HDPE		LDPE/MDPE		Total PE	
	Million pounds		(Thousand tonnes)			
Japan	2454	(1113)	3687	(1672)	6141	(2785)
Western Europe	6476	(2937)	11,565	(5245)	18,041	(8182)
USA	8333	(3779)	10,831	(4912)	19,164	(8691)
Canada	609	(276)	1129	(512)	1738	(788)

Source: *Modern Plastics International, 21* (1) (Jan. 1991): pp. 51–65.

4.1.2 Polypropylenes

There are three predominant types of virgin PP currently manufactured commercially: homo-polypropylene (H-PP), copolymer polypropylene (C-PP), and so-called rubber or impact-modified polypropylene (I-PP). Both H-PP and C-PP are typically manufactured in a single polymerization reactor. In the case of C-PP, one or more additional α-olefins are incorporated in the polymer. The co-monomer may be incorporated either randomly or in a blocked fashion along the polymer chains. This imparts improved ductility to PP. I-PP can be manufactured by either compounding H-PP with an impact modifier or by cascading H-PP into a secondary reactor, where an impact-modifying polymer is produced. Typically, this impact-modifying polymer is a propylene ethylene copolymer. I-PP has improved ductility over both H-PP and C-PP grades.

As is the case with PE, the melt viscosity of PP is related to the molecular weight characteristics of the polymer. The melt flow rate (MFR) of PP is a measure of this characteristic. (This measurement is made in accordance with ASTM Method D-1238 and has the same units, g/10 min, as for PE.) PP is similar to PE in the range of other material characteristics important to commercial uses.

Table 4.2 depicts the 1989 production of PP in Japan, Western Europe, the United States, and Canada [5]. In 1989, an estimated 15% of PP consumption in the United States was used in packaging.

Table 4.2   Polypropylene consumption–1990

Country	Consumption	
	Million pounds	(Thousand tonnes)
Japan	4586	(2080)
Western Europe	8125	(3685)
USA	6670	(3025)
Canada	370	(168)

Source: *Modern Plastics International, 21* (1) (Jan. 1991): pp. 51–65.

4.2   Reprocessing Technologies

The basic features of commercial processes for reclaiming polyolefins are straight-forward. The incoming items are inspected and processed to remove macrocontaminants, reduced to a manageable particle size, washed to remove surface contaminants, and dried. The clean, dry particles can be used "as is" or be pelletized.

Each commercial reclaiming company has its own variation of the above-described technology—major differences are the order of the processing steps and the design of equipment to perform each step. Currently, all commercial reclaiming companies and reclaiming process technology vendors consider the details of their reclaiming equipment to be, to some degree, proprietary. To date, none of these reclaiming technologies is protected by patent; therefore, participants are reluctant to release details that may compromise their competitive positions in this fledgling industry. As a result, discussion of reclaiming equipment and process details in this chapter will be, by necessity, sketchy.

Equipment design and process details for reclaiming polyolefin items are determined, in part, by the physical form of the feedstock. Accordingly, this section of the chapter has been organized to describe recycling of the following: rigid containers; films, including polyolefin-coated substrates; foamed sheet; wire and cable jacketing; and battery cases.

4.2.1   Rigid Containers

A significant portion of the present postconsumer polyolefin recycling industry is rooted in the polyolefin industrial scrap reclaiming business. Therefore, it is instructive to briefly describe this latter activity as an introduction.

Much of the industrial scrap generated during the manufacture of rigid containers is reclaimed in-house. Materials such as runners, sprues, trimmings, and rejected parts are ground in captively operated equipment, mixed with virgin resin, and reused in the parts manufacturing process.

Alternatively, industrial scrap is sold to independent reclaimers who grind this material, and may pelletize it, prior to resale to the original or other manufacturers. Little, if any, cleaning is required to reclaim industrial scrap, since contamination is generally limited to labels. Labels reportedly [6] can be removed by either "wet" or "dry" processes. Wet processes involve the use of agitated, usually hot, water to soften and disperse the label adhesive. Paper labels are pulped by mechanical agitation. This pulp and the polyolefin material may be separated by filtration, buoyancy differences, or a combination of both. With plastic labels, the washed materials are dried and may be separated by air buoyancy differences in a classifying device.

In a dry process, ground scrap may be fluidized or stirred in an upward-flowing stream of hot air. This softens the label adhesive sufficiently to disengage the thin sections of label material from the thicker container-wall sections.

Periodically, industrial scrap may be soiled, including such contaminants as product resins, hydraulic oils, earthen and humic materials, etc. Soil on container scrap is removed by washing the material in water [7]. Dispersion of the soil is enhanced by vigorous agitation and a suitable detergent. Depending on the specifics of the detergent formulation, soil dispersion may be improved by heating the water.

The addition of postconsumer polyolefin-based containers to the pool of scrap polyolefins is a relatively new development. (According to the authors' estimate, in 1987 less than 10 million pounds (4.5×10^3 t) per year of these containers were being recycled. The volumes recycled in more recent years are certainly much larger, but any quantitative estimate would be highly speculative because of the fragmented character of the reclaiming industry and the flexibility and willingness of these reclaimers to handle a variety of raw materials.) This has greatly broadened the range of contaminants that might have to be removed to restore the quality of the material to be comparable to, for instance, reclaimed trim scrap.

4.2.1.1 Postconsumer Containers

Most recyclers of postconsumer polyolefin containers subscribe to the view that these materials have the highest potential value if they are segmented in accordance with the characteristics of the resin in the body of the container [8]. (For instance, one valued segment is composed of one gallon (3.79 liter) and one-half gallon (1.89 liter) PE milk and water jugs. The resin used to produce these extrusion blow-molded containers is very similar, regardless of the manufacturer of the material. This particular material is an unpigmented fractional melt-index homopolymer HDPE, with a density of approximately 0.96 g/cm^3.) As a result, container sorting normally occurs in the recycling infrastructure once or more.

The construction of the overall container system presents new contaminant removal challenges. Items such as decorations, closures, spouts, tamper-evident seals, freshness seals, as well as labels, are routinely manufactured from dissimilar materials. Other challenges to the reclamation process may include residues of the original product in the container; residues of materials resulting from container reuse by the consumer; and incremental (i.e., micro-) contamination from the storage, collection, and distribution infrastructures for postconsumer recyclables.

The first type of postconsumer plastic container to be commercially recycled was the polyethylene terephthalate (PET)-based soft drink bottle. By 1990, the base cup, label, and closure on the plastic soft drink bottle were typically polyolefin-based (although some beverage bottlers continue to use aluminum closure systems). In addition, the closure liner and the label and base cup adhesives are ethylene-based copolymers. As a result, container-related, postconsumer polyolefin recycling began as a coproduct of this activity [9].

The commercially available technologies for reclaiming postconsumer, polyolefin-based containers have been adapted largely from PET soft drink bottle reclaiming systems, described in Chapter 3. The unit operations involved in a representative polyolefin container reclaiming plant are shown diagrammatically in Figure 4.1 and are typically invariant to the type of container being reclaimed, as characterized by resin type, resin grade, or the container end-use.

There is evidence that some reclaiming processes incorporate an additional contaminant removal unit between the sorting and size reduction operations. One of these reportedly water-washes whole containers prior to size reduction [10]. The authors also can envision the use of an inclined rotary or vibratory trommel to separate some contaminants (e.g., loose closures, glass, metal and paper fragments, etc.,) from the containers prior to size reduction.

The principal objective of the size reduction operation is to accommodate the feeding of the reclaimed polyolefin to a melt extruder, either downstream in the reclaiming plant or at a resin processing facility. After size reduction, the mean particle size is typically 0.25–0.5 in^2 (161–322 mm^2). Hereafter, these particles will be referred to as flakes. To produce flake from containers, some advocate combining a shredder and granulator, while others simply use a granulator alone. The shredder-granulator combination is claimed to reduce overall power

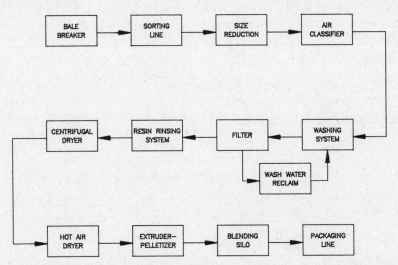

Figure 4.1   Unit operations in a representative postconsumer polyolefin container reclaiming process

consumption and may extend the life expectancy of the knives in the granulator. Both wet and dry granulators are used commercially—at this point, it is uncertain as to which of these is favored [11].

In reclaiming systems that produce dry flake in the size reduction equipment, the process generally includes an air classifier immediately following to remove readily fluidized particles [12]. Contaminants such as labels are at least partially removed by this method. This processing step also results in the removal of "fines" generated during the size reduction operation, resulting in some yield loss of the desired material.

The washing through rinsing systems depicted in Figure 4.1 are regarded as highly proprietary by commercial polyolefin reclaimers and reclaiming process technology vendors. However, one of the available processes has been developed by the Center for Plastics Recycling Research (CPRR) at Rutgers, The State University of New Jersey; therefore, process details are more readily obtainable [13].

The development of the CPRR process was funded, in part, by the Plastics Recycling Foundation (PRF), whose members include manufacturers of resins and plastic products, packaged goods companies, soft drink companies, machinery suppliers, trade associations, and associated companies. The CPRR process was conceived and developed at Goodyear, improved by Owens-Illinois, and, in 1984, was donated to CPRR for further refinement. While the process was initially developed to reclaim PET soft drink bottles, it is readily adaptable to polyolefin containers.

To demonstrate the viability of the process, a 5-million-pound-per-year (2.25×10^3 t) pilot plant was constructed at Rutgers during the latter half of the 1980s. The following is a description of the features of the flake washing through drying processes in this pilot plant— also diagrammed in Figure 4.2. (Readers should note that several process flow paths have been altered by the authors to adapt the system to recover polyolefins.)

Washing of the flaked containers is done in a batch process using a system of three tanks: a primary wash tank, a secondary wash tank, and a detergent solution tank. All three tanks are constructed of carbon steel. The 300-gal (1135-liter) primary wash tank is charged with flake from a screw feeder and hot washing solution from the detergent solution tank. The washing solution is heated in the detergent solution tank by circulation through two electric hair-pin heaters. The typical flake-loading in the washing solution is approximately 0.5 lb/gal (60 g/l).

In the primary wash tank, agitation is provided by a single, clamp-mounted blade mixer. During the wash cycle, the slurried flakes are recirculated by a centrifugal pump. The volume

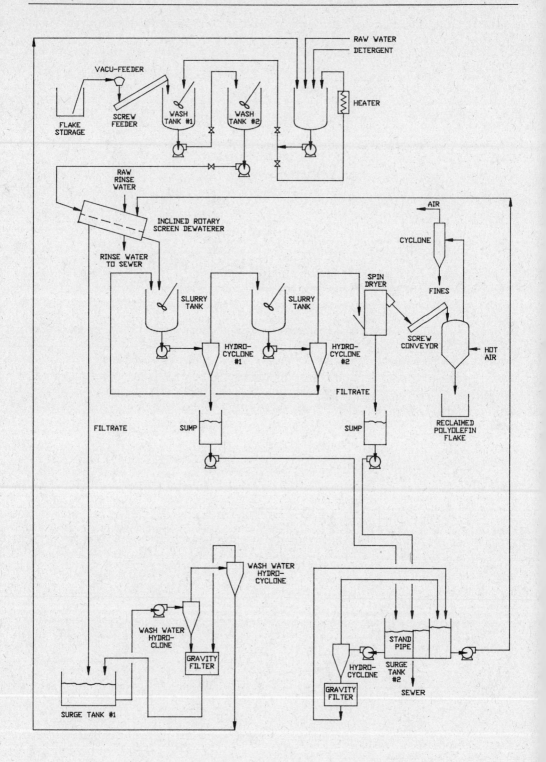

Figure 4.2   CPRR resin reclamation system [13]

of washing solution, the length of the wash cycle, and the temperature of the wash solution are all operator-regulated.

As a result of process research at CPRR, the general set of conditions established to effectively clean flake are wash time (approximately 10 min), wash solution temperature (82 °C), liquid detergent concentration (<5%), and an agitator power of approximately 0.03 hp/gal $(5.9 \times 10^3$ kW/l). Process conditions have not been optimized to specific container types or contaminants. The composition of the liquid detergent is proprietary but can generally be described as a mixture of anionic and nonionic surface active compounds accompanied by conventional inorganic alkalinity builders. Selection of the specific detergent was the result of extensive competitive product evaluations focused on PET reclamation.

Once the wash cycle is completed, the slurry is discharged to the secondary wash tank using a centrifugal pump. Immediately thereafter, another batch of flake and hot detergent solution is charged into the primary washing tank.

The secondary wash tank is fitted with an agitator to maintain the flake in a slurry and to keep contaminants dispersed. This slurry is pumped at a relatively low, constant rate to an inclined rotary screening device. Here, the flake is dewatered while traveling the length of the device. The screen itself is a right cylinder with a 30-in (76-cm) diameter and 10-ft (3-m) length. The screen fits into a stainless steel sleeve, which partitions the screen into two approximately equal lengths. The entire screen rotates at approximately 20 rpm. The inlet end of the steel reel is 30 mesh and passes the contaminated wash solution. In the outlet section of the rotary screening device, fresh water is sprayed on the flakes to remove additional contaminants. This rinse water is sewered.

The contaminated wash solution from the inlet section of the rotary screening device is recovered in a surge tank. Here, the contaminated washing solution is pumped to a series of two hydrocyclones. The bottom and top outlet streams of these hydrocyclones are passed through a filter medium to remove debris, respectively. The filtrate is returned to the surge tank. The bottom outlet stream from the second hydrocyclone is returned to the detergent solution tank, where additional washing solution is made as needed.

The output stream of flakes from the rotary screening device is discharged into a resin-rinsing system. Here, the flakes are slurried in water in a carbon steel tank fitted with an agitator. Using a centrifugal pump, the slurry is discharged into a series of two hydrocyclones. The polyolefin-rich, top outlet stream from the first hydrocyclone is piped to a secondary slurry tank. The slurry is pumped into the inlet nozzle of the second hydrocyclone. The polyolefin-rich, top outlet stream from the second hydrocyclone is discharged into the inlet chute of a centrifugal dryer.

Rinse water is recovered from both the hydrocyclones and centrifugal dryer. The water recovery system also is fitted with a filtration system to remove contaminants and resin fines. Additional moisture is removed from the centrifugally dried flakes by processing the material through an electrically heated counterflow hot-air dryer. The moist air discharged from the dryer is routed through a cyclone separator to remove any particulate matter. In a commercial plant, the hot-air drying system would typically include provisions for energy recovery, although the CPRR system does not have this feature.

Currently, the reclamation system at CPRR can produce only flaked resin. Because of the design of polyolefin resin handling and processing equipment, fabricators prefer to purchase pelletized raw materials [14]. Therefore, a commercial polyolefin resin reclaiming plant typically would include extrusion and pelletizing equipment. This processing offers an opportunity to further enhance the quality of the reclaimed resin. Some of the quality enhancements that can be made include contaminant removal by melt filtration, odor reduction by venting (under vacuum), homogenization by melt-mixing, and stabilization by antioxidant addition. In addition, as a result of being in the melt phase, other functional additives may be incorporated and uniformly dispersed to suit end-use markets—items such as pigments, reinforcements, and processing aids come to mind as examples.

Increasingly, equipment suppliers are offering an array of polymer melt filtration devices [15–17]. Currently, conventional shuttle, rotary, pull-through, and back-flushing screening devices are available. Melt filtration is deemed to be an essential precursor to end-use applications, such as thin-wall extrusion blow molding, to minimize blow-outs. Some of the available screening devices include control systems to automatically refresh the filtration medium—several of these are based on a response to extruder head pressure.

To assist in homogenizing the melt viscosity of the reclaimed polyolefin, high (>30) length-over-diameter extruders are favored. In addition, bulk blending can help to homogenize the pelletized material [18].

Rights to the CPRR plastics reclaiming process are available for a relatively small, one-time licensing fee [19]. The license includes a detailed description of the equipment and process, as well as hands-on training in CPRR's pilot plant.

The capital and operating costs of a 20-million-pound-per-year (9×10^3 t) resin reclaiming plant based on CPRR's technology have been published [20]. The full cost of the reclaimed resin, including raw material purchasing, plant operating costs, depreciation, sales and marketing costs, and overhead costs, is estimated to be 20¢/lb (44¢/kg) (in flake form). This estimate is based on a raw material cost of 6¢/lb (13¢/kg), delivered to the reclaiming plant. The authors estimate that the cost of reclaiming polyolefin containers should be somewhat lower, because aluminum-specific removal equipment is not required. (Some PET soft drink containers have aluminum caps that contaminate the flake, necessitating investment in specific equipment and operating supplies to remove this contaminant.) The cost estimate of reclaimed resin using the CPRR process is consistent with a DeWitt & Company industrywide survey [21], which reported that to sort grind, wash, and dry postconsumer plastic containers costs in the range of 12 to 19¢/lb (26 to 42¢/kg). To the authors' knowledge, no reports estimate the breakdown of costs among these subprocesses.

Worldwide, presently (1991) there are in excess of 15 licenses for the CPRR-developed process [22]. Two of these have constructed plants in the United States—United Resource Recovery, owned and operated by Hancor, Inc., Findlay, OH; and Day Products, Inc., owned and operated by day and Zimmerman, Philadelphia, PA. The Hancor plant is the only one of these designed specifically to recover polyolefin containers. Several firms listed below advertise the availability of engineered systems to reclaim rigid polyolefin containers [23, 24].

Firms Advertising Polyolefin Reclamation Systems

AKW–Hirschau, Germany
Graham Engineering—York, PA
John Brown, Inc.—Providence, RI
M.A. Industries, Inc.—Peachtree City, GA
Pure Tech International—Pine Brook, NJ
SEPCO–Spokane, WA
Sorema–Como, Italy
wTe Corporation–Bedford, MA

The typical capacity of single-train polyolefin reclaiming systems available from one of these companies is in the range of 1,000 to 5,000 lb/h (4.5 to 22.7×10^2 kg/h). Also, several of these vendors offer equipment and construction, installation, and start-up services

In 1990 a minimum of twelve plants with postconsumer, rigid polyolefin container reclamation capabilities were in commercial operation in the United States using primarily self-designed and engineered systems [25–32]. Companies operating these plants are listed below. The total, theoretical operating capacity range exceeds the supply of postconsumer polyolefin containers; however, much of this capacity is adaptable to other types of plastic containers.

U.S. Postconsumer Polyolefin Container Reclaiming Plants

Clean Tech—Dundee, MI
Eaglebrook Plastics—Chicago, IL
Envirothene—Los Angeles, CA
Graham Recycling—York, PA
M.A. Industries—Peachtree City, GA
Partek—Vancouver, WA
Plastics Recycling Alliance—Chicago, IL and Philadelphia, PA
Polymer Resource Group—Baltimore, MD
Resource Recycling—Tallulah, LA
United Resource Recovery—Kenton, OH
Wellman—Johnsonville, SC
Wheaton Recycling—Millville, NJ

Several new plants or plant expansions were announced [33–38] by virgin resin producers (DuPont in partnership with Waste Management, Inc.; Phillips in partnership with Partek, Quantum Chemical, and Union Carbide) and others (Eaglebrook, Midwest Plastics, and Orion Pacific) and were in various stages of development in 1991. These operations will likely add another 100–200 million pounds ($45–91 \times 10^3$ t) per year of reclaiming capacity. Some of these plant announcements indicated that an outside engineering and construction firm was to be used to supply the reclaiming hardware, including John Brown to Quantum Chemical and Sorema to Union Carbide.

Available reports of existing and planned polyolefin container reclamation systems indicate that each has unique design features. Nevertheless, these systems retain the basic processing steps outlined in Section 4.2. Figures 4.3a through 4.3g show the principal process features of a number of these systems [39–46].

At this time, commercial reclaimers represent multimillion-pound-per-year markets for the following grades of postconsumer polyolefin containers: unpigmented, extrusion blow-molded liquid food (milk and fruit juice) and water bottles; and pigmented and unpigmented extrusion blow-molded HDPE household chemical and health care product bottles. In some cases, suppliers commingle these groups of containers with each other or with other types of plastic containers, particularly PET soft drink bottles.

Reclaimers prefer to receive postconsumer polyolefin containers in baled form to improve shipping efficiency and to provide the opportunity to inspect and remediate, as necessary, raw material quality [47, 48]. While grinding provides similar benefits to baling in shipping efficiency, no commercial technology is currently available to separate types and grades of polyolefins in flake form.

The quality of postconsumer polyolefin containers is highly variable. Several of the reasons for this are the inconsistency of the instructions given to consumers related to the preparation and segmentation of plastic containers for recycling and consumer noncompliance with these instructions. Also, the continuing education of consumers in recyclables preparation techniques is lackluster. The inadequacies of consumer education are exacerbated by the collection and sorting infrastructures. To illustrate a few examples: commingled recyclable collection systems frequently result in high contamination of the plastics component with glass fragments from breakage [49]; extended outdoor storage of containers, particularly unpigmented containers, results in measurable photooxidation and material property deterioration [50]; and rusting steel baling wire discolors unpigmented bottles as a result of unprotected outdoor storage [51, 52]. In recognition of these and other quality challenges, the Institute of Scrap Recycling Industries has established a committee to codify the raw material quality requirements of the postconsumer plastics recycling industry [53].

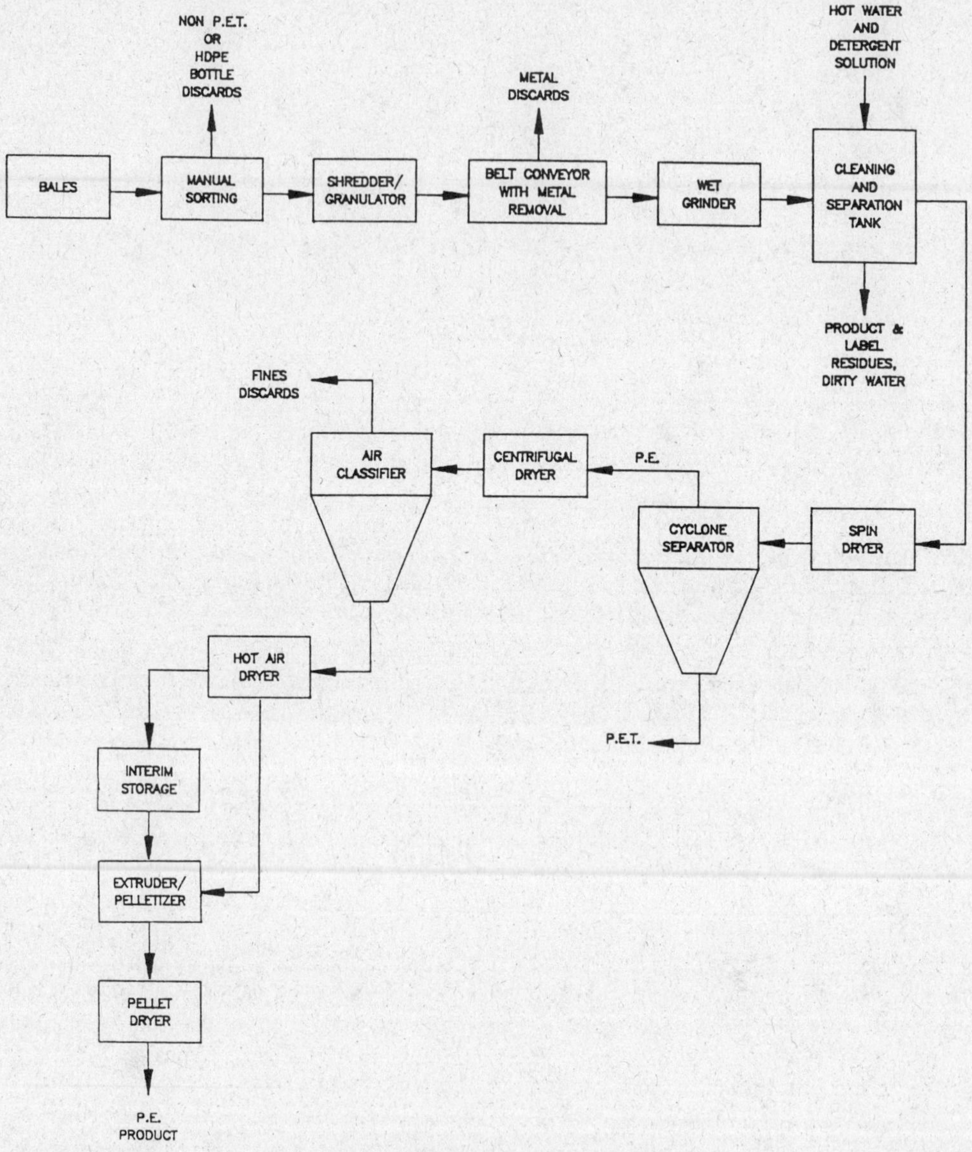

Figure 4.3a   Clean Tech reclaiming process [39]

Currently, polyolefin container reclaimers who receive baled stock usually must sort these materials to achieve reclaimable streams with viable (as measured by size and selling price) end markets. The commercial "state of the art" in sorting is explained herein. First, polyolefin or commingled plastic containers are debaled using a mechanical rake. The debaled containers are conveyed to an inspection station where individuals may either "positively" or "negatively" sort the containers. In a positive sorting operation, the desired containers are removed from the conveyor and fed to downstream processing equipment. In a negative sorting operation, undesired materials are individually removed from the conveyor. The material remaining in the conveyor is then fed to downstream processing equipment.

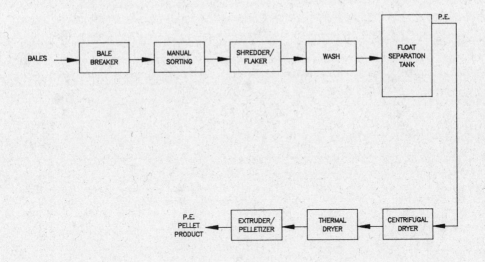

Figure 4.3b   Graham Recycling reclaiming process [40]

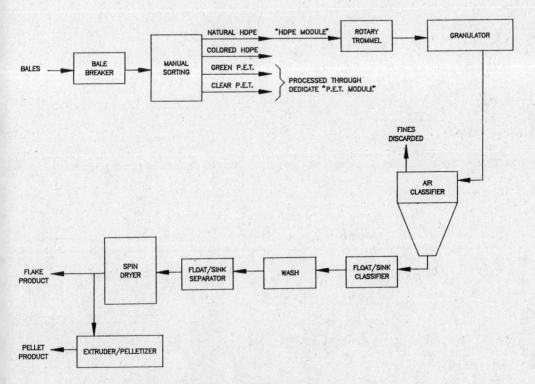

Figure 4.3c   M. A. Industries reclaiming process [41]

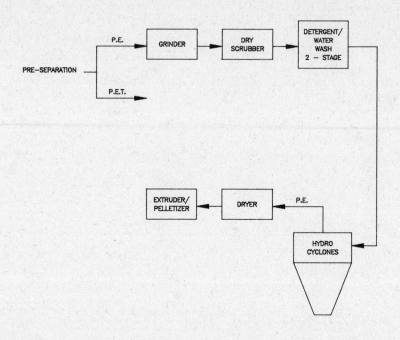

Figure 4.3d   Partek reclaiming process [42]

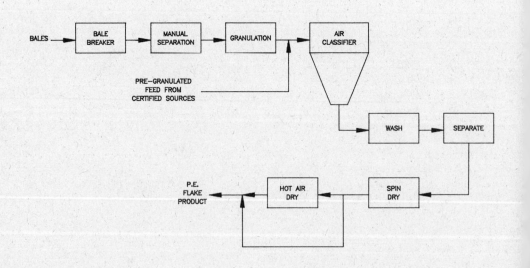

Figure 4.3e   Plastics Recycling Alliance reclaiming process [43]

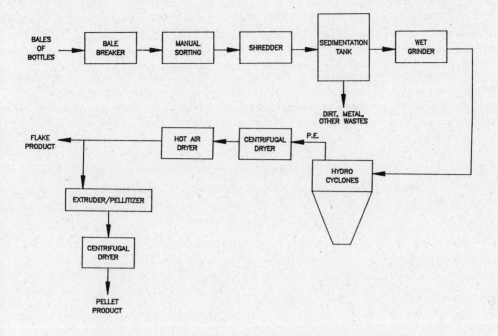

Figure 4.3f   Polymer Resource Group reclaiming process [45]

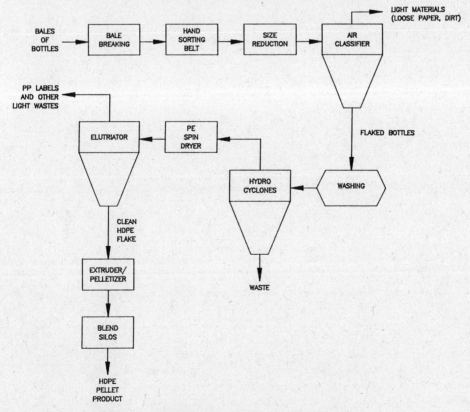

Figure 4.3g   Quantum Chemical reclaiming process [46]

The use of positive or negative sorting techniques is largely dependent on the composition of the baled stock. Commingled baled stock is invariably sorted positively. For example, stock containing predominantly both unpigmented and pigmented extrusion blow-molded HDPE containers would be positively sorted into these two streams. Sorting trials at Rutgers [54] and reports from commercial operations [55] indicate that an individual worker can sort at a rate of 3,000 to 4,000 containers per hour. (For 1-gal (3.79 liter) PE milk bottles, this sorting range translates to approximately 400–500 lb/h (181–227 kg/h).) When baled stock is purchased as being homogeneous, these materials would be typically debaled and sorted negatively to reject noncompliance items. In principle, this minimizes the costs associated with the sorting operation.

As a result of increasing collection of pigmented extrusion blow-molded HDPE containers, the motivation to sort these containers into color-segmented streams is also increasing. For instance, marketers of liquid household chemical products, such as laundry detergent, fabric softener, etc., are expressing an eagerness to recycle the HDPE containers for these products in a so-called "closed-loop" fashion—namely, taking resins reclaimed from postconsumer containers back into the same end market [56].

Infant markets are available for segmented streams of injection-molded HDPE containers [57] and PP-based multilayer food bottles [58]. In many cases, unpigmented PP-based food bottles are accepted as part of the HDPE liquid food container stream, because bottle manufacturers have determined that the inclusion of low levels of PP bottles does not unduly alter the performance of the recovered resin [59, 60]. (It is estimated that unpigmented PP-based bottles represent less than 5 wt % of unpigmented polyolefin liquid food containers.)

Increased postcollection mechanization of household recyclables sorting, including polyolefins, holds the promise of improving feedstock quantities and reducing overall recycling costs. Research on mechanized plastic container sorting is underway at both academic institutions and private companies. For example, Frankel et al. [61] describe an experimental device which sorts plastic bottles into six streams: polyvinyl chloride (PVC) (transparent); PVC (pigmented); PET (transparent, flint); PET (transparent, green-dyed); PE and PP (unpigmented); and all others. The sorting technology is based on the visible light transmission properties of the containers and their back-scattering characteristics after X-ray irradiation. It is the authors' expectation that mechanical plastic container sorting devices will be commercially available before the end of the 1990s.

4.2.1.2 Postcommercial Containers   Other rigid polyolefin containers have been recycled in the United States on a demonstration or experimental basis. For example, E. I. DuPont de Nemours has sponsored a pilot program to collect empty, triple-rinsed HDPE pesticide containers and to recover the resin from them [62]. Reclamation trials indicate that, while resin recovery is technically feasible, equipment modifications are needed to vent or contain volatiles during the washing process [63].

As an outgrowth of the plastic drum reconditioning business, recycling of 55-gal (208-liter) and other sizes of PE drums have been shown to be feasible. Reuse of PE drums is not always possible because of permeation of product contents into the material and the potential contamination of subsequent fills. In fact, the recycling of plastic drums is, in some cases, prohibited by law, including drums that have contained Environmental Protection Agency (EPA) Resource Conservation and Recovery Act (RCRA) "P-list" materials (see Title 40 Code of Federal Regulations (CFR) Section 261.33 for definitions) or Department of Transportation (DOT) "Poison B Liquids" (see Title 49 CFR 173.343 for definitions). In addition, empty drums that have contained DOT "hazardous materials" (see Title 40 CFR 172.101 for definitions) must be handled according to U.S. EPA regulations (see Title 40 CFR 261.7). For containers that do qualify for recycling, the Plastic Drum Institute of the Society of the Plastics Industry, Inc. (SPI), has developed a specific plastic drum purging procedure, which precedes granulation [64].

4.2.2 Films

4.2.2.1 Polyolefin-based Films Industrial monolayer polyolefin film scrap is routinely recycled by manufacturers, either in-house or through sales to others. For example, it is estimated that there is a 300-million-pound-per-year (136×10^3 t) merchant market in clean PE blown film scrap in the United States. By far the largest purchaser of this scrap is Mobil Chemical, which consumes approximately 100 million pounds (45×10^3 t) per year. The majority of this material is used to manufacture trash can liners [65, 66].

At this writing, postconsumer recycling of monolayer polyolefin films is a start-up activity. The emphasis is on collecting grocery sacks, dry cleaning bags, and, in isolated instances, merchandise bags, using specially designed collection containers located in retail stores [67]. These collection efforts are being sponsored principally by bag manufacturers and bag distributors. Some of the firms involved in these programs include Mobil Chemical, Sonoco Products, Polypak, Vanguard Plastics, and Challenge Bag.

The collected materials are baled and forwarded to film plants for recovery. At this point, the bags are hand-inspected to remove readily distinguishable contaminants, such as cash register receipts. The bags are then densified and used as a blend component to make new film products [68].

The first dedicated postconsumer polyolefin film reclaiming line in the United States has been announced by Union Carbide Chemicals and Plastics Company, Inc. [69], and will be part of a plastics recycling plant to be located in Piscataway, New Jersey. The process technology for this film reclaiming system has not been revealed, although plant start-up was scheduled for the end of 1991.

While PE dominates the polyolefin film market in the United States, production of biaxially oriented polypropylene-based (OPP) film is growing rapidly [70]. Most OPP film is used in packaging applications—snack food pouches, cigarette package overwrap, and cookie tray overwrap are a few examples.

OPP films range widely in complexity—from monolayer, to multilayer, to multilayer-multiresin, to multilayer-multimaterial—as determined by the demands of the end-use application. Reportedly, the majority of industrial OPP film scrap, including edge trim, web breakers, etc., is recycled in-house, as blend component in a core layer [71]. Multilayer-multimaterial OPP film scrap, on the other hand, is usually sold to reclaimers [72]. In many cases, aluminum is the other material in these films. The reclaimer may or may not recover the aluminum component, depending on the market value. The recovered PP generally finds its way into injection-molding markets [73].

Some quite complex, multilayer, PP-based films are being reground, with coextrusion of the immediate regrind being used as a distinct layer or layers. For example, HPM (Gilead, OH) recovers scrap from a PP-based polyvinylidene chloride (PVDC) barrier film. Regrind is incorporated into two 30% layers in a seven-layer structure, as shown in Figure 4.4. This processing system was developed by Dow Chemical Company and uses a "double wave" screw design in the regrind extruder to achieve effective mixing at low temperatures [74].

No postconsumer, polyolefin-based, multilayer films are being commercially recycled as such. An incidental amount of these materials may find its way into "mixed" plastics recycling; however, in view of the low volume of these materials, there is no compelling motivation to establish a collection and separation infrastructure.

4.2.2.2 Polyolefin-coated Materials Polyolefins are also used as a film coating on other material substrates, principally wood pulp-based materials. For example, paperboard products used for liquid packaging are either lined or coated with polyolefins for moisture resistance. PE extrusion-coating resins dominate the paper and paperboard coating markets.

Recycling of polyolefin-coated paper and paperboard has its roots in the recovery of plant scrap [75]. Driven by the value of the long-fiber wood pulp in coated paperboard, plant scrap

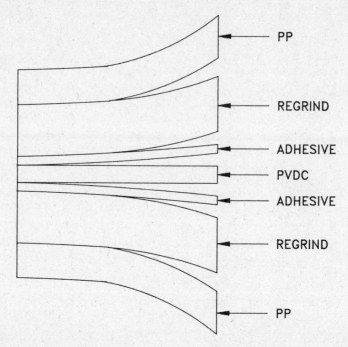

Figure 4.4   Polypropylene-based multilayer film structure incorporating mixed polypropylene/polyvinylidene chloride regrind [74]

recycling has historically paid little attention to the value of the byproduct PE. While contaminated with pulp fiber, this by-product stream is now beginning to find a market in profile extrusions.

At this time there is no United States market for postconsumer, extrusion-coated paperboard packaging. However, there have been recent indications of postconsumer recycling initiatives by some companies with a vested interest in virgin paperboard production and paperboard-based container manufacturing [76, 77].

4.2.3 Foams

Foamed polyolefins are used predominantly in cushioning and in thermal and electrical insulating applications. Both physical and chemical blowing agents are used to produce these foams [78].

To produce extruded polyolefin foams using a physical blowing agent, the agent and the polyolefin resin are fed (under pressure) to an extruder where, on exiting the die, the agent vaporizes, forming gas cells. Alternately, a chemical blowing agent, which is intimately mixed with the polyolefin melt, thermally decomposes and liberates the cell, forming gas in the extruder.

Polyolefins also can be made into expanded foams. Mixtures of polyolefin and blowing agent are partially expanded at elevated temperature to form a bead. These beads are, in turn, poured into a mold where reheating causes further bead expansion and fusion at the bead boundaries, resulting in a solid foam structure.

To produce polyolefin foams with smaller, more uniform cells, as well as with good dimensional stability, particularly at elevated temperatures, processes involving both resin cross-linking and foaming have been developed. Here, mixtures of polyolefin and chemical blocking agent are extruded below the decomposition temperature of the blowing agent. The extrudate is either chemically or radiation-cross-linked prior to reheating to activate the blowing agent. In an alternate process, the cross-linked extrudate is saturated with high-pressure at 10,000 psi (69 MPa) nitrogen, heated, and decompressed to form the foam stock.

Extruded and expanded foam scrap is routinely recovered in manufacturing plants. However, because of the low bulk density of these materials, feeding an extruder can be problematic. One alternative to overcome this is to use a cram feeder, and the other is to use a friction densifier during granulation, which collapses the foam structure. Discarded polyolefin foam from some postcommercial sources also is recycled [79].

Typically, cross-linked polyolefin foams are not recycled because of the difficult melt processing characteristics of these materials. A small amount of this material is used in proprietary, specialty end-uses, while most is simply landfilled [80].

4.2.4 Wire and Cable Jacketing

Thermoplastic polyolefins and PVC are the dominant materials used to insulate and sheath metallic wire conductors. Lesser amounts of fabric, cross-linked polyolefin, and rubber are used for this application [81].

Because of the high value of copper and other metallic conductors, manufacturing wire and cable scrap has historically been recycled. In cases where uncontaminated thermoplastic polyolefins can be recovered, jacketing material is recycled in-house, particularly by large wire and cable manufacturers. Otherwise, contaminated scrap is reclaimed by independent commercial ventures specializing in wire and cable scrap recovery. There are approximately a half-dozen firms involved in this business in the United States. Some of these same firms also are involved in recovering valued materials from aged and obsolete wire and cable discarded by electrical utilities and communication firms [82]. To the authors' knowledge, there is no current effort to reclaim polyolefin jacketing from wire and cable arising from demolition or construction of office or commercial buildings or residences.

Some wire and cable reclaiming processes are described in the literature. Bevis et al. [83] discuss a so-called rising current separator, which uses a series of specially designed, agitated water tanks to effect the separation of the polyolefin, PVC, and copper into their respective streams. Downstream processing equipment is described that further refines the two thermoplastic streams. An overview of the process is illustrated in Figure 4.5. Bevis et al. indicate that this separator system is commercially available from Reconstituted Plastics Ltd. (Denton, Manchester, UK).

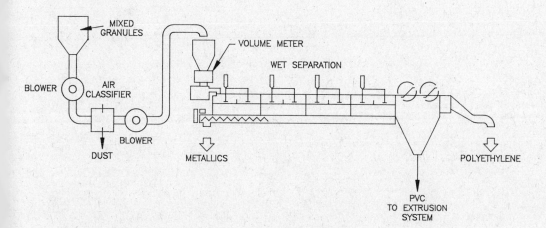

Figure 4.5   "Rising Current" separator for wire and cable reclaiming [83]

Investigators at the Bureau of Mines describe the use of an electrostatic separator to segment jacketed wire into plastic and metal streams [84]. Using a vibratory feeder, uniform particles of wire and jacketing are fed onto a grounded drum rotating in a highly charged electrostatic field. The plastic fraction is pinned to the drum and is removed by a fixed wiper blade, while the metal fraction is thrown from the drum. Multistage electrostatic separators can further improve the quality of the metal-thermoplastic separation.

Plastic Recovery Systems (Toledo, OH) recently reported on the development of a system to recover PE and PVC from wire scrap using modified flotation techniques [85].

4.2.5 Polypropylene Battery Cases

In 1990, an estimated 100 million pounds (45×10^3 t) of virgin PP were used to mold lead-acid battery cases in the United States [86]. Because of enhanced impact properties, rubber modified grades of PP (i.e., I-PPs) are used exclusively in this application.

The first commercial PP battery case was introduced in the 1960s to compete with vulcanized rubber cases. While both PP and vulcanized rubber cases continue to be produced, PP now holds virtually a 100% share of the automotive and truck battery markets [87].

Due to the value of lead and the safety hazards presented by lead-acid batteries to automotive scrap recovery operations, there is a well-established battery recycling industry in the United States. In fact, it is estimated that in excess of 60% of lead production in the United States is derived from recycling batteries.

Because of either mandated deposits or a combined land disposal ban and mandated delivery/take-back provisions on automotive batteries in selected states, consumers are increasingly returning spent lead-acid batteries to automotive parts retailers. The states with mandated provisions for battery return are given in Table 4.3. This provides an additional supply mechanism for the lead-acid battery recycling industry [88].

Table 4.3   States with legislated provisions for lead-acid battery return (1990)

Mandatory Deposit	Land Disposal Ban Mandatory Deliver/Take-Back
Maine	California
Minnesota	Florida
Rhode Island	Hawaii
Washington	Illinois
	Iowa
	Louisiana
	North Carolina
	Oregon
	Pennsylvania
	Wyoming

The PP battery case represents approximately 7 wt % of the recoverable materials in spent and used automotive and truck batteries [89]. To recover PP as a separate stream, whole lead-acid batteries are mechanically crushed and fed to an aqueous flotation tank. Thereafter, the recovered case material is reduced to a smaller particle size, further purified, including neutralization of sulfuric acid residues, modified with additives as required by new end-use applications, and melt-processed into pelletized resin [90, 91]. Four of the largest PP battery case reclaimers in the United States are M. A. Industries, Inc. (Peachtree, GA), Exide Corp. (Reading, PA), Witco Corp., Richardson Division (Indianapolis, IN), and Johnson Controls, Inc. (West Union, SC).

4.3   Effects of Multiple Processing on Polyolefin Resins

During the melt processing of polyolefin resins, changes to molecular structure occur [92–95]. High temperatures, an oxidizing environment, and mechanical shear cause chain-scission reactions. This can lead to either cross-linking reactions, resulting in higher molecular weight species, or chain termination reactions, yielding lower molecular weight species. The change in molecular weight distribution, and concomitant oxidation are manifested in changes in mechanical and melt rheological properties of the resin. Each exposure of the resin to melt-processing conditions is considered a heat history.

The behavior of virgin resins exposed to multiple heat histories is not generally a good predictor of the behavior of postconsumer recycled resins—the effects of compounded additives and contaminants can be marked [96–98]. However, the majority of the research on repeated melt processing of polyolefins has been done on well-defined virgin materials.

4.3.1   Polyethylenes

Ryborz [99] demonstrated that resin processing temperature is a key factor in the flow characteristics of a material after each heat history. Using HDPE of weight average molecular weight (M_w) 91,000–96,000 in an injection molding process, Ryborz showed that the shear viscosity of the resin after eight heat histories was much more reduced at a processing temperature of 300 °C as opposed to 275 °C, as shown in Table 4.4

Table 4.4   Effect of multiple heat histories on the melt flow characteristics of polyethylenes

Polyethylene	Flow Exponent* at 225 °C
"Virgin" Resin	3.93
Processed at 275 °C, 8 cycles	3.64
Processed at 300 °C, 8 cycles	3.17

* Slope of a full logarithmic plot of shear rate versus shear stress

Shenoy et al. [100] proposed a method of predicting the dependence of the melt viscosity of reprocessed generic resins on shear rate using only knowledge of the melt flow index of the resins. They showed that a constant temperature plot of melt viscosity times melt flow index versus shear rate over melt flow index yielded a master curve for all grades of resin in a generic family, including the LDPE and PP families.

Information on the characteristics of contaminated, reprocessed PE is less abundant in the literature. In a controlled experiment, Sikora and Bielkinsk [101] examined the mechanical properties of reprocessed film grades of PE containing up to 5 wt % of mixtures of common agricultural fertilizers and humic soil. Not surprisingly, the tensile strength, elongation at break, and Young's modulus of the reprocessed PE decreases as the level of contamination rises. The moisture content and particle size of the contaminants had markedly less influence on the results. The authors also described the effect of one- and two-year aging on test specimens molded from virgin, production scrap, and reclaimed postconsumer PE. While the precise environmental exposure of the aged test specimens was not detailed, the reclaimed PE shows a more marked time-dependent decline in elongation at break than the other two resins. Presumably, this effect is due to trace contaminants in the reclaimed material.

Gibbs [102] describes an evaluation of commercially available extrusion blow-molding grades of postconsumer recycled homopolymer and copolymer HDPE, derived from milk and household liquid chemical packaging, respectively. The study focuses principally on the question of the acceptability of these materials to produce commercial quality, monolayer, blow-molded containers. He observed significant lot-to-lot variation in the density, MI, environmental stress crack resistance (ESCR), and mechanical properties of the reclaimed materials, as shown in Tables 4.5 and 4.6. Table 4.5 also shows that the mechanical properties of a composite sample of these postconsumer recycled resins are comparable to those of (reprocessed) virgin resins, with the exception of elongation at break. Also, the ESCR of the composite recycled sample of the postconsumer copolymer materials is comparatively reduced, as shown in Table 4.6. It is presumed that these differences can be attributed to contaminants in the reclaimed resins.

Table 4.5   Physical and mechanical properties of postconsumer recycled and virgin homopolymer HDPE used in milk packaging

Property	Commercial Recycler A 7 Lots	Commercial Recycler B 5 Lots	Composite * Recycled Sample	Virgin Sample	Reprocessed Virgin Sample
Melt index (g/10 min)	0.44–0.57	0.40–0.66	0.53	0.77	0.79
Density (g/cm^3)	0.957–0.963	0.960–0.964	0.961	0.963	0.961
Flexural Modulus (10^3 psi)	210–236	207–223	223	219	219
(MPa)	1450–1630	1430–1540	1540	1510	1510
Izod Impact, notched (ft-lb/in)	2.1–3.2	2.7–3.9	2.7	2.4	1.7
(J/cm)	1.1–1.7	1.5–2.1	1.5	1.3	0.9
Tensile Strength, at yield (psi)	4248–4422	4263–4365	4336	4292	4350
(MPa)	29.3–30.5	29.4–30.1	29.2	29.6	30.0
Tensile Strength, at break (psi)	1203–3176	2654–3857	3190	2204	2494
(MPa)	8.3–21.9	18.3–26.6	22.0	15.2	17.2
Elongation at break (%)	58–332	21–115	103	555	613
Tensile Impact (ft-lb/in^2)	58–77	70–76	72	82	85
(kJ/m^2)	123–162	148–159	152	172	178

* Composite of all lots from both Recyclers A and B

In Gibbs's characterizations of the processibility of postconsumer recycled HDPEs, he notes that contaminants lead to "numerous blow holes in thin-walled containers." He also notes a marked difference in the visco-elastic behavior of the recycled resins, compared to virgin resins, as measured by die-swell. The die-swell of the recycled copolymer resins increased compared to a virgin resin control, while that of the recycled homopolymer resin decreased relatively, indicating the possibility of shear modification or another form of degradation in the recycled resins. In addition, the processing of the postconsumer recycled resins produced much more "smoke" than the virgin resin counterparts.

Gibbs's study also describes the ESCR and mechanical and odor properties of blow-molded containers produced from the recycled resins, as shown in Table 4.7. While the column crush characteristics provided by the postconsumer recycled resins are competitive to virgin materials, the odor and ESCR characteristics are not. Again, these differences are probably caused by contaminants in the postconsumer recycled resins.

Table 4.6   Physical and mechanical properties of postconsumer recycled and virgin copolymer HDPE used in liquid chemical packaging

Property	Commercial Recycler A 7 Lots	Commercial Recycler B 5 Lots	Composite * Recycled Sample	Virgin Sample	Reprocessed Virgin Sample
Melt index					
(g/10 min)	0.52–0.66	0.61–0.67	0.62	0.30	0.42
Density					
(g/cm^3)	0.962–0.964	0.964–0.971	0.965	0.954	0.954
Flexural Modulus					
(10^3 psi)	144–212	184–193	189	186	189
(MPa)	992–1460	1270–1330	1300	1280	1300
Izod Impact, notched					
(ft-lb/in)	1.5–2.4	1.1–1.3	1.7	2.4	2.2
(J/cm)	0.8–1.3	0.6–0.7	0.9	1.3	1.2
Tensile Strength, at yield					
(psi)	3466–4365	3828–4089	4017	3843	3814
(MPa)	23.9–30.1	26.4–28.2	27.7	26.5	26.3
Tensile Strength, at break					
(psi)	1203–3176	3031–3582	2755	2306	2538
(MPa)	8.3–21.9	20.5–24.7	19.0	15.9	17.5
Elongation at break					
(%)	58–464	16–55	144	675	694
Tensile Impact					
(ft-lb/in^2)	49–97	36–60	71	100	90
(kJ/m^2)	102–204	76–126	149	210	189
Bent Strip ESCR					
(h to failure)	4.4–12.0	7.3–8.0	7.9	24.0	24.0

* Composite of all lots from both Recyclers A and B

Table 4.7   Column crush, environmental stress cracking resistance (ESCR) and odor characteristics of postconsumer recycled and virgin high-density polyethylene containers

Polymer	Column Crush [a] Lb	(N)	Top-Load ESCR, F50 [a] (h at 50 °C, 67N)	Bottle ESCR, F50 [a] (days at 66 °C)	Odor [b] Amplitude
Homopolymer					
Virgin	72	(318)	–	–	1.5
Reprocessed Virgin	70	(311)	–	–	1.5
Recycler A	68	(303)	–	–	0
Recycler B	59	(264)	–	–	0
Copolymer					
Virgin	54	(238)	15.0	12.4	1.5
Reprocessed Virgin	54	(238)	31.1	10.0	2.0
Recycler A	61	(273)	4.4	1.0	0.5
Recycler B	55	(246)	6.0	1.0	0.5

[a] 16 oz Boston round bottle; target weight 25 g
[b] Rating scale: 0 = worst; 3 = no odor

4.3.2 Polypropylenes

Numerous authors have discussed how the molecular weight reduction occurs during the processing of virgin PP and other resins. According to Shenoy et al. [103], these "can be characterized in terms of parameters like normal stress difference [104], die-swell ratio [105], and shear viscosity [106–110] as a function of temperature and shear rate." In another example, Schott and Kaghan [111] found that six passes through a commercial screw extruder at 260 °C decreased the M_w of a PP sample from 270,000 to 80,000.

In experiments to compare and contrast the mechanical properties and processing characteristics of virgin and uncontaminated, recycled PP, including structures containing the barrier resin ethylene-vinyl alcohol copolymer (EVOH) and compatibilizing adhesive resins, it has been shown that recycled, commercial grades of H-PP, C-PP, and I-PP are virtually indistinguishable from the parent virgin resin after one additional heat history [112]. These data are shown in Table 4.8.

Table 4.8   Mechanical properties and flow characteristics of virgin and recycled PP

Property	H-PP (MFR ≈ 5)		C-PP (MFR ≈ 2)		I-PP (MFR ≈ 4.5)	
	virgin	once recycled	virgin	once recycled	virgin	once recycled
Tensile Strength, at yield						
(psi)	5020	5030	4260	4130	3630	3900
(MPa)	34.6	34.7	29.4	28.5	25.0	26.9
Tensile Strength, at break						
(psi)	4460	3800	2780	2720	3230	3360
(MPa)	30.7	26.2	19.2	18.7	22.3	23.2
Flexural Modulus, 1%						
(10^3 psi)	278	279	174	172	201	216
(MPa)	1915	1922	1199	1185	1385	1488
Flexural Modulus, 2%						
(10^3 psi)	253	257	163	160	183	195
(MPa)	1743	1771	1123	1102	1261	1344
Izod Impact, notched, 23 °C						
(ft-lb/in)	0.6	0.5	1.5	1.4	2.1	1.8
(J/m)	32	27	80	75	112	96
Izod Impact, unnotched, -18 °C						
(ft-lb/in)	2.3	2.3	2.5	3.3	11.0	11.0
(J/m)	123	123	133	176	587	587
Heat Deflection Temperature						
66 psi (°F)	212	221	165	178	207	207
0.5 MPa (°C)	100	105	74	81	97	97
Spiral Flow, 205 °C						
(in)	11.7	11.25	9.5	9.1	11.0	11.0
(mm)	297	286	241	231	279	279
Spiral Flow, 232 °C						
(in)	13.5	13.5	10.75	10.75	12.75	13.0
(mm)	343	343	273	273	324	330
Spiral Flow, 260 °C						
(in)	15.75	15.25	11.75	12.0	14.5	14.75
(mm)	400	387	298	305	368	375

4.4   End Uses

4.4.1   Polyethylenes

Of the approximately 35 million pounds (16×10^3 t) per year of injection molding-grade HDPE recovered as a coproduct of soft drink bottle reclamation, much of this is used to make new soft drink bottle base cups [113]. Reportedly, the reclaimed resin can be commingled with virgin resin to a minimum of 25% with no apparent sacrifice in the performance of the base cup [114].

In 1989, several major consumer product companies created a new market for postconsumer recycled, blow-molding grades of HDPE by committing to incorporate this material in their packaging [115, 116]. Several other companies have since followed suit [117–122]. At this time, the targeted incorporation level is approximately 25% recycled resin, with commitments to go to higher levels as postconsumer recycled resin availability and quality (and cost) permit. The bottles affected by these announcements are for packaging liquid chemicals used in the household, including detergents, hard-surface cleaners, fabric softeners, motor oils, etc. Obviously, the principal challenge here is to produce containers that are functionally and sensorially indistinguishable from all virgin resin bottles.

Currently, two different techniques are being used to produce recycled content HDPE bottles for nonfood applications. In the first, pelletized, reclaimed resin is commingled with virgin resin prior to extrusion to produce a "monolayer" bottle [123]. Depending on the grade of the recycled resin component, this can result in several limitations. For example, if fractional MI homopolymer HDPE is used as the blend component, the ESCR of the resulting container is reduced. However, the resulting bottle usually has to be black to hide the rainbow of pigments in the feedstock [124]. Finally, regardless of the origin of the recycled blend-stock, odorants in the recycled resin can be apparent to those handling the resulting containers [125].

In one approach to overcome the above-described ESCR limitation, a blend of high-ESCR virgin HDPE and 25% postconsumer recycled homopolymer HDPE was first introduced by Solvay Polymer Corporation [126]. Solvay claims that the ESCR performance of the blended resin is competitive with conventional copolymer HDPE used in liquid household chemical packaging applications [127].

The above-described bottle color limitation can be overcome, in part, by color- or brand-sorting of containers prior to reclamation, as discussed briefly in Section 4.2.1.1. However, the supply of postconsumer plastic bottles is not, as yet, large enough, nor is sorting technology sufficiently advanced, to make this relatively cost-effective.

The other technique being used to produce recycled content HDPE bottles is coextrusion blow molding. Reportedly, from two- to four-layer bottles are being produced, at least on an experimental basis, to overcome sensory and functional objectives to recycled content bottles [128–130]. The most popular coextruded structure finds recycled homopolymer HDPE, combined with trim scrap, sandwiched between two layers of (pigmented) virgin copolymer HDPE. The core recycled layer provides column crush strength, while not being directly exposed to a harsh chemical environment or the sensory evaluations of the consuming public.

Since coextrusion blow molding machines are relatively rare, costly, and operated by few and generally large container manufacturers, it is unclear which, if either, of the above-described fabrication techniques will dominate the postconsumer recycled content HDPE bottle market in the long run.

In another high-valued application, postconsumer recycled homopolymer HDPE has been evaluated in three-layer blown film for diaper packaging [131]. Here, the cleanliness of the reclaimed resin is extremely critical to prevent pin-holing and to maintain bubble stability during processing.

Many niche end-use applications for recycled polyolefins have been reported [132]. Custom Pak (Clinton, IA) blow molds HDPE duck nests containing up to 15% postconsumer recycled liquid food containers [133]. Santana Plastic Products (Scranton, PA) compression molds sheeting from recycled HDPE bottles for use in divider walls in public lavatories [134]. Embrace Systems Corp. (Buffalo, NY) manufactures a laminated, spun-bonded HDPE and LDPE film from recycled PE for use as a wind barrier in building construction [135].

4.4.2 Polypropylenes

From a postconsumer recycling perspective, the largest volume (yet least visible) recycled polyolefin resin is derived from battery cases. Of the estimated 100 million pounds (45×10^3 t) per year of PP reclaimed from these cases, approximately 40 million pounds (18×10^3 t) are used to manufacture new battery cases. The disposition of the remainder of this material is not well documented—anecdotally, the material is used in such applications as solid wheels (for lawn mowers, etc.,) decorative shutters, and other injection molding applications [136].

4.5   Global Activities

This section focuses on polyolefin recycling activities in Western Europe, Canada, and Japan. As an introduction, the use of polyolefins for consumer product packaging and the availability of this packaging for recycling in Japan and Western Europe differs from the situation in the United States, while the situation in Canada is very similar [137]. In Canada, one notable exception is milk packaging. The PE film milk pouch is used widely, as opposed to the ubiquitous HDPE milk jug in the United States.

In Japan, comparatively little HDPE is blow-molded into containers for consumer product packaging. PE is used widely for packaging film applications and, as a result, few postconsumer polyolefin items are recycled as materials. However, postindustrial and postcommercial polyolefin items are reclaimed. Recycling via thermal decomposition into useful products such as gasoline or directly as fuel for waste-to-energy also occurs.

In Western Europe, PE is used principally in film applications. (Bottles and other rigid containers for consumer product packaging are most often made from PVC and PET, rather than HDPE.) To spur postconsumer recycling, it is expected that legislation common to all countries in the European Economic Community will be in place by 1992. In the meantime, most organized postconsumer polyolefin recycling occurs in Germany and the United Kingdom, with operations scattered in other countries. Because of the high level of activity in the United Kingdom, this country will be addressed separately from continental Western Europe.

4.5.1   United Kingdom

Data reported by the British Plastics Federation (BPF) show PE film to be the most widely recycled plastic material in the United Kingdom [138]. Approximately 132 million pounds (60×10^3 t) per year of primarily postcommercial and used agricultural films are reclaimed. British Polythene Industries (BPI) reprocesses the bulk of this at plants in Derby and Strathclyde, including all film from the BPF-cosponsored curbside "recycling cities" projects in Sheffield, Cardis, Dundee, and the rural county of Devon [139]. These BPI plants are Alida Reclaimers and Anaplast Ltd., respectively. Reportedly, $5.2 million (£3 million) has been invested in these operations. Alida Reclaimers produces pelletized material, while Anaplast produces PE film with recycled content [140]. In an example end market, BPI expects to supply an estimated 10 million recycled content plastic carrier bags to the Sainsbury grocery store chain each week [141].

Another PE film reclaimer, Polylina Ltd. in London and Middlesex, works with grocery and drug stores countrywide to take back PE carrier bags and convert these into can liners. In addition to BPI, Second Life Plastics, Ltd., reclaims PE agricultural film from collection programs across the country [142].

According to a 1990 BPF directory [143], a total of 24 firms are engaged in reclaiming PE film in the United Kingdom. Besides agricultural films, can liners, and carrier bags, other end uses for reclaimed PE are industrial sheet and profile extrusions. One of the better known companies involved in manufacturing profiles is Superwood, Inc.

Rigid PE items in the above-mentioned "recycling cities" programs are taken by Dow Chemical. Meanwhile, Procter & Gamble is involved with the Newcastle local authority in collecting and reclaiming HDPE containers from approximately 2,000 households. BPF reports that a total of 41 other firms are involved in reclaiming rigid PE items [144]. Most (34) of these firms produce only pellets or granules, while the remainder are integrated into manufacturing end-use products. End uses for the reclaimed resins include pipe, moldings, building components, industrial sheet, extruded shapes, and highway, safety, and street fixtures.

The largest reclaimer of rigid polyolefin items in the United Kingdom is Cookson. At a plant in Northumberland, Cookson Industrial Materials, Ltd. reclaims PP battery cases. Cookson Plastic Products, Ltd. in Cheshire washes, compounds, and processes 22 million pounds $(10 \times 10^3$ t) per year of used polyolefin items into molded horticultural products, such as plant pots and seedling trays [145]. Cookson also produces a container with recycled HDPE content for Johnson Wax's "Shake 'n Vac" carpet cleaner. This is one of the first examples of postconsumer reclaimed HDPE going into consumer product packaging in the United Kingdom. Besides Cookson, 42 companies are reported to be involved in reclaiming PP [146]. One firm, Mayer Cohen Industries, Ltd. of Gwent, recovers PP film. Of those companies reclaiming PP, most (37) produce only pellets or granules. A total of 55 million pounds $(25 \times 10^3$ t) per year of PP is recycled in the United Kingdom.

4.5.2 Continental Western Europe

Western Europe plastics industry organizations working in the field of plastics recycling include the recently formed Plastics Waste Management Institute (PWMI) of the Association of Plastics Manufacturers in Europe (APME). Another group, RECOUP, was formed in 1990 by an alliance of container manufacturers and raw material suppliers to promote recycling of postconsumer plastic containers by funding collection programs, conducting research to improve the efficiency of collection, acting as a clearinghouse for information on plastics recycling, and guaranteeing an outlet for collected plastics [147]. In 1988, the Italian Institute of Waste Material Management was formed by Italy's major plastics manufacturers and suppliers. The nonprofit association disseminates information on the industrial, environmental, and social aspect of solid waste management, with an emphasis on plastic wastes [148].

Generally, Western Europe polyolefin recycling activities have concentrated on PE agricultural and commercial films. A majority of the polyolefin reclaiming industry is located in Germany, Holland, and Italy. These countries also have the highest levels of activity in developing resin reclamation systems.

In Germany, a novel process has been designed to recover PE and PP from household recyclable wastes [149]. The technology, developed by AKW Apparate and Verfahren GmbH, uses hydrocyclones to separate polyolefins from denser plastic materials. Household members are requested to separate recyclables, including plastic, glass, metal, paper, wood, textile items, etc., from compostable wastes. After collection, the recyclable fraction is then segmented further by hand sorting and magnetic separation at a central processing plant. The plastic items are baled and delivered to plastics processing plants. Here, the bales are shredded, and the resulting hand-sized pieces are taken through a magnetic separator. Following a settling tank, which allows mineral contaminants to be removed, the material is fed to a wet-mill granulator. This step flakes and prewashes the plastics before they enter another wash tank. From this second wash tank, the slurried flake is pumped to hydrocyclones, which separate the plastics into two fractions: "lights" (materials less dense than water) and "heavies" (materials denser than water). The lights consist of PE and approximately 5% PP. The polyolefin flakes are dewatered in a small blending silo and then fed, using a cram feeder, into a vented extruder equipped with a screen-pack. The exudate is pelletized. The pellets are cooled in a water bath, dewatered in a centrifuge, stored in a silo, and finally bagged for sale.

According to a 1990 listing [150], there are 19 commercial PE reclaimers in Germany. One of these firms, The Polymer Resource Group, a joint venture between Germany's AKW and U.S. mineral importer ITC, Inc., has three plants located in Germany and Switzerland. The plants have a combined capacity of 66 million pounds (30×10^3 t) tonnes per year.

A number of other German companies offer engineered polyolefin reclaiming systems. For example, Sikoplast of Siegsburg advertises the availability of equipment to reclaim PE from agricultural film, PP from strapping, and OPP from food packaging film. In another example, PP battery cases are the feedstock for a reclaiming system developed by Herbold GmbH of Meckesheim. In this process, the battery cases are first wet-granulated. PP is separated from contaminants in a two-stage wet separation process, based on density differences. Granules of PP produced by this system are used in applications such as battery housings [151].

Several German virgin resin manufacturers have announced polyolefin recycling initiatives [152–154]. BASF of Ludwigshafen has demonstrated a technology for reclaiming PE from agricultural and packaging films. BASF, Bayer, and Hoechst have formed a joint venture company, EWK, to promote plastics recycling in Germany and to develop resin reclaiming processes and end-use applications for reclaimed resins. Hoechst plans to build a $7.3 million (DM12.5 million) polyolefin recycling unit at its Knapsack, Germany, plant. Expected to be operational in 1992, the plant will process up to 11 million pounds (5×10^3 t) tonnes per year. The major feedstocks for the plant will be PP from automobile, appliance, and packaging wastes.

Elsewhere in Western Europe, reclamation systems for recovering plastics have been developed. While these systems are aimed primarily at PET and PVC recovery, they have applicability to polyolefin recovery as well. For example, the PET reclamation technology used by Reko in Holland has application in reclaiming polyolefins. Sorema in Como, Italy, has been supplying plastics recycling equipment since 1976. Approximately 170 lines have been installed at 35 facilities in Europe, South America, Asia, and the United States, with the bulk of these in Europe. Worldwide, Sorema-designed plant capacity exceeds 992 million pounds (450×10^3 t) per year. Sorema's systems are individualized to meet customer needs and reclaim a wide variety of plastic materials and items from a broad spectrum of sources. In polyolefins, Sorema has designed systems for soiled PE agricultural films, HDPE chemical drums, PP battery cases, used garbage bags, and blow-molded containers [155, 156].

Commercial polyolefin reclaimer Wavin in Holland uses recovered PE agricultural film, along with other polyolefin waste, including cable sheathing and industrial scrap, to manufacture drink crates. Raw material quality is gauged with a patented infrared detector to determine the degree of material aging prior to reclamation. The scrap is granulated, washed, dried, and compounded using additives appropriate to the degree of aging of the material. This process reportedly yields crates equivalent in quality to those produced from virgin resins [157].

Other Western European reclaimers include Relatex Company in Wildon, Austria, where Machinenfabrik Andritz AG has built a PE film granulation plant [158], and PolyRecycling in Weinfeldern, Switzerland, with 44 million pounds (20×10^3 t) per year of polyolefin reclaiming capacity [159].

4.5.3 Canada

Several groups in Canada have formed to promote and develop a recyclables collection infrastructure that includes plastic items. Of note are Ontario Multi-Materials Recycling, Inc., and the Collecté Seléctive Québec. Additionally, the Environment and Plastics Institute of

Canada is cooperating with the Ontario Ministry of the Environment in a collection study involving the City of Peterborough, Ontario.

In Canada, postconsumer polyolefin items are collected in both curbside and drop-off programs. Approximately one-half of the current 350 curbside recyclables collection programs accept plastic items. HDPE containers are among these items. Several curbside programs also accept PE bags [160].

A number of retail stores have set up merchandise bag or grocery sack drop-off programs, the most notable being a collaboration by numerous supermarkets in Nova Scotia and New Brunswick. Here, used carrier bags are turned over to PCL Industries in Oakville, Ontario, and Eastern Packaging in St. John, New Brunswick, and reclaimed and remanufactured into new bags [161].

In a different approach, convenience store operator Becker's Ltd., in Toronto, Ontario, places a deposit of 25¢ per container on HDPE milk jugs. The returned jugs are processed by the Plastics Recycling Alliance [162].

According to the 1990 Canadian Recycling Directory, there are currently 17 companies in Canada with resin reclamation capability, including washing capacity [163].

4.5.4 Japan

It appears that very little household-derived postconsumer plastics are recycled in Japan. According to a report by the Japanese Plastic Waste Management Institute [164], there is only one municipal plastic waste collection in the country, and it collects less than 2 million pounds $(0.9 \times 10^3$ t) per year of material.

Feedstocks for Japanese polyolefins recycling operations include manufacturing scrap, film wraps used in product distribution systems, agricultural films, and wastes from electrical and communication networks. Currently, polyolefin reclamation facilities produce about 265 million pounds $(120 \times 10^3$ t) per year of PE and 154 million pounds $(70 \times 10^3$ t) per year of PP. Some portion of these volumes include postindustrial reclaimed resins [165].

In the recycling process development area, Teijin Engineering Company of Osaka has been active in developing hydrocyclone-based resin separation systems using a variety of working fluids. Teijin has demonstrated the separation of polymers with density differences as small as 0.1 g/cm^3. Rates of up to 1322 lb/h (600 kg/h) have been achieved [166].

Fuji Recycle Industry K. K. reportedly has built a 110-lb/h (50-kg/h) pilot plant that thermally decomposes PE into gasoline-rich oil [167]. The reaction is conducted over a zeolite-based catalyst (Mobil ZSM-5) to improve the yield of desirable products, including gasoline (50–60%) and kerosene (20–30%). The coproducts are C-22 and lower hydrocarbons. In a similar vein, a pilot plant built in Aioi is capable of processing 88 lb/h (40 kg/h) of PE into 106,000 gal/yr (400 kl/yr) of oil. The system reportedly can tolerate a few percent of PVC with the polyolefin feed.

4.6 Future Outlook

It is likely that the supply of postconsumer plastics in the United States, including polyolefins, will increase dramatically during the closing decade of the twentieth century. Fourteen states have mandatory refuse recycling goals, ranging from a minimum of 15% to a maximum of 40% of the municipal solid waste (MSW) stream. Only one of these states has passed the implementation deadline for its mandate. An additional nine states have voluntary refuse recycling rate objectives of between 10 and 40%. Again, these voluntary goals are to be reached before 2000 A.D. [168, 169]. These 23 states include a majority of the U.S. population.

In the absence of governmental influence, there is, and will continue to be, a rising environmental ethic among the American population. In part, this will manifest itself in increased voluntary recycling, including the recycling of polyolefin packaging.

Assuming that the above-described mandatory and voluntary recycling goals will be pursued with cost minimization as a factor in determining which recyclables will be collected, polyolefin beverage and household chemical bottles have excellent prospects of being included in this mix of materials. These plastic containers are composed of a limited array of relatively homogeneous and high-volume resins and have viable domestic and international end-use markets after being reclaimed.

The quality issue looms as one downside to the optimistic prediction for the future of postconsumer polyolefins (and other plastics and other materials) recycling depicted above. Improvements are clearly needed in household materials preparation and collection systems to obtain lower levels of contamination, including product and related resins, extraneous contaminants, and intermaterial contamination.

Highly segmented household source separation and segmented collection also is an alternative to achieve feedstock quality improvements. In the authors' opinion, this is neither economically viable nor practical, particularly in view of the public education effort required and the collective discipline needed thereafter.

Contaminants of all types ultimately limit the range and size of markets for reclaimed materials. If quality improvements of the type described above are not made, demand for these materials will slacken in the face of increasing supply. This will be reflected initially in decreasing values and eventually in no additional markets, regardless of prices. Some state governments, however, have shown a penchant for proposing or mandating recycled content in consumer products, thereby "guaranteeing" an end market for recycled materials. These proposals naively ignore the quality issues that a free market economy would naturally correct.

Another point of concern is that curbside collection of household, source-separated recyclables is not generally cost-competitive with the alternate refuse management techniques, including landfilling and incineration (with energy recovery). While waste disposal facility siting delays and increasingly stringent environmental performance and closure standards for these facilities are driving disposal costs up [170, 171], there is no short-term prospect that these costs will ascend to those of curbside recycling as currently practiced. How local, state, and federal government will react to these realities, particularly in the face of existing mandatory recycling goals, is unclear. There are some indications that, through innovation, the cost of recyclables collection can be reduced and may become competitive with the disposal alternatives. Obviously, such developments would obviate this cost-competitiveness concern.

Enhancements and refinements of currently available polyolefin reclamation systems may result in improved reclaimed resin quality, from both fundamental and consistency perspectives. For example, alternate washing techniques, fluids, and conditions appear to be fruitful process development areas, as well as multistep washing processes. Also, the melt processing step in reclamation plants appears to offer an opportunity for further resin quality improvement.

Several fundamentally different polyolefins recycling technologies may be attractive. For example, Nauman and Lynch [172, 173] describe an experimental process in which thermoplastics are selectively dissolved at increasingly higher temperatures and, subsequently, filtered and reclaimed by flash devolatilization of the solvent. However, neither the resolution of this technique nor the quality of the resultant resins have been explored completely. A detailed description of this process is given in Chapter 9.

In another example, Battelle Laboratories has developed an experimental process to pyrolyze PE, yielding up to 50 wt % ethylene [174]. The recovered ethylene could be purified to produce polymerization-quality monomer.

There are numerous reports of polyolefins being treated by high-temperature processing to produce hydrocarbons [175, 176–184]. One of these is briefly described in Section 4.5.4. The economics of these largely exploratory processes have not been described in the literature but may be viable. Finally, it has been suggested that waste polyolefins would make a suitable (co-) feed for synthesis gas (CO) plants [185]; however, there are no reported feasibility demonstrations.

References

[1]   *Modern Plastics Encyclopedia 90,* Mid October Issue, *65*(11) (Oct. 1989).

[2]   "Olefin Polymers." *Kirk-Othmer Encyclopedia of Chemical Technology.* 3rd ed. New York, NY: John Wiley and Sons, 1981, Vol. 16, pp. 385–469.

[3]   "Petrothene Polyolefins...a processing guide." 5th ed. Cincinnati, OH: USI Chemicals, Division of National Distillers and Chemical Corp., 1986.

[4]   *Modern Plastics International 20*(1): p. 31–45.

[5]   Reference 4.

[6]   Bitar, V., Partek, Vancouver, WA. Personal communication.

[7]   Stephens, A., Eaglebrook Plastics, Chicago, IL. Personal communication.

[8]   "Plastic Prices, The Markets Page." *Recycling Times* 2(15) (17 July 1990): p. 5.

[9]   "Plastics Recycling: A Strategic Vision." Piscataway, NJ: Plastics Recycling Foundation and the Center for Plastics Recycling Research, Rutgers, The State University of New Jersey.

[10] Koester, L., AAC Development Corporation, Omaha, NE. Personal communication.

[11] Fernandes, J., Center for Plastics Recycling Research, Rutgers, The State University of New Jersey, Piscataway, NJ. Personal communication.

[12] Schrieber, G. F., Plastics Recycling Alliance, Chadds Ford, PA. Personal communication.

[13] "Technology Transfer Manual #1—Plastic Beverage Bottle Reclamation Process." Sidney Rankin, ed. Piscataway, NJ: Center for Plastics Recycling Research, Rutgers, The State University of New Jersey, Sept. 1987.

[14] Charnas, D. "Short Supply of Recycled Resins to Keep Prices Up." *Plastics News* (17 Sept. 1990): p. 12.

[15] *Modern Plastics, 66*(10) (Oct. 1989): p. 32.

[16] Vernyl, B. "Entrepreneurs Dominate Equipment Sector." *Plastics News* (17 Sept. 1990): p. 10.

[17] Fleischer, R., Quantum Chemical Corporation, Cincinnati, OH. Personal communication.

[18] Reference 17.

[19] "License Agreement—Plastic Beverage Bottle Reclamation Process." Piscataway, NJ: Center for Plastics Recycling Research, Rutgers, The State University of New Jersey.

[20] Reference 13.

[21] *Chemical Week, 146*(17) (2 May 1990): p. 36.

[22] Wenzel, J., Director, Center for Plastics Recycling Research, The State University of New Jersey, Piscataway, NJ. Personal communication.

[23] *Modern Plastics, 67*(4) (April 1990): pp. 19, 20.

[24] "Sorema Expands Recycling Machine Market." *Plastics News* (10 Dec. 1990): p. 12.

[25] *Resource Recycling, X*(1) (Jan. 1991): p. 40.

[26] Rattray, T., Procter & Gamble Company, Cincinnati, OH. Personal communication.

[27] "M. A. Industries Offers Recycled HDPE." *Plastics News* (19 June 1989): p. 11.

[28] Rowand, R. "Plastipak Starts Own Recycling Operation." *Plastics News* (4 Dec. 1989): p. 4.

[29] Reall, J. "Philadelphia Site for Recycling Facility." *Plastics News* (18 Dec. 1989): p. 2.

[30] "SGC Building Recycling Plant." *Plastics News* (5 March 1990): p. 2.

[31] "Wheaton to Process PC, PVC." *Plastics News* (20 Aug. 1990): p. 5.

[32] "California Firm Planning Recycling Plant." *Plastics News* (24 Sept. 1990): p. 2.

[33] *Modern Plastics, 67*(7) (July 1990): pp. 11, 12.

[34] "Union Carbide Will Open Facility to Recycle Film and Rigid Plastics." *Plastics News* (5 Feb. 1990): p. 2.

[35] Ackermann, K. "John Brown Develops PET, HDPE Recycling System." *Plastics News* (19 March 1990): p. 5.

[36] "Two Recycling Ventures Announced." *Plastics News* (1 Oct. 1990): p. 2.

[37] "Sorema Expands Recycling Machine Market." *Plastics News* (10 Dec. 1990): pp. 3, 12.

[38] Charnas, D. "Quantum to Build $5 Million Recycling Plant." *Plastics News* (11 Feb. 1991): p. 18.

[39] *Packaging Digest,* 27(2) (Feb. 1990): p. 62.

[40] *Packaging Digest,* 27(12) (Nov. 1990): p. 98.

[41] "The Efficient M. A. Way." Peachtree City, GA: M. A. Industries, Inc.

[42] "PE Recycler Gears Up for Post-Consumer Sourcing." *Plastics Today* (6 Aug. 1989): p. 1.

[43] *Packaging Digest,* 26(5) (May 1989): p. 6.

[44] *Plastics News* (30 April 1990): p. 5.

[45] Bader, J., Polymer Resource Group, Baltimore, MD. Personal communication.

[46] Grubba, M. J., Quantum Chemical Corporation, Cincinnati, OH. Personal communication.

[47] Traweek, C. H., United Resource Recovery, Inc., Kenton, OH. Personal communication.

[48] Flemming, R., DuPont Company, Wilmington, DE. Personal communication.

[49] Merriam, C. N., Center for Plastics Recycling Research, Rutgers, The State University of New Jersey, Piscataway, NJ. Personal communication.

[50] Norwalk, S., Union Carbide Chemicals and Plastics Company, Danbury, CT. Personal communication.

[51] Reference 48,

[52] Reference 50.

[53] Dinger, P., Council for Solid Waste Solutions, Washington, DC. Personal communication.

[54] Fernandes, J. R. "Experiments on Manual Sorting." Piscataway, NJ: Center for Plastics Recycling Research, Rutgers, The State University of New Jersey, May 1990.

[55] Reference 53.

[56] "Ink Maker Starting to Recycle Bottles." *Plastics News* (1 April 1991): p. 8.

[57] *Recycling Times,* 2(1) (5 Jan. 1990): p. 2.

[58] *Plastics News* (1 Oct. 1990): p. 2.

[59] *Plastics News* (24 Dec. 1990): p. 2.

[60] "Specifications for Reclaimed Post-consumer High Density Polyethylene Bottle Scrap." Toledo, OH: Owens-Illinois (as of 16 July, 1990).

[61] Frankel, H., J. R. Fernandes, and D. R. Morrow. "Research and Development for Sortation, and Development for Sortation and Material Handling in Pilot Plant No. 3." Progress Report No. 47, July 1989–June 1990. Piscataway, NJ: Center for Plastics Recycling Research, Rutgers, The State University of New Jersey.

[62] "Pesticide Container Recycling Considered." *Techpak* (11 Sept. 1989).

[63] Rankin, S., Center for Plastics Recycling Research, Rutgers, The State University of New Jersey, Piscataway, NJ. Personal communication.

[64] "Report on Plastic Drum Recycling." Washington, DC: The Plastic Drum Institute of the Society of the Plastics Industry, Inc.

[65] Barrett, R. J. "Mobil Chemical's Solid Waste Solutions." Conference Proceedings—Recyleplas III, May 25–26, 1988, pp. 144–148.

[66] *Modern Plastics,* 65(8) (July 1988): p. 14, 15.

[67] *Resource Recycling, IX*(11) (Nov. 1990): p. 24.

[68] Reference 67.

[69] *Modern Plastics,* 67(3) (March 1990): pp. 16, 20.

[70] Reference 4.

[71] *European Plastics News, 17*(21) (Jan. 1990): p. 18.

[72] Galecki, G., Quantum Performance Films, Streamwood, IL. Personal communication.

[73] Koehn, H., Quantum Performance Films, Streamwood, IL. Personal communication.

[74] *Modern Plastics, 64*(11) (Nov. 1987): pp. 35–36.

[75] Speed, D., AAA Recycling, Wilmer, AL. Personal communication.

[76] "Paper Firm Recycling Milk, Juice Cartons." *Plastics News* (22 Oct. 1990): p. 2.

[77] Lauzon, M. "Superwood to Recycle Multimaterial Boxes." *Plastics News* (11 June 1990): pp. 1, 20.

[78] "Desk-top Data Bank: Foams 1978." International Plastics Selector, Inc., pp. 462–467.

[79] Renner, M., Richter Manufacturing, Pomona, CA. Personal communication.

[80 Schwaber, D. W., Monarch Rubber, Baltimore, MD. Personal communication.

[81 Brady, T. "Commercial Recovery of Mixed Industrial Plastic Waste." SPE RETEC New Developments in Plastic Recycling, Charlotte, NC, 30–31 Oct. 1989.

[82] Bostrum, D., Quantum Chemical Corporation, Cincinnati, OH. Personal communication.

[83] *Conservation and Recycling, 6*(1 & 2) (1983): pp. 3–10.

[84] Jensen, J. W., J. L. Holman, and J. B. Stephenson. "Recycling and Disposal of Waste Plastics" in *Recycling and Disposal of Solid Wastes*. Yen ed. Ann Arbor, MI: Ann Arbor Science, 1974, pp. 219–249.

[85] "Collection is the Big Problem Facing PVC." Earth Day Supplement, *Plastics World* (22 April 1990): pp. 55–59.

[86] Champi, C., Himont USA, Inc., Wilmington, DE. Personal communication.

[87] Woods, R., Quantum Chemical Corporation, Cincinnati, OH. Personal communication.

[88] *Resource Recycling, IX*(2) (Feb. 1990): p. 21.

[89] Lamm, K. and R. Pfaff. "A New Way to Reprocess Polypropylene Waste From Used Starter Batteries." Paper presented at Recycle '89 Forum, Davos, Switzerland, 10–13 April 1989.

[90] Reference 86.

[91] Reference 89.

[92] Braun, D. *Kunstoffe, 55* (1965): p. 473.

[93] Knappe, W. and U.G. Kress. *Kunstoffe, 53* (1963): p. 346.

[94] Vogel, H. *Kunstoffe Plastics, 22* (1976): p. 11.

[95] Schott, H. and W.S. Kaghan. "Extrusion and Applied Rheology," SPE RETEC Preprints, 1962, p. 48.

[96] *Plastics Technology, 37*(1) (Jan. 1991): pp. 26, 28.

[97] Dietz, S. "The Use and Market Economics of Phosphite Stabilizers in Post-Consumer Recycle." Recycle '90 Forum and Exposition, 30 May 1990, Davos, Switzerland.

[98] Pearson, L. T. and R. Todesco. "Effect of Co-Additives on the Process and Long Term Stability of Polyolefins." SPE Polyolefins VII International Conference, 24–27 Feb. 1991, Houston, TX.

[99] Ryborz, D.H. *Kunstoffe, 69* (1979): p. 134.

[100] Shenoy, A. V., D. R. Saini, and V. M. Nadkarni. *Polymer, 24,* (1983): p. 722.

[101] Sikora, R. and M. Bielinski. *Kunstoffe, 78*(4) (1988): p. 335.

[102] Gibbs, M. *Plastics Eng., 46*(7) (July 1990): pp. 57–59.

[103] Reference 100.

[104] Rokudai, M. *J. Appl. Polym. Sci., 23* (1979): p. 463.

[105] Rokudai, M. and T. Fujiki. *J. Appl. Polym. Sci., 26* (1981): p. 1343.

[106] Reference 104.

[107] Wolczak, Z. K. *J. Appl. Polym. Sci., 17* (1973): p. 153.

[108] Csupor, I. and T. Toth. *Int. Polym. Sci. Tech., 7* (1980): p. T16.

[109] Springer, P.W., R.S. Brodkey, and R.E. Lynn. *Polym. Eng. Sci., 15* (1975): p. 588.

[110] Poller D. and A.M. Kotliar. *J. Appl. Polym. Sci., 9* (1965): p. 501.

[111] Reference 95.

[112] "Recycling Multilayer Multimaterial Polypropylene Containers." EVAL Company of America, Lisle, IL.

[113] Bennett, R. A. "New Product Applications and Evaluation, and Continued Expansion of a National Database on the Plastics Recycling Industry." Progress Report No. 41, July 1989 to June 1990. Piscataway, NJ: Center for Plastics Recycling Research, Rutgers, The State University of New Jersey.

[114] Huges, C., Plastics Division, Wellman, Inc., Johnsonville, SC. Personal communication.

[115] Reference 56.

[116] Reall, J. "PA Firm to Produce Recycled HDPE Containers for Motor Oil." *Plastics News* (6 Nov. 1989): p. 20.

[117] *Modern Plastics, 66*(6) (June 1989): pp. 170, 172.

[118] *Packaging Digest, 27*(11) (Oct. 1990): p. 128.

[119] Ackermann, K. "Jennico Increases Use of Recycled Resin." *Plastics News* (5 Nov. 1990): p. 6.

[120] *Packaging Digest, 26*(12) (Nov. 1989): p. 82.

[121] *Modern Plastics, 67*(8) (Aug. 1990): p. 127.

[122] Reference 114.

[123] Reference 116.

[124] Gibbs, M., Quantum Chemical Corporation, Rolling Meadows, IL. Personal communication.

[125] Rattray, T., Procter & Gamble Company, Cincinnati, OH. Personal communication.

[126] "Soltex Resin Has 25 Per Cent Recycled HDPE." *Plastics News* (27 Aug. 1990): p. 11.

[127] Polka, A., Solval Polymers, Houston, TX. Personal communication.

[128] Reference 124.

[129] *Modern Plastics, 67* (1) (Jan. 1990): pp. 47–49.

[130] *Modern Plastics, 67* (10) (Oct. 1990): pp. 41–42.

[131] Rojas, A., Sengewald USA, Inc., Marengo, IL. Personal communication.

[132] "Public-Restroom Panels." *Plastics News* (17 Sept. 1990): p. 25.

[133] "Duck House Provides Molded-In Comforts." *Plastics Today* (24 June 1989): p. 18.

[134] Product Literature, Santana Solid Plastic Products, Scranton, PA.

[134] Lauzon, M. "Plastics Insulation Maker Plans Ambitious Expansion." *Plastics News* (19 Feb. 1990): p. 15.

[136] Reference 86.

[137] Reference 4.

[138] British Plastics Federation, UK Directory, 1990.

[139]   Gibson, P., British Plastics Federation, London, England. Personal communication.

[140] "Output at BPF Reclamation Centre Set to Double." British Plastics Federation Press Release, 15 Feb. 1991.

[141] "The Plastics Fix." *Ethical Consumer* (Feb./March 1991).

[142] Reference 141.

[143] Reference 138.

[144] Reference 138.

[145] Reference 139.

[146] Reference 138.

[147]   Noone, A. "Recycling of Used Plastic Containers—Progress and Plans." Paper presented at conference January 30–31, 1991, Birmingham, UK.

[148]   Wolpert, V. M. "Recycling of Plastics: International Techno-economic Report." February 1989, p. 139.

[149]   von Schoenberg, A. "Coburg's Plastics Recycling Plant." Warmer Bulletin, Number 24 (Spring 1990): p. 19.

[150]   "Kunstoffe Recycling Liste." *Verband Kunstoffezeugende Industrie* (June 1990).

[151]   *Plastics Machinery & Equipment, 19*(3) (March 1990).

[152]   Shanley, A. "Recycling's Continental Touch." *Chem. Business* (Sept. 1989): p. 15.

[153]   "Green Power Drives Recyclers." *Chem. Week* (21 Nov. 1990): p. 24.

[154]   *European Plastics News, 17*(11) (Dec. 1990): p. 39.

[155]   *Resource Recycling, IX* (6) (June 1990): p. 42.

[156]   Gazzoni, D. Sorema, Como, Italy. Personal communication.

[157]   *European Plastics News, 15*(12) (Dec. 1988): p. 17.

[158]   Wolpert, V. M. "Recycling of Plastics: International Techno-economic Report." February 1989, p. 137.

[159]   Wolpert, V. M. "Recycling of Plastics: International Techno-economic Report, Vol. 2." March 1990, p. 307.

[160]   Press release of the Environment and Plastics Institute of Canada, 6 Feb. 1991.

[161]   Reference 160.

[162]   Reference 160.

[163]   "Canadian Plastics Recycling Directory '90," Canadian Plastics (July/Aug. 1990).

[164]   "A Guidebook to Reclaimed Plastics Molded Products," Plastics Waste Management Institute of Japan.

[165]   Reference 164.

[166]   *European Packaging Newsletter and World Report, 22*(10) (Oct. 1989): p. 4.

[167]   "Useful Oil Reclaimed from Waste Plastics (Polyolefins)," *Japan Chem. Week* (31 May 1990): p. 6.

[168]   Glenn, J. "The State of Garbage in America," *BioCycle* (April 1990): p. 34.

[169]   "Recycling in the States. Update 1989." Washington, DC: National Solid Waste Management Association.

[170]   Fitzgibbon, J. T., Foster Wheeler USA Corporation, Wexford, PA. Personal communication.

[171]   "NSWMA Annual Tip Fee Survey 1988." Washington, DC: National Solid Waste Management Association.

[172]   Lynch, J. C. and E. B. Nauman. "Application of Compositional Quenching Technology to the Recycling of Polymers from Mixed Waste Streams." Progress Report No. 39, July 1988 to June 1989. Piscataway, NJ: Center for Plastics Recycling Research, Rutgers, The State University of New Jersey.

[173]   Nauman, E. B. and J. C. Lynch. "Application of Compositional Quenching Technology to the Recycling of Polymers from Mixed Waste Streams." Progress Report No. 50, July 1989 to June 1990. Piscataway, NJ: Center for Plastics Recycling Research, Rutgers, The State University of New Jersey.

[174]   Battle Institute, Columbus, OH. Confidential communication.

[175]   Reference 167.

[176]   Meszaros, M., Amoco Chemical Company, Naperville, IL. Personal communication.

[177]   Rowatt, R. J., Chevron Chemical Company, Kingwood, TX. Personal communication.

[178]   Barrett, R. J., Mobil Chemical Company, Pittsford, NY. Personal communication.

[179]   Stapp, P. R. and H. H. Wandke, National Institute for Petroleum and Energy Research, IIT Research Institute, Bartlesville, OK. Personal communication.

[180]   Madorsky, G. *J. Res. Nat. Bur. Std., 42* (1949): p. 499.

[181]   Takesue, T. *Petrol. Petrochemical International, 12*(4) (April 1972): p. 36.

[182]   Bacci, L. "Mechanisms of the Thermal Degradation of Polyolethylene."   Ph.D. Thesis, U. of Maryland, 1970.
[183]   Luff, P. and M. White. *Vacuum, 18*(1968): p. 137.
[184]   Segwick, P.D. *J. Polym. Sci.—Polym. Chem. Ed., 19* (1981): pp. 2007–2015.
[185]   Reference 86.

5 Polystyrene (PS)

G. A. Mackey, R. C. Westphal, R. Coughanour

Contents

5.1   Introduction 111

5.2   Industry 111

    5.2.1   Manufacturers 111

    5.2.2   Processes 111

    5.2.3   Products 112

5.3   Sources and Volumes 113

    5.3.1   Packaging and Disposables 113

        5.3.1.1   Injection Molding 113

        5.3.1.2   Sheet Extrusion 114

        5.3.1.3   Foam Extrusion 114

        5.3.1.4   Expandable Polystyrene 115

    5.3.2   Nonpackaging 115

5.4   Reprocessing Technologies 116

    5.4.1   Feedstock 116

    5.4.2   Wash-Dry-Pelletize Process 117

    5.4.3   Densification Process 117

    5.4.4   Densification Equipment 118

5.5   Recycled Products 121

    5.5.1   Properties 121

    5.5.2   Densified Product Applications 123

    5.5.3   Nondensified Product Applications 124

        5.5.3.1   Void Fill 124

        5.5.3.2   Thermal Insulation and Construction 124

5.6   Economics 125

    5.6.1   Food Service Material 125

    5.6.2   Durable Goods Packaging 126

5.7   The National Polystyrene Recycling Company 127

5.8   Global Activities 127

     5.8.1   Japan 127

     5.8.2   Europe 127

5.9   Future Outlook 128

References 128

Polystyrene (PS) is a thermoplastic resin that is used in a wide variety of applications. Typically, the market is divided into four major product segments: injection-molded solid, extruded solid, extruded foam, and expandable bead. These product categories are further segmented into many end-use applications. Of the 68 billion pounds (30.8×10^6 t) per year produced in the United States, about 1 billion pounds (454×10^3 t) are used in single-use disposable applications. It is this application that has drawn the attention of public concern and that is targeted as the raw material stream for the emerging recycling activity. This chapter outlines the typical applications of PS products and identifies various recycling streams available for reprocessing. Some description of process equipment used for recycling is included along with some discussion of end markets for recycled materials. Activity in the legislative area or in hydrocarbon pricing can have significant impact on the future of plastic recycling, and it is hoped that this chapter will form a useful background in evaluating the impact of these forces.

5.1   Introduction

PS was first commercially produced in the United States in 1938, about eight years after its introduction in Germany. Earlier British patents issued in 1911 describe PS resin, and literature references to PS go back to the mid-nineteenth century [1].

Foam PS was introduced in the early 1940s, and impact grades (PS enhanced with rubber to improve impact strength) were commercialized in the United States after World War II [2]. Domestic sales of crystal and impact grades have grown from an estimated 265 million pounds (120×10^3 t) in 1950 to over 5 billion pounds (2.3×10^6 t) in 1990. The large growth rate of PS is due to a combination of versatility, easy processing, and low production cost.

5.2   Industry

5.2.1   Manufacturers

Crystal, impact, and expandable polystyrene (EPS) types of PS are manufactured by fifteen producers, who currently have a total nameplate capacity of 6.4 billion pounds (2.9×10^6 t). About 5.8 billion pounds (2.6×10^6 t) of this is represented by eight major PS producers who comprise the National Polystyrene Recycling Company (NPRC), a limited partnership formed to develop recycling technology and establish PS recycling centers throughout the United States. Suppliers and capacities are shown in Table 5.1 [3].

5.2.2   Processes

The two common methods of polymerizing styrene are the batch-suspension process and the mass-continuous technique. The batch-suspension system utilizes a stirred reactor in which styrene monomer is dispersed in water, with a suspending agent utilized to maintain dispersion of the monomer droplets in the water phase [4]. Free radical initiators are used to both accelerate the polymerization and to control the molecular weight. The resulting polymer resembles granules of sugar and is washed, dried, and extruded into pellets suitable for use by converters.

EPS is manufactured using the batch suspension system; however, a volatile hydrocarbon blowing agent is added during the polymerization. The EPS bead does not undergo the extrusion/pelletization step.

The mass-continuous method pumps styrene monomer through a series of heat exchangers until it is converted to a specified solids level [5]. The unreacted monomer is then stripped off and recycled to the beginning of the process.

Table 5.1   U. S. PS capacity–1990

Supplier	Million Pounds			(Thousand Tonnes)		
	PS	EPS	Total	PS	EPS	Total
Huntsman	1155	130	1285	(524)	(59)	(583)
Dow Chemical	1255	20	1275	(569)	(9)	(578)
Polysar	805	—	805	(365)	—	(365)
Fina	640	—	640	(290)	—	(290)
Arco	100	445	545	(45)	(202)	(247)
Mobil	505	—	505	(229)	—	(229)
Chevron	480	—	480	(218)	—	(218)
Amoco	310	—	310	(141)	—	(141)
BASF Wyandotte	—	175	175	—	(79)	(79)
Scott Paper	—	120	120	—	(54)	(54)
American Polymers	70	—	70	(32)	—	(32)
Dart	70	—	70	(32)	—	(32)
Kama	65	—	65	(29)	—	(29)
A&E Plastics	55	—	55	(25)	—	(25)
Total	5510	890	6400	(2499)	(403)	(2902)

Note: Used with permission from *Modern Plastics*

Impact PSs are produced by polymerizing styrene monomer in the presence of an elastomer. Polybutadiene rubber is the preferred elastomer. Block copolymers of styrene-butadiene are sometimes compounded with PS to enhance impact strength.

The suspension technique creates a more versatile product, while mass-continuous results in a more efficient operation. Suspension methods today are usually restricted to specialty products such as EPS or to materials which will undergo subsequent compounding such as precolored materials, flame retardant modified grades or color concentrates.

5.2.3 Products

General purpose, or crystal, PS is a clear rigid material with a specific gravity of about 1.05, a tensile strength of 8000 psi (55 MPa), and a Vicat softening point up to 108 °C, but with a relatively low Izod impact strength of 0.2–0.5 ft-lb/in (11–27 J/m) of notch. Typical weight average molecular weights (M_w) will range from 200,000 to over 300,000. Physical properties are dependent upon molecular weight, molecular weight distribution, and the use of internal additives.

Crystal PS products, which are transparent, can be injection molded or extruded. Common injection-molded applications include audio equipment dust covers, clear audio tape cassette cases, windows for video cassettes, boxes for compact discs, office supplies, computer disc reels, tumblers, flatware, housewares, display cases, and medical applications such as bottles, petri dishes, and pipettes.

Impact grades are opaque but have substantially higher impact strengths compared with crystal PSs. The usual grades of medium impact have notched Izod impact strengths from 0.6 to 1.0 ft-lb/in (32 to 53 J/m) notch, high impacts range from 1.0 to 3.0 ft-lb/in (53 to 150 J/m) notch, and extra high impact grades of 3.0 to 5.0 ft-lb/in (160 to 267 J/m) notch are available. The discrete rubber phase not only improves the impact strength but also increases the elongation at rupture up to 60% or better. Clarity and stiffness are reduced accordingly. Both impact and crystal PS degrade readily in the presence of ultraviolet light; therefore, they are not suitable for use where sunlight exposure occurs unless protected with coatings or suitable additives [6].

Impact PS injection-molded applications include electronic appliance cabinets, business machine housings, video cassettes, small appliances, smoke detectors, and furniture. Extruded and thermoformed items include refrigerator door liners, luggage, horticultural trays, and some food packaging, such as dairy and yogurt containers.

Extruded non-foamed PS sheet is used in glazing and decorative panels. Oriented polystyrene (OPS) film and sheet is used in food packaging, such as cookie trays, document wrap, blister pack, and salad containers.

Extruded foam PS sheet will range from 3.0 to 10.0 lb/ft^3 (48 to 160 kg/m^3) and is thermoformed into egg cartons, meat and poultry trays, food service trays, and fast food packaging. Thicker foam board for insulation can be produced at densities of 1.3–5.0 lb/ft^3 (21–80 kg/m^3). Protective jackets for glass bottles provide an additional application for foam sheet and provide an example of source reduction by allowing glass companies to manufacture bottles with thinner walls.

EPS is made from PS resin granules impregnated with a blowing agent [7]. After a steamed pre-expansion process that produces foamed beads, the beads are put into a mold again, where they are heated with steam and expand further until they fuse together, forming the finished product. The small cellular foam structure gives the material a natural white color. The finished products, ranging in density from 1.0 to 10.0 lb/ft^3 (16 to 60 kg/m^3), are used in a wide variety of applications, such as insulation board for the construction industry, molds for metal castings, flotation devices, and packaging materials.

Table 5.2 [8] summarizes the PS market by product type.

Table 5.2　PS markets by product type

Product	Million Pounds	(Thousand Tonnes)
Molded parts	1879	(852)
Extruded sheet	1415	(642)
Extruded foam sheet	737	(334)
EPS for molding	659	(299)
Export	196	(89)
Other	251	(114)
TOTAL	5137	(2330)

5.3　Sources and Volumes

5.3.1　Packaging and Disposables

The volume of PS used in packaging and other products did not increase from 1989 to 1990. This is due to improved resin compositions and processing technologies that allowed for thinner wall sections and lighter weight parts. In addition, rising concern over municipal solid waste (MSW) disposal problems slowed down conversion from permanent ware to disposables.

Methods used for converting PS to packaging and disposable products are injection molding, solid extrusion, foam extrusion, and EPS molding of cups and containers. Each of these methods requires specific equipment. Each method also presents its own set of problems for recycling. These problems and the techniques used to solve them are discussed in Section 5.4.

5.3.1.1　Injection Molding　Injection-molded items used in the packaging and disposable markets are shown in Table 5.3 [9]. Injection-molding methods are used when more intricate designs are required or heavier wall sections are needed. Tumblers and beverage glasses with wall sections of up to 0.030 inch (0.76 mm) made from PS represent traditional applications.

Table 5.3   Injection-molded solid PS packaging/disposable products–1990

Product	Million Pounds	(Thousand Tonnes)
Rigid packaging	85	(39)
Closures	96	(43)
Produce baskets	24	(11)
Tumblers/beverage glasses	90	(41)
Flatware, cutlery	100	(45)
Dishes, cups, bowls	58	(26)
Blow-molded items	10	(51)
TOTAL	463	(210)

In recent years, however, much of this volume has been converted to extruded/thermoformed styrene-butadiene block copolymer material, which results in lighter weight items. Flatware and cutlery are injection-molded using either crystal or medium-impact PS. Because of its use in cafeterias and take-out services, rubber-modified PS is a significant component in the recycle stream.

5.3.1.2 Sheet Extrusion   The major markets and 1990 volumes for standard extruded sheet in the packaging market is shown in Table 5.4 [10]. The equipment used for producing sheet is similar to the first stage of a tandem extrusion line (described below) used for foam sheet. However, no foaming agent is added and, instead of extruding into the secondary stage of a tandem line, the melted plastic is forced through a flat-faced die. Extruded sheet is then thermoformed on conventional equipment. Impact PS is the usual material of choice since the reinforcing rubber results in excellent elongation properties and uniform draw on forming. Crystal PS compounded with 40–60% loadings of styrene-butadiene block copolymers has seen increased use since this results in a transparent item instead of the usual translucent appearance resulting from other rubber additive systems.

Table 5.4   Extruded PS solid sheet packaging/disposable products–1990

Product	Million Pounds	(Thousand Tonnes)
OPS film, sheet	245	(111)
Dairy containers	150	(68)
Vending and portion cups	260	(118)
Lids	110	(50)
Plates and bowls	45	(20)
TOTAL	810	(367)

A more rapidly growing market is OPS [11]. In this process, high molecular weight crystal PS is extruded and biaxially oriented as it exits the die. Strength properties are related to the extent of orientation. This material is thermoformed on conventional equipment and used in applications such as lids and packaging where clarity is important. Prepackaging of salads in take-out restaurants and delicatessens has contributed to this growth rate.

5.3.1.3 Foam Extrusion   Extruded foam fabricated products and 1990 volumes of PS used in producing these are shown in Table 5.5 [12].

Table 5.5   Extruded PS foam sheet packaging/disposable products–1990

Product	Million Pounds	(Thousand Tonnes)
Food trays	191	(86)
Egg cartons	60	(27)
Single service plates	140	(64)
Hinged containers	125	(57)
Cups fabricated from sheet	40	(18)
Miscellaneous sheet	31	(14)
TOTAL	587	(266)

Extruded PS foam is produced by extruding high molecular weight crystal PS with a nucleating agent and a foaming agent. Foaming agents commonly used are hydrocarbons or a hydrochlorofluorocarbon (HCFC). At one time, the nonhydrogen chlorofluorocarbon version (CFC) was used, but this has been discontinued because of the potential adverse effect of CFCs on ozone levels in the stratosphere. By early 1990, the food service and packaging industry was no longer using any of the fully halogenated CFCs. Research is continuing to develop even more options in foaming agent selection.

A tandem extrusion line is commonly used to produce foam sheet with the foaming agent added between the two extruders [13]. Material is extruded through an annular die, and foaming occurs as the material exits the die. The sheet is then cooled and aged prior to thermoforming. Sheet thickness and density is varied to meet end-use requirements. The low-density foam produced results in excellent thermal insulation properties. This, along with the rigidity of PS, resistance to moisture absorption, and good printability, make it the ideal material for certain food packaging. Thermoformed trays and hinged containers usually range in thickness from 0.080 to 0.150 inch (2.0 to 3.8 mm), and the densities vary from 3.0 to 10.0 lb/ft^3 (48 to 160 kg/m^3).

5.3.1.4 Expandable Polystyrene
EPS beads are used to fabricate a wide variety of finished foamed products for the packaging/disposable market. Small beads are expanded to produce cups and containers for hot liquids and beverages; medium beads are used for shape-molded packaging for the medical, electronics, and transportation industries; and large beads are expanded for loose-fill packaging applications. The volume breakdown for these products is given in Table 5.6 [14].

Table 5.6   EPS foam packaging/disposable products–1990

Product	Million Pounds	(Thousand Tonnes)
Shapes for packaging	103	(47)
Cups and containers	153	(69)
Loose-fill	75	(34)
TOTAL	331	(150)

5.3.2 Nonpackaging

Applications other than packaging and disposable food service form the largest percentage of PS usage. These include appliances, construction, electronics, furniture, toys, and housewares.

However, even these items will reach the MSW stream at some point. Whether it is a simple appliance with a five-year life span or PS foam insulation that may be in use for over one hundred years, at some point it will have served its purpose and end up as a candidate for recycling. If the durable good is made from PS alone, it can obviously be recycled in the same manner as packaging. Typically, durable goods are made up of several materials and may include metals or glass, which means the article must be disassembled. Recycling of PS separated from multimaterial items is unknown to the authors at this time.

5.4 Reprocessing Technologies

5.4.1 Feedstock

PS resins and their various fabricated forms have already been described. These fabricated forms are characteristic of the products that appear as raw materials in a recycling effort. The kinds of organizations which provide quantities of PS packaging products capable of recycling are military bases, cafeterias, community drop-off centers, stadiums and special events, elementary through university level schools, fast food restaurants, airline meal service, manufacturing or retail business packing, and material recycling facilities (MRFs). The actual types of articles received are illustrated in Figure 5.1.

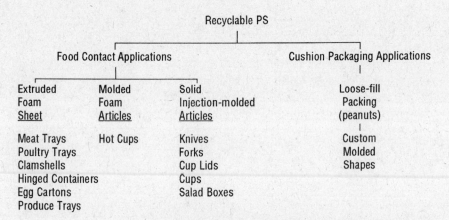

Figure 5.1 Recyclable PS products

These feedstocks may represent used EPS protective packaging foam material, scrap from curbside pickups or neighborhood drop-off collection points, food packaging, and trays and salad containers from school or institutional cafeterias.

Food service packaging, whether it originates from curbside pickup, fast food restaurants, or school cafeterias, is fairly consistent in nature since the virgin feedstock is the same in all cases. However, the mixture of items in the waste will vary. Foam items represent the bulk of the stream by volume. Compatible solid PS items that may be present will include solid thermoformed lids, clear salad trays, thermoformed cold drink cups, injection-molded flatware, and EPS foam hot drink cups. Lids, cutlery, and some cups contain rubber that will enhance the physical properties, although the concentration is too low to have a noticeable effect on notched Izod impact strength.

The dry EPS foam from shape moldings used for packaging and cushioning of items during shipping usually represents a relatively low molecular weight PS. EPS grades modified with flame-retardant additives and used in applications such as building insulation and loose-fill packaging must be handled carefully because of the temperature sensitivity of the additive systems. It is preferred to recycle this material into nondensified products.

The above PS products could provide four grades of recycled resins for sale if the collected materials were separated into the following four types:

1. Extruded foam sheet—typically a high M_w resin of $> 300,000$ with a low melt flow rate (MFR) of < 2 g/10 min;
2. Molded foam cups—a high M_w resin of about 250,000 with a MFR of 5–10 g/10 min;
3. Solid injection-molded or thermoformed sheet—generally made of rubber-modified resin with M_w of 250,000 with a MFR of 5–10 g/10 min (OPS sheet salad container resin does not contain rubber but has a M_w of 300,000 and a MFR of < 2 g/10 min.); and
4. Custom-molded shapes and loose-fill—generally with M_w of 250,000 with a MFR of >10 g/10 min.

It is unlikely that the mix of incoming materials could be separated as indicated, but it is possible to separate the low M_w packing foams from the types used in food service. The better these subdivisions can be maintained, the more specific and uniform the end product becomes.

5.4.2 Wash-Dry-Pelletize Process

Food service postconsumer PS is available from fast food restaurants, schools, business cafeterias, etc. This feedstock, when separated for a recycling plant, should not contain any nonrecyclable solid-waste material. Some nonPS constituents, such as aluminum foil container lids, polypropylene (PP) straws, and napkins, are to be expected.

A typical plant places the incoming feedstock, which usually arrives in baled form, on a conveyor or screening table where obvious contaminants are removed by hand-sorting and screening. The PS is then granulated and thoroughly washed with water containing a detergent. A slurry of granulated PS fluff and water flows onto a vibrating screen where the water is separated from the PS fluff. Fresh rinse water is sprayed onto the fluff as it is conveyed across the vibrating screen. The wash water is typically pumped through a strainer to remove the solid particles and any pulped paper, and then it is pumped to water treatment facilities. The PS fluff is conveyed to a centrifugal drier where the surface moisture is removed. The fluff at this point still contains a significant amount of moisture that must be removed. This can be done by actually squeezing the water from the fluff, or through the use of convection heat drying, vacuum-vent extrusion, or some economical combination of all of the above. Some novel ways have been used to remove up to 50% of the water from the fluff, such as a screw press continuous pressure system or the batch ram compactor system. Although high in capital cost, a multiple vacuum-vented set of extruders for dewatering, along with a melt filtering system, may be the most economical of the current technologies.

If dewatering is done by extrusion, a single-screw or twin-screw extruder may be used. Single-screw extrusion requires preliminary drying and results in a high level of contaminants since any suspended solids remaining in solution are left on the product during the drying stage. Properly designed twin-screw extruders provide much cleaner products since most of the remaining wash water is expressed from the feedstock in the compression zone of the extruder and steam-stripped in the next zone before devolatilization occurs.

The latter technique not only results in a product with a substantial reduction in black specks but also one that is more odor free. Odors may be a potential problem since incomplete washing of oil and grease from food contamination may result in the formation of trace amounts of aldehydes with noticeable odors. Twin-screw extrusion also provides a method for adding other materials, such as rubber modifiers, before pelletization to improve physical properties.

5.4.3 Densification Process

There are a number of options for densifying EPS. Actual densification of EPS is usually accomplished by total or partial collapse of the cell structure rather than by compaction alone, which eliminates the air between the foamed beads but only partially from the cells.

One of the principal options is to use heat to soften the polymer and collapse the cell structure. The heat can be applied from external sources or generated by friction. The final product can be anything from a solid block of PS to extruded pellets. Another approach generally applicable to the lower density molded EPS foams rather than the higher density extruded food service foams is to densify through the use of pressure. The most important issue in foam densification is to accomplish the desired density without degrading the polymer. EPS dry foam will result in a clear, transparent pellet, while feedstock obtained for fast food restaurants usually contains pigmented clamshells. Printed containers also contribute to coloration on extrusion.

5.4.4 Densification Equipment

In this section, an attempt is made to briefly describe the various equipment currently available for densification. Selection of the type of densification equipment for any given situation depends on the volume to be processed, cleanliness of the PS foam, and final disposal of the densified product. For example, it may not be desirable to densify postconsumer food service foam since this would result in trapping the contamination internally in the polymer. In this instance, compaction may not be desirable until the polymer is thoroughly washed. The densification systems described below are currently being used or offered for sale in the marketplace.

Repro Machine Industries of Japan offers several models of densifiers. In the FT-type machine, the EPS foam is shredded and pushed into an electrically heated melting chamber. Molten polymer is pushed out of the chamber by rotary blades into rectangular block molds. Cooled polymer in the form of a solid block can be ground and subsequently extruded and pelletized. Typical polymer degradation is a 40% molecular weight loss. Capacity is 200 lb/h (90 kg/h).

The FS-type machine (Figure 5.2) [15] operates on the same principle as the FT-type, except that the molten plastic is extruded through a T-die and special cooling rolls and then cut into pieces. The shredding and melting chamber are mounted directly above the extruder, and the solid-block-forming apparatus is eliminated.

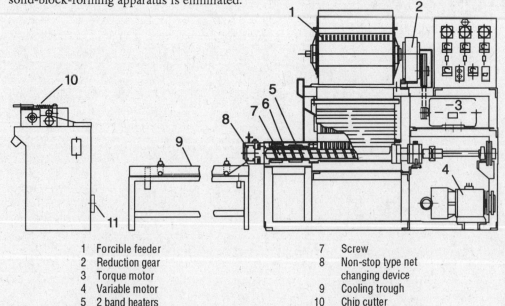

1	Forcible feeder	7	Screw
2	Reduction gear	8	Non-stop type net
3	Torque motor		changing device
4	Variable motor	9	Cooling trough
5	2 band heaters	10	Chip cutter
6	Extruder cyclinder	11	Variable speed handle

Figure 5.2   Repro Machine FS-type

The Sentinel EPS Recycler, Figure 5.3 [16], was basically designed for densifying steam-chest-molded EPS. The EPS particles are heated by rotating blade shearing forces and are compressed into closely spaced 300 °C contact heating panels. The melt leaves the chamber and drops into block molds where it is cooled. As with the Repro Machine, the polymer is degraded in molecular weight to somewhere in the range of 60% of the original value.

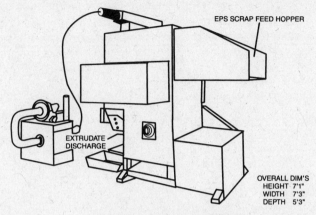

Figure 5.3   Sentinel EPS Recycler

The Instamelt System [17] consists of a rotary extruder combined with a melt pump to boost the pressure for pelletizing. The low density of the EPS feed makes this system relatively high-cost for the volume processed. Also, the use of a melt pump with possible contaminated feed could lead to serious mechanical problems. There is, however, minimal degradation of the polymer.

The AP-Shrinker (Figure 5.4), manufactured by Sekisui of Japan [18], feeds chunks of EPS foam onto a conveyor belt and subjects the pieces to infrared rays generated by special ceramic

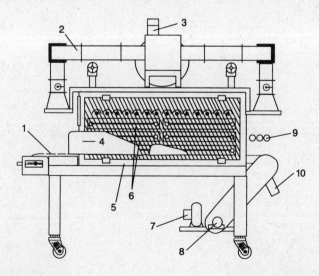

1	Conveyor belt	6	Far-infrared heater
2	Exhaust unit	7	Cooling blower
3	Exhaust blower	8	Drive motor
4	PS scrap	9	Cooling air pipe
5	Heater furnace	10	Scraper

Figure 5.4   Sekisui AP-Shrinker

heaters. There is no flame involved, and the heater's surface temperature is a relatively low 470 °C. The electromagnetic wave concept is highly energy efficient, as the wavelengths are particular to those most easily absorbed by the foam PS. Some polymer degradation occurs as a result of local overheating of the densified polymer. The capacity of most existing shrinkers is less than 100 lb/h (45 kg/h).

The Glenro Densifier [19] was designed specifically for the NPRC plants to densify EPS postconsumer packaging materials from the nominal 1–3 lb/ft^3 (16–48 kg/m^3) to 9–12 lb/ft^3 (144–192 kg/m^3). This partially densified material is then pelletized on an extruder specially designed for these same NPRC plants. This densifier system consists of a Weisser granulator where large EPS molded foam parts are shredded into small particles. The small particles are placed on a continuous moving belt and are subjected to rays from gas infrared heaters. The densified particles are then fed to the extruder system for pelletizing. Some degradation (20% molecular weight loss) of the polymer occurs during the densification. Capacity of the Glenro unit is 1000 lb/h (454 kg/h).

In the Buss-Condux Inc. Plastcompactor [20], the EPS foam to be densified is first shredded into chips or flakes. This material is sintered by frictional heat between a rotating and stationary disc. The clearance between the discs is adjustable to control the amount of heat generated. The sintered material is cut into granular size determined by the mesh in the grinding screen. Granulated product is either used as-is for injection molding or repelletized in an extruder. Capacity is 180–200 lb/h (82–91 kg/h).

Another system is the Valex Converter [21] or Walex System [22]. In this system (Figure 5.5), PS foam is shredded in a dosing roll, and the ground EPS foam is metered into a heated rotary drum through rollers controlled by an adjustable speed motor. The molten PS is pushed through a die head over cooling rollers and into a granulator. Heat history on the polymer is a potential problem. Capacity is 100 lb/h (45 kg/h).

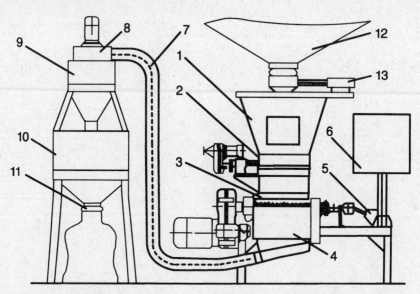

1	Input hopper	7	Pipeline
2	Dosing roll infinitely variable	8	Conveying blower
3	Roll extruder	9	Cyclone separator
4	Granulator	10	Pellet container
5	Hot air blower for internal	11	Sack-filling position
	heating of roll	12	Storage silo
6	Control cabinet	13	Shut-off slide

Figure 5.5   Walex System

The Gelimat system, manufactured by Draiswerke, Inc. [23], consists of a horizontal chamber with a central rotating shaft and staggered mixing blades at various angles. High tip-speed on the rotating blades results in thermo-kinetic heating of the batch. The cycle ranges from 10–38 seconds and the batch is discharged through a bottom door. These batches, each weighing about 4 lb (1.8 kg), are then fed to an extruder for pelletization. Capacity is 500 lb/h (227 kg/h).

Another densifier, the Toronita [24], uses a horizontal ram to compress molded foam with a density of 1–5 lb/ft^3 (16–80 kg/m^3) to a density of 18–25 lb/ft^3 (288–400 kg/m^3). The final product is a compressed rectangular block of densified PS weighing approximately 40–50 lb (18–23 kg) that can be easily ground and then extruded into pellets. The main advantage of the Toronita Densifier over other densifiers is the fact that densification is accomplished with hydraulic pressure only, and no heat is generated or applied to the system. Therefore, there is no polymer degradation and no air emissions from the system. Capacity of the 40T is 100–150 lb/h (45–68 kg/h). Other models both larger and smaller are available.

The Mobil Chemical Company [25] manufactures a Mobil Thermoplastics Densifier. This electrically powered densifier supplies variable heat from 149 to 204 °C, which collapses the air cells of the foamed packages and allows the material to flow into a solid mass with a density of 35 lb/ft^3 (570 kg/m^3) or greater. The inert block can be safely stored and is ready for recycling. This densifier has been designed to be used by fast food restaurants, institutions, schools, etc. Since contaminated particles are trapped within the polymer structures, the blocks would be ground and processed by a recycler with special emphasis on the filtration of foreign materials. Capacity is less than 50 lb/h (23 kg/h).

Densification equipment manufactured by MK Recycling Inc. [26] consists of a 15 hp (11 kW) extruder designed to feed PS fluff. The product is densified to a pellet form in a water-cooled pelletizer. The extruder is not vented; therefore, the feed must be very low in blowing agent-content to make a satisfactory product. Approximate capacity is 150 lb/h (68 kg/h).

Many extruder manufacturers provide the type of equipment necessary to feed foamed EPS directly and produce a final pelletized product. These extruders are generally of a tapered screw design to facilitate feeding of the low-density foam. Extruders are also vented to remove any residual blowing agent in the final product. Capital costs are generally high due to the size of the equipment required to achieve a decent production rate better than 400 lb/h (181 kg/h). One such system is marketed by Reifenhauser [27].

Another system, the EREMA-RM [28], provides a direct transformation from plastic foam waste to final pellets. The material is fed from a conveyor to a shredding drum, which chops the material and generates the force necessary for feeding an extruder screw. The extruder screw plasticizes the material, which is subsequently passed through a screen changer and into a hot die-face pelletizer and then a pelletizer cooler and drying device. The final product is a cylindrical pellet. This system is capable of processing 500–600 lb/h (227–272 kg/h) of foam PS.

5.5   Recycled Products

5.5.1   Properties

Physical properties of a typical sample of product representing PS food service items that were recycled into pellets by a commercial operation specializing in postconsumer PS recycling were evaluated. Properties are recorded in Table 5.7 and designated as "Foamsheet Recycle." For comparative purposes, samples of in-house regrind from a converter producing foam trays also was obtained. This is designated as "Foamsheet Regrind." Also listed are properties of a commercial resin of similar molecular weight to the foam sheet recycle. These data [29] are based on compression-molded test specimens. The presence of an internal plasticizer and a

Table 5.7   Comparison of recycled PS foam sheet, regrind PS foam sheet, and a virgin PS injection-molding grade

Property	Foamsheet (Recycle)	Foamsheet (Regrind)	A Typical Injection-Molding Grade of Commercial Resin
Melt Flow Rate, g/10 min	5.3	6.4	5.0
Vicat Softening Point, °C	101	106	107
Heat Deflection Temperature			
66 psi, °F	187	180	196
(0.45 MPa, °C)	(86)	(82)	(91)
Yield Tensile Strength			
psi	4300	5711	6440
(MPa)	(29.6)	(39.3)	(44.4)
Ultimate Elongation, %	3.2	1.2	1.6
Tensile Modulus			
10^3 psi	422	479	420
(MPa)	(2900)	(3300)	(2894)
Flexural Modulus			
10^3 psi	428	506	425
(MPa)	(2949)	(3486)	(2928)
Flexural Strength			
psi	8010	7990	10,000
(MPa)	(55.2)	(55.1)	(68.9)
Izod Impact Strength at 23 °C			
ft-lb/in, notch	0.4	0.2	0.2
(J/m), notch	(21)	(11)	(11)
Density, g/cm^3	1.05	1.05	1.05
ZnSt, ppm	1512	1253	1200–1500
M_w	240,500	212,800	240,000
M_w/M_n	2.5	2.5	2.5
Volatiles, %	0.86	0.51	0.20
Black Specks, %	0.56	0.01	0.00
Granulation, g/100 pellets	2.9	3.8	2.5
Styrene, ppm	143	399	500
Ethyl Benzene, ppm	12	78	50

small amount of rubber reduces the tensile strength and thermal properties but improves the elongation and impact strength. Although the residual styrene monomer content is lowest on the recycled sample shown in the table, total volatiles were highest. The volatility test is based on the weight loss after heating the sample for 60 min at 200 °C in a vacuum oven [30]. These volatiles include traces of various decomposition products (aldehydes, ketones, other aromatics) and moisture. This is typical for a sample processed on a single-screw extruder, while twin-screw technology results in improved devolatilization.

Recent experiments conducted by the NPRC illustrate the molecular weight changes that occur during densification and densification-pelletization of EPS cup and packaging scrap material. These changes are shown in Table 5.8 [31]. The recycled product may be upgraded during the reclaiming operation by compounding additives into the system during the extrusion stage. These additives could be flame-retardant or rubber modifiers. One experiment showed that a 10% loading of a rubber modifier increased the notched Izod impact strength from 0.3 to 1.8 ft-lb/in (16 to 96 J/m) and reduced the flexural modulus from 430,000 to 380,000 psi (2962 to 2618 MPa) [32].

Table 5.8 Effects of processing on EPS scrap molecular weight

| | | EPS Cup Scrap | |
Molecular Weight	Undensified	Densified	Densified/Pelletized
$M_w \times 10^3$	292	273	203
$M_n \times 10^3$	104	102	77
M_w/M_n	2.8	2.7	2.6
MFR (g/10 min)	—	—	8.4
		EPS Packaging Scrap	
$M_w \times 10^3$	222	195	166
$M_n \times 10^3$	83	67	66
M_w/M_n	2.7	2.9	2.5
MFR (g/10 min)	—	—	14.0

In general, physical properties of a number of recycled products tested are similar to standard commercial injection-molding-grade materials. However, because of the presence of pigments, end-use applications are restricted to items where color or color variation is acceptable.

5.5.2 Densified Product Applications

End-use items made from postconsumer recycled PS are numerous. Existing Food and Drug Administration (FDA) regulations make it difficult to use postconsumer scrap for food packaging applications. One supplier of PS foam egg cartons has received an FDA letter of "no objection" for use in egg cartons—assuming the applicable requirements are met [33].

Office items such as desk trays, waste paper baskets, cardfile boxes, and rulers are ideal items for recycled PS since these items are usually darker colors, which makes color matching easier. Some of the original novelty items, such as combs, key chain tags, and yo-yos, have expanded into more sophisticated items. Coat hangers and picture frames are frequently made from recycled material, as well as flower pots. Other horticultural items include nursery trays extruded and thermoformed from blends of recycled PS mixed with rubber modifiers or virgin impact PS. In the home entertainment field, video and audio cassettes are potential applications. The heavy-duty reusable serving tray used by fast food restaurants traditionally has been molded from PP. However, one molding company [34] has produced these from blends of virgin impact and recycled scrap PS. These trays meet the requirements of major fast food restaurants.

Foam applications include the use of recycled scrap in producing loose-fill for packaging. A major foam converter also has used recycled products in making a fan-folded, foam-core insulation board. Other foam insulation board companies are also investigating optimum levels of recycled PS that can be incorporated with virgin material to produce insulation with suitable compressive strength, density, and cell size.

Some converters are successfully utilizing recycled PS at a 100% level, while others prefer processing 50:50 blends with virgin. Nearly every application can tolerate at least a 10% loading of recycle with virgin material with no property or processing problems.

Recycled material may be used as the base for compounding color concentrates. The higher flow, lower molecular weight reclaimed EPS scrap is particularly suited for this. Converters operating coextrusion lines have also utilized this lower cost product by coextruding it into an inner layer in which any color inconsistencies are concealed by the exposed outer layers of virgin feedstock. The availability of recycled PS from postconsumer sources will actually encourage the development of new uses.

5.5.3 Nondensified Product Applications

Given the unique physical properties of shredded foam, one may conclude that if PS foam were merely cleaned and then shredded there could be economic value to such a product. The applications for recycled PS foam would fall into two categories: (1) void fill material and (2) thermal insulation/construction applications.

5.5.3.1 Void Fill   For a number of years, scrap PS foam has been ground (using typical plastic scrap grinders) to some specific particle size and used in floral vase stuffing, lawn furniture fill, large doll stuffing, and occasionally in plant nurseries as a soil lightener to improve aeration. Another development in this area has been the use of shredded foam as a drainage medium for ground water. It has been used in a loose form in the bottom of trenches or large landscape planters to allow water to drain freely away. More recently, a refinement on that application has appeared in the form of a loose-weave cloth tube filled with shredded foam that is used in place of perforated drainage tile [35].

5.5.3.2 Thermal Insulation and Construction   In the early 1960s, the BASF Corporation performed extensive work on a foamed PS bead product that could be added to concrete mix to form a lightweight insulating concrete. This application used virgin prefoamed beads that were treated with an epoxy/cement coating to keep the beads from floating and separating from the concrete mix while it was being prepared. One targeted application for this type of concrete was for use as highway road beds, railroad beds, and beneath airport runways to prevent frost damage in seasonal frost areas or permafrost damage in the Arctic. It was also proposed that the material be used for lightweight fill for road beds or other applications where light weight but low load-bearing requirements were important [36].

Now that PS foam is available as an inexpensive recycled raw material, foam-filled concrete applications are receiving more attention. In California, Rastra Building Systems, and in Florida, Mandish Research International, are promoting PS foam-filled concrete applications. In the Rastra system, foam-filled concrete panels are prepared which then become the form work for a poured-in-place concrete wall [37]. The poured-in-place concrete becomes the structural member while the foam-filled concrete form work is left in place and becomes the thermal insulation for the wall system. The Mandish systems [38] use the foam-filled concrete in more typical lightweight concrete applications such as building facades and architectural details.

Another construction-related application of recycled PS foam involves the use of the foam particles as the primary insulating medium in a phenolic resin encapsulated foam board. In this patented process [39], the foam particles are coated with a phenol-formaldehyde or melamine-formaldehyde resin, placed in a mold, and cured by the application of dielectric heating. The resultant product is a good thermal insulation board with a very high fire resistance. The board is particularly well-suited for roof insulation application because of fire performance and high service temperature potential.

Other applications that have been developed for recycled PS foam include a filter medium and an oil spill clean-up material. As a filler medium, the PS foam must be ground to a very small particle size using cryogenic grinding. This particular technology is being developed by Matcom Inc. and is covered by several U. S. patents [40–43]. Using finely ground PS foam to soak up oil spills on water has only been tried a few times with limited success. The biggest drawback is transportation of the very low-bulk-density foam material.

5.6 Economics

The collection system developed for any recyclable product generally becomes the economic factor. For beverage containers, the deposit laws in some states drive the consolidation of containers so that sorting and collection is not an economic problem. For nondeposit-type items, the sortation and collection become a more complex problem.

5.6.1 Food Service Material

Since most PS food packaging is used in a manner that causes it to be commingled with other waste upon disposal, one needs to add a sortation step in the collection process. In the case of fast food restaurants, customers are asked to sort their waste when they clear their tables. Trash receptacles are provided to accept PS articles, and a separate container is designated for paper and food waste. This sortation by the customer is rarely perfect and generally results in a bag which is 50% recyclable and 50% nonrecyclable by weight. (Of course, one must remember that one 2-ounce (56.7 g) ice cube weighs as much as 15 hamburger clamshells.) Since a recycling plant cannot accept a 50:50 mix of recyclables/nonrecyclables without being permitted as a waste transfer site, a second sortation must take place. This second sortation is normally done by the hauler who picks up the bags from the restaurants, separates the PS, and bales and ships it to the recycling plant.

In the case of school cafeterias, commercial cafeterias, or the like, customers are normally asked to remove the loose waste from their trays and deposit the soiled trays in a rack upon exiting. Flatware and cups are deposited in separate receptacles. This type of separation has been found to be very effective, and material collected this way does not need a second sortation.

PS products from a curbside program are normally collected along with newspaper, glass, steel, aluminum, and other plastics. These materials are then further separated into their individual stream at the MRF. PS foam, which is readily identified by appearance, is normally hand-sorted and then baled and shipped to a recycling facility.

In all of the above cases, the recycle facility avoids the cost burden of the collection costs by utilizing any or all of these three means: (1) the generator must pay part or all of the collection costs; (2) collection of the material is piggy-backed on the collection of another material so the costs are shared; or (3) each pick-up must be of substantial volume so as to make it cost-effective on a unit cost basis. Each of these alternatives are described in more detail below:

1. The cost of disposal that the generator saves by recycling should certainly enter the economic model, and, more significantly, it should be a value calculated in volumetric, not weight terms. Trash containers at restaurants are deemed ready to haul away when they are physically filled, not when they reach some specific weight. In one U. S. survey, it was estimated that the average cost to a business owner to have a dumpster removed, emptied, and replaced was $10/yd^3 ($13/m^3). PS food packaging, when lightly compacted into a trash bag, has a bulk density of about 2 lb/ft^3 (32 kg/m^3). On this basis, the

restaurant owner is paying $0.185/lb ($0.40/kg) to dispose of PS packaging, which is obviously saved if the restaurant recycles. Another consideration in the economic equation is the cost of alternate materials. Many times, the replacements for the PS foam will be so expensive that a "recycling charge" can still make economic sense.

2. Collection systems are most cost-effective when more than one recyclable is being picked up. In the case of restaurants, the corrugated paper is often of enough value to warrant a pickup. If PS packaging can be collected simultaneously at cost, the cost benefits accrue to both materials. In some deposit states, the collection and return of used beverage containers can be combined with the collection of separated PS packaging materials. It is unlikely, however, that waste packaging could be backhauled on food shipments because of possible contamination.

3. Every pickup must contain an economically worthwhile amount of PS packaging. An average fast food restaurant may dispose of about 100 lb/week (45 kg/week) of PS packaging. In that case, a once-a-week pickup would be appropriate. In the case of a school or cafeteria program, about 2,000 meals would have to be served to approach the 100 lb (45 kg) of recyclables. As the weight per pickup drops below 100 lb (45 kg), the cost per pound rises nonlinearly, so it becomes cost-ineffective to look at very small pickup quantities.

When all of the above are brought into play, a very cost-effective program for all parties can be established.

5.6.2 Durable Goods Packaging

PS foam that is used in the packaging of durable goods is somewhat easier to collect than food service materials. This kind of packaging material may be contaminated with nonrecyclable items such as paper or other plastics, but these contaminants are normally easy to remove and in small quantities. This type of waste stream is usually generated at electronic, appliance or department stores, or assemblers of a variety of durable goods. Some quantities may be available from foam manufacturers as well, but, unless the product has been used in an end-use application, it may be classified as plant scrap.

The type of PS foam used in this application is usually molded-type with an absolute foam density of approximately 1.25 lb/ft^3 (20 kg/m^3). When typical shapes are collected loose in bags or in bins, the average bulk density is less than 1 lb/ft^3 (16 kg/m^3). An average 40-ft (12 m) tractor trailer can hold 3,000 ft^3 (86 m^3) and may only hold 2,000–3000 lb (907–1361 kg) of product.

In order to improve transportation economics, some form of densification is required. Typically, a compression type of baler is required to make the product easier to handle and increase the load factor. Unless the baler device is capable of exceeding 100 psi (0.69 MPa) at the face of the compression, the resultant bales will only achieve about 2 lb/ft^3 (32 kg/m^3) bulk density. Machines that can achieve 400–500 psi (2.7–3.4 MPa) may crush the foam cell structure and thereby achieve bale densities of 17–25 lb/ft^3 (272–400 kg/m^3). At this density, a 40-ft (12 m) trailer can be loaded to maximum weight rather than volume. When material is prepared in baled form and is free of contaminant, the product will have a positive economic value at the recycler's dock. This value is in addition to the avoided cost of landfilling. When these two values are added together and combined with a collection system for other packaging recyclables such as corrugated paper, the economic value can become significant.

The process to wash-dry-densify-pelletize or merely densify-pelletize is relatively straightforward. The actual costs will depend on labor, utilities, and cost of capital as one would expect; however, scale of operation will have a very significant impact. Since these economics are so variable, it would be difficult to cite examples here, but let it suffice to say that a profitable business can be established.

5.7 The National Polystyrene Recycling Company

The NPRC is a $16 million effort [44] between eight corporations: Amoco Chemical Company, ARCO Chemical Company, Chevron Chemical Company, The Dow Chemical Company, Fina Oil and Chemical Company, Huntsman Chemical Company, Mobil Chemical Company, and Polysar Inc. The NPRC has an initial goal to facilitate the industry's efforts to recycle 25% of the PS produced in the United States for food service and packaging applications each year. By the start of 1991, the NPRC will have established regional recycling facilities in Los Angeles, San Francisco, Chicago, and Philadelphia, in addition to an existing facility in the Boston area. The NPRC also has been instrumental in the facilitation of a PS recycling plant in Portland, Oregon. As these plants are brought up to capacity, markets for the resin produced are being developed in major applications, such as building insulation products, office accessories, and refuse containers, as well as packaging systems. The creation of a large, high-quality, dependable stream of recycled PS stimulates additional applications that can benefit from the recycle origins. Additional facilities beyond those already listed are in the planning stage.

5.8 Global Activities

While it is not the purpose of this review to be global in scope, a few comments on PS recycling activities observed outside of the United States seem to be in order. Since the entire area of plastic recycling is getting increased attention, one may expect the situations to change rapidly.

5.8.1 Japan

Although the Japanese seem to be active in the design and manufacture of thermal and radiation densifiers, there have been no reports of activity surrounding washing systems. It appears that the Japanese equipment used to densify foam plastic has been developed for cushion packaging from products that do not require cleaning. The densification equipment which seems most prevalent in Japan uses conductive heating systems to soften and densify the foamed plastic. This type of densification exposes the polymer to a nonuniform heat history and results in nonuniform properties in the final product. If the purpose of densification is to reduce the volume of disposal or to facilitate the use of the end product as a fuel source, the heat conductive densification may be quite effective. Certain product applications with less severe property requirements for the resin may allow the inclusion of a small percentage of nonuniform recycled resin, but the value of the resin will be low. Because of the different solid waste handling methods in Japan, it is difficult to forecast the future direction of the PS recycling efforts in that area.

5.8.2 Europe

Although there is more equipment developed in Europe to wash plastics prior to recycling, none of it has been developed specifically for foamed products. Most washing equipment has been developed for plastic films, although it can be modified to handle foamed articles. Densification systems in Europe seem to center about conductive heating systems and frictional heating via extrusion or intensive grinding. The latter methods tend to produce a more uniform product, albeit somewhat degraded, and are the types of systems capable of higher throughput rates. Although no food service recycling of PS foam has received any publicity in Europe, there is an active recycling program for PS foam cushion packaging in Germany. The Storopak Company in Metzingen, Germany, offers an "Umwelt service" [45] that is designed to recover PS foam materials used to package fragile items. The foam is returned to control collection stations where periodic pickups are made, and the foam is returned to the manufacturers. This foam is typically clean and does not require washing. The returned foam is ground to a small particle size and either added back into new molded foam articles or thermally densified to a solid resin pellet. The system was first introduced in 1986 but seemed to receive more attention in 1989.

5.9   Future Outlook

The future of PS recycling is dependent on many factors, all of which impact the underlying economics facing such an activity. The major elements that affect this effort are: (1) public attitude toward reusable items versus disposable items, (2) legislative mandates on product usage and composition, (3) solid waste disposal concerns, and (4) hydrocarbon pricing and availability. If one were to assume an extreme polar position on any or all of the above issues, the future direction would change dramatically. As it stands today, PS recycling is expected to be a significant activity through the 1990s and PS, along with other plastics, classed with glass, paper, and metal as a recyclable commodity.

References

[1]   Pohleman and Echte. "Fifty Years of Polystyrene." *Polym. Sci. Overview*. ACS Symposium Series, Vol. 175, 1981.

[2]   Reference 1.

[3]   *Modern Plastics*, 68(1) (1991): p. 122.

[4]   "Polystyrene." *Modern Plastics Encyclopedia* (1986).

[5]   Ames, J. L., International Award Address, *Polym. Eng. & Sci.* (Jan. 1974).

[6]   "Polystyrene." *Modern Plastics Encyclopedia* (1979–86).

[7]   Swett, R. M. "Polystyrene." *Modern Plastics Encyclopedia* (1986–87).

[8]   *Modern Plastics*, 68(1) (1991): p. 123.

[9]   Reference 8, p. 117.

[10]   Reference 9.

[11]   *J. Commerce* (9 May 1988).

[12]   *Modern Plastics*, 68 (1) (1991): p. 117.

[13]   Eaton, Christopher. "Foam Extrusion." *Modern Plastics Encyclopedia* (1986–87).

[14]   Reference 12.

[15]   Uniglobe Kisco, Inc., 709 Winchester Avenue, White Plains, NY 10604.

[16]   Package Industries, Sentinel Products Division.

[17]   Instamelt Systems, Inc., 300 N. Marienfeld, P.O. Box 1714, Midland, TX 79702.

[18]   Texmax Inc., 3001 Stafford Drive, P.O. Box 668128, Charlotte, NC 28216-8128

[19]   Glenro Inc., 39 McBride Avenue Extension, Patterson, NJ 07501-9997.

[20]   Buss-Condux, Inc., 2450 Delta Lane, Elk Grove, IL 60007.

[21]   Valex Converter, RAPAC, P.O. Box 428, Industrial Park Drive, Oakland, TN 38060.

[22]   Walex Roll Extruder, Storopack, Inc., 4758 Devitt Drive, Cincinnati, OH 45246.

[23]   Draiswerke, Inc., 3 Pearl Court, Allendale, NJ 07401.

[24]   Toronita Corporation, 1760 Toronita Street, York, PA 17402.

[25]   Mobil Chemical Company, Solid Waste Management Solutions, 1159 Pittsford-Victor Road, Pittsford, NY 14534.

[26]   MK Recycling, Inc., P.O. Box 658, Cape Girardeau, MO 63702.

[27]   Reifenhauser-Van Dorn Co., 35 Cherry Hill Drive, Danvers, MA 01923.

[28]   Katema Process Equipment, P.O. Box 1406, El Cason, CA 92022.

[29]   *Styrene Polymers*. 2d ed. John Wiley, 1989.

[30]   Sugden, J.,; NPRC internal correspondence. 15 October 1990.

[31]   Coughanour, R., NPRC internal report. 10 September 1991.

[32]   Westphal, R., NPRC internal report. 10 November 1990.

[33]   *Food Chemical News* (26 March 1990).

[34] Traex Mfg., NPRC internal correspondence. 15 September 1990.
[35] EEZZZ Lay Drain. Product literature. Pisqah Forest, NC.
[36] Styropor Division, BASF Corp., Parsippany, NJ.
[37] Rastra Building Systems, Los Angeles, CA.
[38] Mandish Research International, Mims, FL.
[39] Mosier, B. U. S. Patent 4 714 715, 1987.
[40] Klein, M. U. S. Patent 4 200 679, 1980.
[41] Klein, M. U. S. Patent 4 207 378, 1980.
[42] Klein, M. U. S. Patent 4 304 873, 1981.
[43] Klein, M. U. S. Patent 4 427 157, 1984.
[44] *Chicago Tribune* (7 June 1989): p. 3.
[45] Storopack GmbH, Postfach 1333, D7430 Metzingen, Germany.

6 Polyvinyl Chloride (PVC)

W. F. Carroll, Jr., R. G. Elcik, D. Goodman

Contents

6.1	Introduction	133
6.2	PVC Production and Markets	133
	6.2.1 Suspension PVC	134
	6.2.1.1 Rigid Formulations and Applications	134
	6.2.1.2 Flexible Formulations and Applications	134
	6.2.2 Plastisol PVC	135
	6.2.3 Copolymers	136
6.3	Disposal Streams for PVC	136
6.4	Incineration	137
6.5	Recycling of Rigid PVC Products	139
	6.5.1 PVC Bottles—History and General Considerations	139
	6.5.1.1 Programs and Developments—Europe	139
	6.5.1.2 Programs and Developments—United States	141
	6.5.2 Calendered Sheet	143
	6.5.3 Building Products	144
6.6	Recycling of Flexible PVC Products	144
	6.6.1 Automotive Waste	144
	6.6.2 Wire and Cable Jacketing	145
	6.6.3 Packaging Film	145
	6.6.4 Agricultural Film	146
6.7	Recycling of Plastisol PVC Products	146
	6.7.1 Bottle Closures	146
	6.7.2 Flooring	146
6.8	Future Outlook	147
	Acknowledgments	147
	References	147

Polyvinyl chloride (PVC) is the second largest volume thermoplastic in the United States. It is extremely versatile and may be compounded for rigid or flexible construction or packaging products. PVC recycling is in its infancy in virtually all postconsumer applications, although reuse of postindustrial scrap is time-honored. New techniques for sortation of PVC from other polymers have recently been developed. Coupled with company repurchase of postconsumer material, the potential for viable recycling is great.

6.1 Introduction

At this writing, discussion of total PVC recycling is a bit premature. Major applications for PVC are building products of long life span; packaging constitutes only a small part of end-product usage of PVC, yet packaging is the initial hotbed for recycling activity. To address the packaging issue, programs for collection and reprocessing of PVC bottles are now underway in earnest; but industry also perceives the future need for recycling other products as they are removed from service.

6.2 PVC Production and Markets

PVC homopolymer has the Chemical Abstracts Service (CAS) identification number [9002-86-2], and the basic structure $(CH_2CHCl)_n$, where n is typically 300–1700 repeating units [1–3]. In fact, PVC is best characterized by molecular weight, as it determines most physical properties of the polymer. Molecular weight is also a primary distinction among different commercial grades of PVC.

In 1990, U. S. production of PVC resins was approximately 9 billion pounds (4.05×10^6 t), the second largest of the major polymers in production volume. This represented a 7% increase in production over 1989. For simplicity, we will categorize the uses of PVC into rigids, flexibles, and plastisols. End uses and volume for PVC (1990 data) are summarized in Table 6.1 [4].

Table 6.1   U. S. PVC sales volume, 1990, by application

Resin Application	Million Pounds	(Thousand Tonnes)
Pipe, Tube and Fittings	3777	(1700)
Siding and Accessories	860	(387)
Packaging Film and Sheet	480	(216)
Wire and Cable	410	(185)
Flooring	246	(111)
Window Profile/Gutters/Skirts	401	(180)
Bottles	223	(100)
Other Calendering/Extrusion	1234	(555)
Other Coating/Molding/Sealants	728	(328)
Export	937	(421)
TOTAL	9296	(4183)

Source: *Modern Plastics*

PVC is never fabricated as the neat polymer, yet it is an excellent base with which to create literally thousands of diverse products. A variety of additives are required to enhance the performance of products made from PVC. Heat stabilizers, which prevent dehydrohalogenation, subsequent color formation, and loss of physical properties, are always used. Other additives such as processing aids, impact modifiers, and lubricants are used in rigids. Plasticizers are materials which impart flexibility to PVC and are used mostly in flexibles, although low levels of some plasticizers add to stability and processibility.

6.2.1 Suspension PVC

Over 85% of PVC production is by suspension polymerization of vinyl chloride monomer (VCM) via free radical addition. Suspension-grade PVC consists of relatively porous particles of about 130 µm in diameter [5].

6.2.1.1 Rigid Formulations and Applications   Rigid PVC products differ in molecular weight requirements. Bottles and sheet, which obtain most of their toughness from impact modifier and require exceptional processibility, are typically toward the low end of the molecular weight scale. Most commodity products that depend upon the properties of the base polymer itself are of intermediate molecular weight. A typical formula for rigid PVC is shown in Table 6.2.

Table 6.2   Typical formulation for rigid PVC

Ingredients	Amount (phr)
PVC	100
Impact Modifier	0–15
Processing Aid	0–3
Stabilizer	0.4–3

Most rigid PVC is extruded into pipe and conduit, siding, window profiles, gutters, downspouts, and sheet. Rigid PVC is also injection-molded into pipe fittings and electrical and appliance applications, calendered for blister pack, or extrusion blow-molded into bottles. Rigid applications account for the majority of PVC uses.

Each of the additives shown is more costly than PVC itself and is not used up in initial processing. Packaging formulations are rich in these additives; pipe and siding formulations are not. Packaging, then, is an excellent candidate for recycling as it has enhanced raw material recovery value (to justify collection) and commands higher prices (to defray the cost of processing).

PVC bottles are made by extrusion blow-molding, which has relatively low equipment cost but also relatively low productivity. PET bottles, on the other hand, are fabricated by injection stretch blow molding. Equipment and molds for this process are expensive but offer good economies of scale for large runs of identical bottles. The combination of its unique moldability, crystal clarity, and intermediate price has given PVC a niche in the bottle market, but unlike polyethylene terephthalate (PET) soft drink bottles, PVC bottles come in many different shapes and colors so that they are not readily identifiable and separable manually.

6.2.1.2 Flexible Formulations and Applications   With the addition of plasticizer, tough, rigid PVC becomes flexible and pliant, and a whole new range of applications can be achieved.

Flexible PVC is calendered for pool liners, auto upholstery, handbags, films, tapes, labels, shower curtains, shades, and wall covering. It is extruded into electrical wire and cable, garden hose, blood bags, and medical tubing. A typical formulation for flexible PVC is shown in Table 6.3.

Table 6.3   Typical formulation for flexible PVC

Ingredients	Amount (phr)
PVC	100
Plasticizer	25–100
Stabilizer	1.5–5.0
Lubricant	0.5–1.0
Filler	0–30

Choice of plasticizers varies according to the end-use application. Monomeric diesters of phthalic, adipic, and azelaic acid and triesters of trimellitic acid are used because they impart different temperature and physical performance to products. Polymerics have low volatility and migration but are more difficult to blend with resin and are more expensive. Flexible PVC, therefore, is a much less homogeneous set of products than rigid PVC, and recovery of these products will yield a widely variable recycle stream unless collected by application.

The molecular weight of PVC used for flexibles is typically intermediate to high, as these applications require toughness and abrasion resistance yet must retain a soft feel. Products fabricated with very high molecular weight resin have properties of thermoplastic elastomers [6].

6.2.2 Plastisol PVC

Plastisol PVC, which accounts for about 10% of PVC production, differs from suspension PVC in particle size and structure. Dispersion resins are synthesized in a latex by emulsion or microsuspension polymerization and mixed with plasticizer to form a fluid "paste." Plastisol resin particles range in size from 0.1 to 50 µm and are also generally more isotropic and less porous than suspension resin. When suspension resins are mixed with plasticizer, PVC is the continuous phase and plasticizer is absorbed within the porous particles. In plastisol pastes, however, plasticizer is the continuous phase and PVC is the disperse phase.

This inversion creates a different set of handling properties and end-use applications. Plastisol is processed as a liquid, and rheological properties of this liquid must be controlled. This is done by managing the type and level of resin surfactant, size and distribution of particles, and types and amounts of additives. A typical plastisol formulation is found in Table 6.4.

Table 6.4   Typical formulation for plastisol PVC

Ingredients	Amount (phr)
PVC	100
Plasticizer	30–400
Stabilizer	0–4
Diluent	0–20
Pigment	0–20
Filler	0–100

The molecular weight range for these resins is intermediate to high, with the average over all products being higher than the other two classes of resins. Plastisol pastes are roto- or slush-molded into toys, sports equipment, auto dashboards and door panels, footwear, and traffic cones. They are dip-molded for medical gloves or tool handles. By various coating processes, they are used for sheet vinyl flooring, carpet backing, table cloths, and wall covering. Plastisols can be sprayed as automobile undercoating or spun as cap liners for bottles.

6.2.3 Copolymers

Vinyl chloride (VC) can be copolymerized with several other monomers, but only copolymers of vinyl acetate (VAc) or vinylidene chloride (VDC) are made in any commercial quantity. VC/VAc copolymers (CAS #9003-22-0) were initially commercialized as internally plasticized PVC. Copolymers were developed to increase the thermoplastic flow properties of PVC during melt fabrication of products and to improve solubility characteristics in common organic solvents. Poly(VCM-co-VAc) has the basic structure $[(CH_2CHCl)_x(CH_2CHOAc)_y)]_n$, where x is greater than y. Vinyl acetate is typically used as a comonomer at 3–20%.

VCM/VAc copolymer resins generally have lower glass transition temperature, heat distortion temperature, and tensile strength than PVC homopolymer, but they can be used in applications where low melt viscosity, low processing temperatures, freedom from external plasticizers, and good solubility are needed. Product areas for such applications include floor coverings, phonograph records, and protective coatings.

Chlorinated polyvinyl chloride (CPVC) and polyvinylidene chloride (PVDC) are similar in that they both contain an additional chlorine atom in the repeating polymer unit. In the case of CPVC, chlorine is added to the polymer backbone by an additional process after the PVC has been made. Inclusion of this added chlorine on vicinal carbons substantially increases the heat distortion temperature and other properties of the resultant polymer. For example, a pipe made out of CPVC might be able to withstand water temperatures 30 °C higher than a corresponding PVC pipe.

PVDC and PVC are made via similar processes. In fact, VCM and VDC are copolymerized in production of Saran™ (Dow Chemical Company). The polymer or copolymer is more crystalline than PVC and is generally used as a barrier film or coating to prevent permeation of gases and moisture.

6.3   Disposal Streams for PVC

Via end-use analysis, Franklin Associates estimates that 23% of domestic PVC production will eventually enter the municipal solid waste (MSW) stream [7]. By definition, MSW includes disposal from single and multifamily residences, retail and wholesale commercial establishments, institutions, offices, farms, and small factories [8]. While a sizable amount, PVC is only 7% of the total plastics in MSW. By comparison, low-density polyethylene (LDPE), high-density polyethylene (HDPE), polypropylene (PP), and polystyrene (PS) contribute 27, 21, 18, and 16% by weight of the total MSW plastics, respectively. As plastics in MSW account for about 7.3% by weight, PVC is less than 0.5% of the total stream [9]. PVC packaging discarded as MSW includes about 600 million pounds (270×10^3 t) of bottles, closures, wraps and sheets.

Since most PVC is used as building products such as pipe, house siding, window profiles, electrical insulation, and flooring, most PVC disposal will be in the nonmunicipal solid waste stream (non-MSW). By definition, this includes heavy industry generation, construction waste, demolition debris, automotive scrap, sludge, and incinerator residues. Fully one-third of all potential plastic discards destined for non-MSW disposal are calculated to be PVC due to its heavy use in construction-related applications. Franklin Associates estimates that 69% of the 8.3 billion pounds (3.74×10^6 t) produced in 1988 will eventually be disposed of in non-MSW landfills [10].

6.4   Incineration

About 10% of U.S. waste is incinerated, so no discussion of disposal of PVC would be complete without a review of the facts and lore surrounding incineration. All three major arguments made against incineration by the environmental lobby also are made against PVC as part of incinerator feeds:

1.   Critics allege that incineration presents a health risk from airborne toxics, especially 2,3,7,8-tetrachlorodibenzdioxin (also known as TCDD or simply "dioxin"). Since PVC is a chlorinated material, it is said that PVC makes the problem worse.

2.   Critics allege that toxic metals in incinerator feeds are emitted to the air or make ash a hazardous waste. Since cadmium and lead stabilizers are used with some PVC applications, it is said that PVC makes the problem worse.

3.   Critics allege that incineration is expensive, particularly when stack-gas remediation is included in the capital and incremental cost of running the incinerator. Since PVC yields hydrogen chloride as a combustion product, which must be neutralized in the stack gases, it is said that PVC makes the problem worse.

Although each of these questions has been answered separately in the literature, a recent summary has been issued [11]. This study confirms that fear of incineration, and particularly fear of incineration of PVC, is not justified.

The original characterization of dioxin congeners in combustion gases from incinerators was made by Olie, Vermeuler and Hutzinger [12]. The possibility that dioxin may be generated during many types of combustion is now well-accepted. Although dioxin is said to be ubiquitous [13], it also has a history. Hites and Czuczwa found a correlation between dioxin levels in lake sediments and the year the sediment was formed [14]. Dioxins were absent prior to about 1940; they rose steadily until the 1970s and appeared to peak near the end of that decade. Hites and Czuczwa hypothesized that the increase in dioxins paralleled the increase in incineration of chlorinated materials, including plastics such as PVC [15]. The logic is appealing: chlorine is a constituent of dioxin; therefore, to eliminate dioxins from incinerator emissions, eliminate the sources of chlorine.

Discoveries of other modes of dioxin generation present a different picture, however. Rappe and coworkers [16], as well as Ballschmitter and coworkers [17], found dioxins in exhaust gases from automobiles burning leaded gasoline. They hypothesized that chlorinated anti-knock additives were the source of chlorine. Their hypothesis also fits very well with Hites's data, as the use of leaded gasoline with such additives increased up until the 1970s, when gas crises and unleaded fuel made it obsolete. Thus, the historical correlation of dioxin and incineration of PVC is questionable.

Fundamental studies of incineration have been carried out over the past ten years to find the mechanism of formation of dioxins and the contribution of PVC to that formation. Karasek [18] carried out small-scale experiments with elevated levels of PVC in the incinerator feed and found no enhanced production of dioxin. Giugliano [19] increased the PVC loading from the typical 0.5% to 10%, yet observed no increased dioxin production.

Other attempts to add or delete chlorine in incinerator feeds gave varying results; two factors contribute to this poor experimental reproducibility. First, incineration experiments and dioxin analyses at part-per-trillion levels are difficult to perform. Second, small-scale incinerators do not reliably scale up to industrial apparatus.

In our opinion, the premier study of the effects of PVC on incineration off-gases remains the so-called New York State Energy Research and Development Authority (NYSERDA) study [20], conducted in the Vicon Incinerator at Pittsfield, MA. In this study, MSW and PVC-free waste were burned in a full-scale incinerator at varying temperatures, and off-gases were analyzed. The data show no correlation between PVC in the feed and dioxins in the exhaust. In fact, the dioxin emissions were of similar magnitude to the dioxin coming into the incinerator with the feedstock. The main significant variable is incineration temperature, with lower temperature generating higher levels of carbon monoxide and dioxins. Low combustion temperature, however, may not be the only significant variable contributing to production of dioxin during combustion. Rappe [21] reports on a fire in a warehouse containing 200 tons of PVC and 500 tons of carpet that burned out of control (especially as to temperature) for days. Significantly less dioxin was generated in this fire than would be generated during state-of-the-art, Swedish-government-approved incineration of waste containing the same amount of PVC.

Critics point out that a recent Danish study suggests that by doubling the PVC in an incinerator feedstream, a 34% increase in dioxin results [22]. Yet Rappe points out that experimental error associated with dioxin analyses performed in this Danish study is 100%; replicate experiments differ in results by 600% [23]. Clearly, no meaningful conclusion can be drawn from such poor data.

Concern about acidic combustion gases arises from concern about acid rain and incinerator corrosion. Without doubt, increasing the chlorine content of the feedstock increases the amount of hydrochloric acid (HCl) generated in combustion [24], but HCl is a minor component of acid rain, and HCl from incineration is an even smaller contributor (less than 0.3%) [25]. With regard to corrosion from HCl, even incinerator manufacturers now believe it to be a minimal problem and controllable [26].

To minimize emission of HCl and dioxins, equipment for stack gas remediation must be installed [27]. In this way, the small risk posed by an incinerator can be reduced by a factor of ten, with incremental cost about $10/ton ($11/t). As an example, the Flakt Lime system reduces dioxins and acid gases by over 90% [28]. Rappe notes the capability of such systems to reduce dioxin emissions by two to three orders of magnitude [29]. Its cost is nominal, and its use addresses the question of HCl as a combustion product. Since PVC accounts for one-third to one-half of the chlorine content of incinerator feeds, total removal of PVC would reduce HCl emissions about 30–50%; there would still be a need for acid-gas scavenging for the remainder of the HCl.

Incineration qualifies as a version of recycling if the heat released in burning is recovered and put to good use [30]. A new and unique approach to recycling during incineration relates to the salts created when HCl is neutralized in stack gases. In the "Closed Salt Cycle" [31], HCl is recovered with caustic soda to yield sodium chloride. Salt is then electrolyzed to regenerate caustic and chlorine, and the chlorine is reacted with ethylene to produce vinyl chloride monomer. Although this seems expensive at first blush, every step is necessary for other reasons, and the cost of landfilling contaminated salts is avoided. This process has been commercialized in the Iserlohn and Hamburg incinerators [32, 33].

Incinerator ash contains the residues of metals inherent in the feed. Some metals (mercury and lead) may vaporize and issue from an untreated stack. While cadmium and lead stabilizers are occasionally used in some types of PVC, their occurrence is declining. Neither is used in packaging.

A novel approach to removal of heavy metals from ash was proposed by Vogg [34]. In the "3R" system, acid gases including HCl are mixed with water and used to acid-leach metals from ash. Subsequently, solutions are concentrated and metals are recovered from the solutions.

Hospital incineration of infectious and other wastes presents an interesting analysis of relative risks. The hazard associated with transportation and landfilling of infectious wastes is obvious; in recent years, the perception of hazard associated with incineration has increased, and hospitals must reconsider their disposal method of choice [35]. Although the scale of hospital incinerators is different from typical MSW incinerators, and hospital waste streams tend to be enriched in PVC, operation of both types of incinerators for minimal environmental impact is similar [36].

New intermediate-scale incineration studies are underway that will relate small- and large-scale combustor performance under scrupulously controlled conditions [37]. Even under incineration conditions currently in use, however, the health risk to the general public from incinerators is insignificant [38, 39].

6.5   Recycling of Rigid PVC Products

6.5.1   PVC Bottles—History and General Considerations

The history of recycling postconsumer PVC bottles is short. Before there was postconsumer recycling, however, processors were already reusing their postindustrial scrap. For example, as much as 40% of the resin used to mold a typical half-gallon (1.9 liter), handleware bottle is neck flash, handle knockout, or bottom tab. This scrap is fairly pure and is generally ground and reused in production of virgin bottles.

Scrap generated from mislabeling or from discolored or dirty polymer cannot be reused in virgin bottles, and other outlets must be found. Discolored or very lightly contaminated scrap is generally sold to injection molders and extruders. Bottles contaminated with paper labels or bottle contents are problematic for any fabricator and were typically landfilled until very recently. In fact, these bottles are nearly equivalent to true postconsumer bottles in that they must be washed, dried, and possibly repelletized to yield a usable product. Learning to reprocess this fill-line scrap has been of use in deciding how to attack the garbage problem.

For true postconsumer waste, collection is the first stumbling block. No program can begin in earnest unless large quantities of raw material are available, and since it is such a small part of the bottle stream, no large collection program will be organized exclusively for PVC. Eventually, either through political expediency or collection economies of scale, plastic bottles beyond those of beverages will be collected at the curbside. PVC will be included in this collection.

Sortation is difficult, especially as related to PVC and PET. Most packaging plastics are incompatible with one another, but PVC and PET present the worst of the cross-contamination problems. Although PVC and PET are easily separable from HDPE or PP by water flotation, they both have densities of 1.30–1.35 g/cm^3 and are inseparable from one another by normal flotation methods. If PET is contaminated with PVC, PVC will degrade at the high PET processing temperature and produce char. PET does not melt at PVC processing temperatures and must be filtered out.

Once separated, PVC itself does not present any unusual problems for reprocessors. It can be ground, albeit with some production of fines, and regularly is in reuses of mold flash. It can be washed in any unit capable of recovering material with density greater than water. Despite the well-known thermal sensitivity of PVC, it can be remolded; in some cases, addition of stabilizers or lubricants as boosters may be advised. Once again, the reuse of PVC with many heat histories is well-known in the processing of in-plant regrind. There are no unusual emissions created in the recycling process as working with recycled PVC is equivalent to working with virgin polymer, given the additional heat history it has experienced.

At first glance, the pipe market would seem to have the greatest potential for absorption of recycled PVC with the least risk to volume growth. However, use of recycle in certain kinds of pipe may be prohibited by codes that prescribe regrind other than that from the manufacturer's own process [40]. The extent of this restriction is being studied, but nonpressure drain pipe would not be code restricted.

Reprocessing recycled PVC into packaging is the largest challenge, since manufacturing specifications for gloss, color, and clarity are quite stringent. However, environmental lobbyists argue that a product is not truly recycled until it reappears in the same application. Also, due to the cost of recycling, higher priced applications must be found to support this cost if PVC recycling is to be sustainable and profitable. Finally, recycled content may be mandated by legislation. Packaging must be seen as a prime candidate for recycled PVC, even if other, more permanent, applications are technically available.

6.5.1.1   Programs and Developments—Europe   Europeans began the search for effective recycling programs in the middle 1970s with the establishment of GECOM (Groupe d'Etude

pour le Conditionnement Moderne) and GREPP in France. GECOM is an industry group that combines the talents of PVC producers and users and packaging and equipment manufacturers. Under its guidance, various programs related to PVC and PS have been carried out over the years [41], including critical studies of the health effects of vinyl chloride monomer, as well as reviews of developments in PVC packaging. Today, GECOM also acts as the purchaser of waste PVC bottles.

GREPP was an association of bottled water-based interests including Dorlyl, Solvic, BAP, and Rhone-Poulenc. In the late 1970s, GREPP organized a program to collect PVC bottles from households. Collection efficiency was good in the areas served, recovering 20–47% of the theoretical amount of PVC available [42]. GREPP offered a national support price, contracts, and technical assistance to get the program started. Recovered resin was used to fabricate electrical conduit and pipe [43, 44]. Pipe produced in this way was shown to be sensitive to polyethylene (PE), paper, and PS contamination, and the physical properties of the articles were affected by these impurities.

Earliest PVC collection efforts were based on drop-off centers, still an important technique. In Belgium, supermarkets set up receptacles for plastics near similar receptacles for glass [45].

Recently, the French government, environmental groups, and GECOM announced a large-scale recycling effort in southern France called "Operation Pelican." Various collection methods will be tried, many in conjunction with schools as educational programs. Operation Pelican intends to collect 2,500 tons of plastic bottles in 24 municipalities, a capture rate of about 20% for those areas (about 5% of France's total consumption) [46]. In late 1990, an environmental group (France Nature Environment) also had joined the venture [47]. Operation Pelican paid 1200 FF/t as of late 1990 for recovered PVC. Purification of the PVC after recovery will be accomplished by "Micronizing" (see below).

Sortation of PVC depended largely upon manual identification until Govoni, the European Vinyls Corporation (EVC), Solvay et Cie, and EniChem announced the development of an automatic sortation system called "Tecoplast" [48]. This system is part of an integrated reverse-vending-machine operation. At a participating store, a consumer feeds PVC and PET bottles into the machine. Glass and metal are rejected and, in return for the plastic bottles, the consumer receives store credit. In the machine, bottles are flattened by a heated roll, scraped off the roll with a knife, and bagged; available space in the machine is monitored automatically. At the processing plant, bottles are vibrated into cells of a cleated conveyor, then passed single-file beneath X-rays.

A European patent application describes a system that may be functionally similar to the Govoni process [49]. In the claimed device, objects attenuate transmitted X-rays as a function of the composition of the material. PVC is more efficient at attenuation of X-rays than other common packaging plastics and is recognized on this basis. Non-PVC plastics are diverted to a second line, and processing takes place in two different grinding and washing systems. A later United States patent issued to Govoni describes the entire process in more detail [50].

Much development effort has also been given to reprocessing of recycled PVC. The most versatile product is made by fractional pulverization, and Micronyl SA [51], supported by GECOM, has led in this effort. PVC bottles and impurities are passed over magnets, and large waste is removed by sieve. The remainder is dried, lights and heavies are removed, and it is ground to 0.79 inch (20 mm) and screened. After further grinding of PVC particles to about 0.39 inch (10 mm), material is washed and separated by gravity. At this point, glass and PE have been removed, so PVC and some PET remain in the heavy fraction, which is micronized. Aluminum is removed electrostatically. Separation depends on differential grinding of PVC and PET; PET remains in the coarser fraction.

Micronyl starts with nearly pure PVC and returns product fractions of pure PVC (powder), PET and other impurity (flake), and a mixture of the two. The finest particle-size product can be used like virgin resin in many extrusion applications. Solvay has patented a foam-core pipe that encapsulates this type of recycled PVC between two layers of virgin resin [52]. Water bottle regrind is preferred for this process. Other applications for recycled bottles include pallet corners, shoe soles, and flooring underlayment; profile extrusions are under development.

Cep Adour Industry, also with support from GECOM, developed a version of commingled extrusion technology for dirty or contaminated PVC [53]. Feed is desirably at least 50% PVC, but bottle-type formulations are not required. Cep Adour makes various products for agriculture including wine, vegetable, and fruit stakes. Thus, between Cep Adour and Micronyl, the quality of the feed determines the final products made from reprocessed resin.

In 1989, EVC announced a joint venture with PVC Reclamation Ltd. for recycling of all types of plastics. This venture, called Reprise International Plastics Recycling, is based upon both curbside collection and drop-offs and is the largest bottle recycling initiative to date in the United Kingdom. Reprise will mechanically separate PVC, HDPE, and PET; these materials then will be granulated and cleaned. Harvested PVC will be processed neat or blended with virgin in EVC's compounds for cables, wall covering, and window frames [54]. Other plastics will be sold on the open market.

Two supermarket chains, Tesco and Sainsbury, have signed on to the program by sponsoring drop-off sites in their parking lots. Another recycling project, called Operation Recoup [55], sponsored by the plastic bottle industry in the United Kingdom, will collect all types of plastic bottles at the curbside and pass them on to the Reprise venture [56].

Until recently, Germany emphasized incineration as a waste disposal method; however, in early 1990, Deutsche Solvay announced another water-bottle-to-pipe venture. Evian water bottles are sold under a mandatory deposit system in Germany. When they are returned, they go to a central distribution site. After collection they are ground, washed, micronized, and extruded into sewer pipe [57]. A similar program has been organized for Austria by Solvay Oesterreich in Salzburg [58].

Plastic bottle separation proposals for Italy spring from "low-tech" ideas as well [59]. Known as the "Loop" project, PVC and PET semisparkling mineral water bottles would be distributed in crates made of the same plastic. Due to unique shaping of the bottle bottoms, a bottle fits only in its appropriate crate.

There appears to be little or no attempt to reuse PVC in bottles in Europe. This may be because PVC is used predominantly for food packages, and manufacturers are understandably reluctant to use recycle in direct food contact.

6.5.1.2 Programs and Developments—United States

Earliest attempts at recycling vinyl were aimed at postindustrial material. A solvent-based process was reported by Hafner Industries in the 1970s [60]. This process was reportedly poised for commercialization in 1990 if funding could be obtained [61], although it appears not to be forthcoming [62].

Postconsumer recycling projects in the United States began in earnest in the late 1980s. Programs have been organized by The Vinyl Institute (VI), Dow-wTe, Vermont Republic Industries (VRI), Recoverable Resources Boro Bronx/Bronx 2000 (R2B2), Occidental Chemical (OxyChem), BF Goodrich, Georgia Gulf, and Keysor Corporation; some programs are collaborations by two or more of these parties.

In 1987, BF Goodrich distributed fliers identifying PVC bottles to employees and children from a local school. The company paid 5¢ per bottle to all who collected them during the pilot period [63]. Bottles were ground and washed in the laboratory, and regrind was mixed with virgin PVC. This mixture was shown to be extrudable and moldable in the laboratory. In 1988, Goodrich agreed to take postconsumer PVC collected by the Dow-wTe joint venture from its collection point in Akron, Ohio [64]. Although PVC was not specifically requested for this collection, Goodrich accepted that which was received inadvertently.

VRI (St. Alban's, VT) is an "affirmative workshop," that is, a private business employing people with minor mental and physical disabilities. VRI obtains PVC computer chip carrier tubes from local electronics vendors and refurbishes or grinds them [65]. VRI then sells this material, which is functionally equivalent to clean bottle regrind, to profile extruders. In conjunction with OxyChem and Ben and Jerry's Homemade Ice Cream, VRI washes and grinds some HDPE bucket scrap as well [66].

R2B2 collects, sorts, and grinds postconsumer plastics using hard-to-employ urban residents in "community-based recycling." In the mid-1980s, R2B2 was unique in its willingness to sort PVC by hand; it was nearly universally believed at that time that PVC sortation was futile. Operations in at least six more cities were planned as of 1989.

An industrywide attempt to quantify PVC recycling problems was organized by VI in 1988 [67]. The "Delaware Valley Program" depended upon employees of member companies located in the Philadelphia, PA, area and consisted of drop-off centers at each factory or office. The program lasted three months.

VI members learned early in the program that education, even education of employees within the PVC industry, is difficult but critical to success. Five thousand employees in the area represented a potential yield of about 1,000 lb (454 kg) of PVC; about 100 lb (45.4 kg) was obtained. This amount is small but consistent with the 10–15% recovery rate obtained in other U. S. drop-off experiments [68]. Collected PVC was combined with recycled water bottles obtained in Europe, washed at the Center for Plastics Recycling Research at Rutgers University (CPRR), and fabricated into pipe and fittings.

Two other challenges surfaced. As noted by Atochem [69], European recycle contains about 5% PET contamination. Since the ground bottles had been washed but not micronised, this contamination carried through to the final products. Still, functionally acceptable if unattractive perforated drain pipe was obtained commercially using 100% recycled PVC.

In 1989, VI agreed to sponsor sortation scale-up engineering by two organizations. CPRR proposed the use of X-ray fluorescence (XRF) to detect chlorine atoms in bottles, and by inference, PVC [70]. A commercial unit manufactured by ASOMA (Austin, TX) was grafted to a delivery line designed by Pace Packaging (Fairfield, NJ).

XRF spectroscopy depends upon absorption of X-rays from a radioactive source by atoms larger than sodium. From their excited state, these atoms fluoresce at known wavelengths, which can be monitored. In the analytical laboratory, XRF is used for simultaneous multielement detection; in the ASOMA device, X-ray sources and detectors are optimized for detection of a single element: chlorine.

The detector operates in a reflectance mode; samples must pass within ~0.25 inch (~6 mm) of the detector as both the signal and the response decrease as the square of distance. Engineering optimization of the presentation and ejection mechanism continued in early 1990 by both ASOMA and CPRR, but the system operated smoothly at detection rates of multiple bottles per second.

The second system was proposed by National Recovery Technologies (NRT) (Nashville, TN) [71]. This system, although proprietary, also operates via electromagnetic radiation. NRT received a grant from the U. S. Environmental Protection Agency (EPA) for development of a system to remove PVC from whole refuse, thus reducing HCl emissions from incinerators [72]. Detector development continued in early 1990, although the basic ejection mechanism is available commercially as part of metal removal systems marketed by NRT/Magnetic Separations Systems (MSS) [73]. An intermediate-scale prototype was shown in late 1990 [74], which easily achieves multiple bottle-per second sortation with >99% selectivity. A fully-developed commercial-scale unit was to be placed at XL Disposal (Crestwood, IL) in 1991. NRT and EVC have agreed to participate in a joint venture that will expand technology developed by both companies into a sortation line for all plastics [75].

Even outside the PVC industry *per se,* sortation technologies continue to develop. For whole bottles, an optical method was developed by Microfilm Archimed (Repentigny, Quebec) to separate HDPE, PVC, and PET [76]. For flake material, Refakt, the recycling division of Herbold, USA, announced a venture to commercialize hot-belt separation of PET and PVC. In this system, chips of PVC and PET are heated to a temperature where PVC melts but PET does not. PVC chips stick to the heated belt, while PET chips can be thrown off. Residual PVC is then scraped from the belt [77]. A proprietary nondestructive system for separation of PVC and PET or PVC and HDPE flake reached benchtop scale at DevTech Laboratories (Amherst, NH).

In anticipation of a separated stream of bottles, OxyChem began a nationwide buy-back of postconsumer PVC bottles in fall, 1989. OxyChem offered prices for PVC bottles that were competitive with prices paid to recyclers for other packaging plastics and also paid freight costs. Initial minimums were set at 2,000 pounds (907 kg) to allow easy participation by small recyclers [78].

By definition, the program included all bottles that held the product for which they were originally intended. Material was ground and washed in existing plants via tolling arrangements. In the same way that other recycled bottles are processed, it seems necessary to repelletize PVC with melt filtering to remove small bits of paper and PET. Recovery of these bottles in the first year exceeded 400,000 pounds (181×10^3 kg). Commercialization of EcoVinyl™ recycle-content compound occurred in early 1991.

Unlike Europe, there is a significant market for nonfood-grade bottles in the United States. With the advent of recycle-content legislation [79, 80], it is imperative that bottle-grade product be developed, and this was the announced intention of the OxyChem program. Opaque bottles could be used in the inner layer of coextruded bottles, but coextrusion of PVC bottles is extremely rare. Still, new tooling capable of coextruding two-layer PVC bottles has been announced by Willi Mueller KG of Germany [81].

In 1990, BFGoodrich announced the formation of a recycling laboratory to study the technology of recycle PVC cleaning and processing [82]. Although they have not announced a general buy-back, they are recycling bottles collected at drop-off centers in one Wisconsin county in conjunction with Schoeneck Containers, Incorporated [83].

Important results related to sortation were reported by the Goodrich laboratory. First, the relationship of detector distance from sample in X-ray fluorescence detection of PVC bottles and the effect of bottle shape was determined [84]. Also, an exhaustive characterization of material obtained from the Akron MRF was made. As a result, some thoughts on manual sorting and pipe made from such materials, was reported [85].

ICI Australia announced a line of products based on cryogrinding, which may be considered a variation of micronising. Bottles are shredded, cooled with liquid nitrogen, and pulverized. Early products contained 30% recycle content. Revinyl, as the product is known, sold for 10–15% above normal virgin prices at the time of commercialization [86].

Georgia-Gulf has announced formation of a market team to identify markets for recycled PVC [87] and subsequently have agreed to buy clean, ground recycled PVC [88]. Keysor Corporation has announced its intention to buy clean, pelletized PVC bottles [89].

The question of available markets for recycled PVC bottles, at least if cost/price are not considered, appeared to be answered by a 1990 study by Bennett and Lefcheck of the University of Toledo [90]. It showed the demand potential for recycled PVC to be 494 million pounds (224×10^3 t) per year, an excess demand of 287 million pounds (130×10^3 t) over 1989 bottle usage. Markets cited depended heavily on building products applications.

6.5.2 Calendered Sheet

Although blow molding is the largest packaging application for PVC, the fastest growing application is rigid calendered or extruded sheet for blister pack and food packaging [91]. So far, collection programs have included only bottles, and no current municipal programs actively anticipate capture of postconsumer blister pack except as mixed plastics waste. Still, one of the most active postindustrial recycling programs is for blister pack scrap.

When blisters are fabricated, they are cut from a large sheet, leaving "trim" or "skeletons." When ground, skeletons yield a clean, clear, high-impact material that has been a staple of scrap markets for many years. Some fabricators of calendered sheet have demonstrated the use of recycled skeletons [92]. Skeletons are typically extruded into drain pipe or decorative moldings and calendered into sheet for cooling towers or water purification plants. At least one manufacturer, American Mirrex, uses postindustrial recycle in marketing recycle-content compounds [93].

6.5.3 Building Products

Although this category of materials (including pipe, siding, and window profiles) has received little attention in the United States, it has not escaped notice in Europe. Certain environmental lobbyists perceive the need for environmental impact statements concerning disposal of PVC after its useful life. Despite the greater-than-25-year useful life of these products, certain German cities declared themselves "PVC-free" zones with regard to new construction until PVC disposal could be addressed. In one case, the prohibition has been lifted [94], but the situation is far from stable.

Proper disposal of long-lifetime products will be a task unique to the PVC industry. None of the other commodity thermoplastics has as large a fraction of its production dedicated to durables, and this share will be increased by "cascading" of nondurables. Cascading is the practice of removing one product from the waste stream and fabricating it into more durable and presumably more forgiving applications. It is a great temptation to view PVC pipe as an infinite sink for packaging, but cascading simply delays the problem of recycling packaging until recycling durables becomes the imperative.

Fortunately, the need to anticipate the end of the useful lifetime of durable products has not been lost on U.S. manufacturers. In late 1989, the Vinyl Siding Institute and Vinyl Window and Door Division of the Society of the Plastics Industry (SPI) organized a task force to address the need for pre- and postconsumer recycling [95].

Pipe, siding, and window profile cutting ends also are a staple of the plastics scrap brokerage industry, although collection of small, widely dispersed amounts is sometimes a problem. Metals brokers may eventually handle decommissioned siding and pipe as an extension of their aluminum and iron scrap business.

There will come a time, however, when the volume of pipe and siding coming out of service will be significant with respect to the amount going into service. Fortunately, with appropriate cleaning and reprocessing there seems to be no technical barrier to starting this kind of PVC on another 50- to 100-year lifetime as a useful product.

6.6   Recycling of Flexible PVC Products

Unlike rigids, most flexible PVC products are destined for MSW disposal. At first glance, the diversity of products would seem to preclude any viable curbside recycling efforts, and MSW collection programs will probably not include these plastics for some time. Recycling of a few large-volume or easily isolated flexible PVC product groups is possible, however, especially those destined for non-MSW disposal.

6.6.1 Automotive Waste

PVC automotive products enter the waste stream via non-MSW. Vinyl upholstery, crash pads, door panels, body side moldings, and wire insulation are recovered as part of the automobile scrap residue called "fluff." Fluff remains after crushed and shredded automobiles' metal content has been salvaged and consists primarily of waste glass, fiber, plastic, and dirt. The plastic component, which is known to be a large percentage of the total, contains a mix of thermoset, thermoplastic, and foamed materials. The U.S. EPA estimates that between 500 and 850 pounds (226 and 386 kg) of fluff is generated for each of the 10 million cars shredded each year [96]; for 1990 models, about 20 pounds (9 kg) of that was PVC [97].

Previously, automotive fluff was landfilled, but the combined problems of cost, lack of space, and concern for potentially hazardous material is forcing the automobile rendering industry and public officials to seek alternate solutions. The mix of materials in fluff—both thermoset and thermoplastic—makes it a particularly difficult recycling problem. In the 1970s, attempts were made to recover some polymers generically [98]; in the 1990s, the concept of auto disassembly lines with subsequent sortation of materials is being discussed.

6.6.2   Wire and Cable Jacketing

Approximately 500 million pounds (227×10^3 t) of PVC enter the wire and cable insulation market annually. As a result of the maturity of this product, tens of millions of pounds of decommissioned wire and cable scrap enters non-MSW each year as it is removed from service during demolition, remodeling, and rendering of electrical and telecommunication devices. Copper and aluminum are recovered through various forms of chopping operations. The major byproduct is PVC insulation compound mixed with quantities of cross-linked HDPE, paper, cloth fibers, and from 1 to 3% residual metal. The obvious problem is that the waste is a mixture of polymers and metals; the hidden problem is the presence of lead-containing compounds used as thermal stabilizers.

Lead stabilizers are used in wire and cable insulation compounds because they provide excellent protection against thermal degradation during processing but do not generate salts, which can degrade the dielectric properties of the insulation. Lead is a toxic heavy metal specifically targeted for substitution by the U. S. EPA [99]. As a result, landfilling this material has been severely curtailed because of the perceived potential for ground water contamination. Incineration is restricted because of possible air emission of volatile lead-containing compounds and the presence of lead in the ash. Further, the presence of residual copper has been suggested as a catalyst for dioxin formation on flyash [100]

Historically, several systems for recovery of wire insulation have been proposed [101, 102]. In general, these involve solvent or flotation separation or commingled processing. Recently a commercial venture to recover HDPE and PVC from wire scrap was announced by Plastic Recovery Systems (Toledo, OH) pending accumulation of funding. As described [103], the system also involves flotation but appears to use other chemicals to modify the density of water.

Since wire and cable scrap can be treated to remove metal and HDPE (albeit with some difficulty), recycling would appear to be the preferred means of solving this very visible problem. Some companies are known to buy cable scrap in the United States and recycle it into shoe soles in offshore operations. Still, this does not address the issue of ultimate disposal; shoe soles will go into MSW and will surely be landfilled or incinerated. Neither of these methods may be deemed acceptable.

In the 1990s, the PVC industry will need to address the issue of lead removal from this material. Four possibilities exist today: (1) direct reuse as cable sheathing, (2) solvent recovery of polymer with recovery of insoluble lead by filtration, (3) incineration in special apparatus that will allow recovery of lead, or (4) extrusion processing into articles of low surface-to-volume ratio for appropriate landfill. Simply putting the material back into items of commerce is unacceptable, and manufacturers will need to act as the ultimate stewards for their products. At least two companies, BFGoodrich and Vista Chemical, are involved in studies of appropriate disposal strategies for this material [104, 105].

6.6.3   Packaging Film

Flexible PVC packaging films include semirigid tamper-evident films, as well as meat, produce, and consumer food wrap. All are disposed of in MSW. Recycling of those items will occur only if they are included in curbside collection programs, sorted mechanically from whole refuse, or become part of a commingled effort on mixed films. One manufacturer,

Reynolds, has expressed a desire to recycle food-wrap films, but programs in this area will probably not be significant in the short term [106].

6.6.4 Agricultural Film

PVC is used extensively in Japan as agricultural film. As such, reprocessing is accomplished by shredding and washing in the same manner as PE film is recovered in Europe. Mitsubishi Kasei Vinyl leads in this effort [107].

6.7 Recycling of Plastisol PVC Products

6.7.1 Bottle Closures

A serious problem related to recycling plastic bottles is the difference in composition among bottle, label, top, and gasket. In order to create a proper seal, a foamed-in-place or solid gasket is put in the cap or an insert is used to create a mechanical seal. In either case, PVC is a predominant material for these seals.

Table 6.5   Typical sheet vinyl flooring construction

Layer	Materials	Thickness mil	(μm)
Topcoat	Urethane	1	(25)
Wearlayer	PVC	20	(508)
Foam Layer	PVC	30	(762)
Substrate	PVC, "Organic Felt"	25	(635)

For some bottles, this poses a daunting problem because cap and gasket are recovered along with the bottle. Previously, recyclers of PET bottles noted that PVC seals and metal caps contaminate their product. In operations subsequent to washing, metal could be removed but not PVC. By evolution, this problem has largely disappeared as ethylene-vinyl acetate (EVA) copolymers have achieved acceptance as cap liners for soft drink bottles.

6.7.2 Flooring

A much larger and more important use of PVC plastisol resins is sheet vinyl flooring. A typical flooring construction is summarized in Table 6.5. The composite structure is about 75 mil (1900 μm) thick and is predominantly PVC.

Current practice in the United States is to cement flooring in place over wood subfloor or existing flooring. At present, foam and wearlayer PVC cannot be separated from the substrate, and the entire composite cannot easily be removed from the subfloor, so wood or concrete residue may contaminate it. In the future, all of the composite could be made of PVC, with flooring tacked only to the subfloor at its perimeter or not tacked at all (as is done in Europe) so that recovery of the flooring can be accomplished with less effort.

Tarkett/Pegulan has been recycling PVC flooring products for a number of years. Recently, an association between European PVC producers and flooring manufacturers, AgPR, was established to recycle this material on a broader scale. AgPR, with about twenty members, will sponsor collection points for old flooring and will invest in facilities to recover materials of value. This reprocessing plant was scheduled to open in Fall 1990.

In-plant scrap has been reprocessed for a number of years [108]. Cuttings and returned rolls are chopped and then fluxed on a mill. This gray-colored material is calendered and used as backer sheet for more flooring. Presumably, postconsumer flooring will be processed similarly.

6.8 Future Outlook

PVC recycling is underway and will continue to grow. Packaging will be the first target, but building products will not lag far behind if European-style pressure spans the Atlantic Ocean. The potential to upset markets for virgin material is not as strong in the 1990s for PVC as it is for HDPE, but as commodity materials are decommissioned they will displace some virgin production. The time is close when not all new PVC plants will make product by polymerization of monomer.

Acknowledgments

The authors thank OxyChem for allowing the publishing of this chapter. Also, editorial assistance of Dr. Roy T. Gottesman of The Vinyl Institute and Rolf Buehl and Willem Prinselaar of EVC is gratefully acknowledged.

References

[1]   "Vinyl Chloride Polymers" in *Encyclopedia of Polymer Science and Technology*. Norman M. Bikales, Exec. Ed. New York: Wiley-Interscience, 1971.

[2]   "Vinyl Polymers" in *Kirk-Othmer Encyclopedia of Chemical Technology, Third Edition*. Martin Grayson, Exec Ed. New York: Wiley-Interscience, 1983.

[3]   *Encyclopedia of PVC Volumes 1–3*. Leonard Nass and Charles Heiberger, Eds. New York: Marcel Dekker, 1986.

[4]   *Modern Plastics, 67* (1) (Jan. 1990): p. 104.

[5]   See "Particulate Nature of PVC: Formulation Structure and Processing." G. Butters, Ed. London: Applied Science Publishers, Ltd., 1982.

[6]   Attwood, P. C., Brookman, R. S. "Ultrahigh Molecular Weight PVC Resins." *J. Vinyl Tech., 11*(42) (1989).

[7]   Franklin Associates, Ltd. "Characterization of Plastic Products in Municipal Solid Waste." Final Report to Council for Solid Waste Solutions, Feb. 1990.

[8]   Office of Technology Assessment, "Facing America's Trash: What's Next for Municipal Solid Waste?" OTA-0-424. Washington, DC: U.S. Government Printing Office, October 1989.

[9]   Franklin Associates, Ltd., "Characterization of Municipal Solid Waste in the United States, 1960 to 2000, Update 1988." Prepared for U.S. Environmental Protection Agency, Contract No. 68-01-7310. Prairie Village, KS, 30 March 1988.

[10] Reference 7.

[11] "The Environmental Impact of Municipal Solid Waste Incineration." Findings from the International Symposium on Municipal Solid Waste Incineration, Sept. 26–27, 1989. Sponsored by the Coalition on Resource Recovery and the Environment and the United States Council of Mayors. Report available from the Council for Solid Waste Solutions, 1275 K Street, NW, Suite 400, Washington, DC 20005.

[12] Olie, K., Vermeuler, P. L., Hutzinger, O., *Chemosphere, 6* (1977): p. 455.

[13] Tschirly, F. H. *Scientific American, 254* (2) (1986): p. 29.

[14] Hites, R.A., and Czuczwa, J.M. *Environ. Sci. Technol., 20* (1986): p. 195.

[15] Hites, R. A., and Czuczwa, J. M. *Environ. Sci. Technol., 18* (1984): p. 444.

[16] Marklund, S., Rappe, C., Tsklind, M., *Chemosphere, 16* (1987): p. 29.

[17] Ballschmitter, K., Buchert, H., Niemczyk, R., Munder, A. and Swerev, M. *Chemosphere 15* (1986): p. 901.

[18]   Karasek, F. W., Viau, A. C., Guiochon, G. and Gonnord, M.F. *J. Chromatogr. 220* (1983): p. 227.

[19]   Giugliano, M., Cernuschi, S., and Ghezzi, U. *Chemosphere, 19* (1989): p. 407.

[20]   Midwest Research Institute. "Results of the Combustion and Emissions Research Project at the Vicon Incinerator Facility in Pittsfield, Massachusetts." New York State Energy Research and Development Authority Report 87-16. Albany, NY, 1987.

[21]   Marklund, S., Andersson, R., Tysklind, M. and Rappe, C. *Chemosphere, 18* (1989): p. 1031.

[22]   Kreisher, K. R. *Modern Plastics, 67* (1) (June 1990) p. 60.

[23]   Rappe, C. "Critical Evaluation of Three Danish Inquiries and Reports." Manuscript in preparation.

[24]   Reference 20.

[25]   Lightowlers, P. J. and Cape, J. N. "Sources and Fate of Atmospheric HCl in the U.K. and Western Europe." *Atmospheric Environ., 22* (1988): p. 7.

[26]   Leaversuch, R. "Incinerator Operator: Vinyl is Harmless." *Modern Plastics, 66* (11) (Nov. 1989): p. 19.

[27]   United States Environmental Protection Agency. "Municipal Waste Combustion Study: Report to Congress." EPA/530-SW-87-02. Washington, DC, June 1987.

[28]   Flakt of Canada and Environment Canada. "The National Incineration Testing and Evaluation Program: Air Pollution Summary Report." EPS3/UP/2. Ottawa, Canada, Sept. 1986.

[29]   Marklund, S. "Dioxin Emissions and Environmental Immissions. A Study of Polychlorinated Dibenzodioxins and Dibenzofurans in Combustion Processes." Doctoral Dissertation, University of Umea, Sweden, 1990.

[30]   Curlee, T. R. "The Economic Feasibility of Recycling Plastic Wastes: Preliminary Assessment." Oak Ridge, TN: Oak Ridge National Laboratories, 1984.

[31]   "Global Recycling Proposal Modeled on PVC." Arbeitsgemeinschaft PVC und Umwelt e.V. Dusseldorf, FRG, 1989.

[32]   "Datum, Fahten, Meinungen." Arbeitsgemeinschaft, PVC und Umwelt, e.V., 2/90, (30 April 1990).

[33]   Reference 22.

[34]   Vogg, H., Christman, A. and Wiese, K. "Environmentally Compatible Fly Ash from Refuse Incineration Plants Obtained by the 3R Process." *VGB Kraftwerkstechnik, 68* (3) (1988): p. 237.

[35]   Marks, C. H. "Burn or Not to Burn: the Hospitals' Modern Day Dilemma." *Pollution Eng.* (Nov. 1988): p. 97.

[36]   Tessitore, J. L. and Cross, F. L. "Incineration of Hospital Infectious Waste." *Pollution Eng.* (Nov. 1988): p. 83.

[37]   Rappe, C., Marklund, S., and Fangmark, I. Untitled manuscript in preparation.

[38]   Kellermeyer, D. A., Ziemer, S. E., Stewart, S. L. "Health Risk Assessment for the Pinellas County Resource Recovery Facility, St. Petersburg, Florida, Executive Summary." HDR Techniserv, Santa Barbara, CA, 1987.

[39]   Reference 27.

[40]   As an example, ASTM Standard D 2665-87: Specification for Poly(vinyl chloride) (PVC) Plastic Drain, Waste and Vent Pipe and Fittings.

[41]   Atochem. "Recycling of PVC Packaging: the French Approach" and "PVC: Let's Look at it Again." Paris, France, 1989.

[42]   Perin, G. "Recuperation des Bouteilles en Polychlorure de Vinyle." *Materiaux et Techniques* (April 1978): p. 154.

[43]   Burgaud, P. "Recycling of PVC Bottles: Properties of Products Manufactured with Regenerated PVC." *J. Appl. Polym. Sci, Appl. Polym. Symp., 35* (1979): p. 469.

[44]   "PVC Bottles Being Recycled in France." *Plastics World* (May 1979): p. 24.

[45] Derry, R. "Plastics Recycling in Europe." Hartfordshire, Great Britain: Warren Spring Laboratory, Department of Trade and Industry.

[46] "France Launches Demonstration Project to Promote PVC Bottle Recycling." *Materials Reclamation Weekly* (6 May 1989): p. 22.

[47] "Two French Groups Join Forces in Plastics Recycling Venture." *PetroChemical News* (24 Sept. 1990): p. 3.

[48] *Modern Plastics, 67* (3) (March 1990): p. 14.

[49] Giunchi, G. and Piere, G. "Method for the Separation of Manufactured Articles of Plastic Materials and Apparatus Suitable for this Purpose." Instituto guido Donegani. *Eur. Pat. Appl.* 88107970.1, 1988.

[50] Carlo, G. "System for Recovering, Selecting and Recycling Rejected Plastics Containers." U. S. Patent 4 884 386.

[51] GECOM. "La Regeneration des Bouteilles en Polychlorure de Vinyle par Voie Humide" 65 rue de Prony, 75854 Paris cedex 17.

[52] Claerbout, J. "Recylcing of PVC Bottles into Coextruded Pipes" Recycle '88 Forum "Polymers, Processing, Applications, Business Development and Marketing." Davos, Switzerland, 31 May–2 June 1988.

[53] GECOM. "Fabrique a Partir de Dechets Plastiques Melanges." 65 rue de Prony, 75754 Paris cedex 17.

[54] "EVC Enters the Mixed Plastics Recycling Business." Press release. Brussels: European Vinyls Corporation, 1989.

[55] "Operation Recoup to Go Live." *Packaging Week* (28 Feb. 1990): p. 1.

[56] "New U. K. Project Aims to Fill Bottle Recycling Gap." *Plastics Industry Europe, 14* (4) (1990): p. 1.

[57] "Evian faces up to German deposit scheme: minimal effect on sales." *Plastics Industry Europe, 14* (1990): p. 1.

[58] "Recycling System" *Chemical Week* (12 Sept. 1990): p. 42.

[59] Celata, C. "Italian Way to Recycling" *Macplas* (Aug. 1988): p. 24.

[60] Hafner, E. A. "Recycleability [sic]—Another Big Plus for PVC." *Abstracts, Third International Symposium on Polyvinyl Chloride*, 1980, p. 338.

[61] Charnas, D. "Developer Committed to Recycling Process." *Plastics News*(1 Jan. 1990): p. 3.

[62] "Hafner Project Lacks Backers" *Plastics News* (17 Sept. 1990): p. 17.

[63] "Vinyl: Part of the Recycling Solution." Videotape prepared by the Vinyl Institute, Wayne, NJ, 1988.

[64] Summers, J. W., Mikofalvy, B. K., Wooton, G. V. and Sell, W. A. "Recycling Vinyl Packaging Materials from the City of Akron Municipal Wastes." *Society of Plastics Engineers Annual Technical Conference Abstracts*, 1990, p. 1422.

[65] Goodman, D. "'The Mouse That Roared,' Recycling at Vermont Republic Industries." *Society of Plastics Engineers Annual Technical Conference Abstracts*, 1990, p. 1414.

[66] "Finding the Silver Lining." *Plastics World Earth Day Supplement* (22 Apr. 1990): p. 28.

[67] Goodman, D. and Carrol, W. F., Jr. "Recycling PVC Bottles." *Society of Plastics Engineers Annual Technical Conference Abstracts*, 1989, p. 765.

[68] Plastics Recycling Foundation. "Collection Systems Technology Transfer Manual." Washington, DC, 1989.

[69] Reference 32.

[70] Misner, M. "X-ray Technology Separates PVC Bottles" *Recycling Times* (24 Apr. 1990): p. 8.

[71] Salimando, J. "Pulses to Target PVC Containers." *Recycling Times* (30 Jan. 1990): p. 3.

[72] Reall, J. "Grant Awarded for Plastics Recovery Plan." *Plastics News*(28 Aug. 1989): p. 1.

[73] Sommer, E. J., Kenny, G. R., Kearley, J. A. and Roos, C. E. "Mass Burn Incineration with a Presorted MSW Fuel." *JAPCA, 39* (1989): p. 511.

[74]   Lashinsky, A. "NRT Process Sorts PVC from Other Plastics." *Plastic News* (12 Nov. 1990): p. 3.

[75]   "U. S., European Firms Team on Recycling." *Plastics News* (28 May 1990): p. 2.

[76]   Lauzon, M. "Optical Bottle Sorter Promised for 1991." *Plastics News* (8 Oct. 1990): p. 22.

[77]   Toensmeier, P. A. "PVC Melt Point Keys Scrap Separation." *Modern Plastics 67* (6) (June 1990): p. 15.

[78]   "PVC Enters the Recycling Arena." *Plastics Today* (21 Oct. 1989): p. 3.

[79]   Greer, B. "New Laws Likely to Make Recycled Resin More Costly." *Plastics News* (30 Apr. 1990): p. 1.

[80]   State of Wisconsin Act 335. "The Comprehensive Recycling Act of 1990." Signed by the governor on 27 April 1990.

[81]   "Two- and Three-Layer Coex Retrofits for Extrusion Blow Molders." *Plastics Technology* (June 1990): p. 13.

[82]   *Modern Plastics 67* (4) (Apr. 1990): p. 123.

[83]   DeGaspari, J. "Opportunities Take Shape in Recycling." *Plastics Technology* (May 1990): p. 57.

[84]   Summers, J. W., Mikofalvy, B. K. and Little, S. *J. Vinyl Tech., 12* (1990): 154.

[85]   Summers, J. W., Mikofalvy, B. K., Wooton, G. V. and Sell, W. A. *J. Vinyl Tech., 12* (1990): p. 154.

[86]   McIlwaine, K. "ICI Australia Recycles PVC Cryogenically." *Plastics News* (3 Sept. 1990): p. 6.

[87]   "Georgia Gulf Forms New Business Unit." *Plastics News* (4 June 1990): 2.

[88]   Meade, K. "Georgia Gulf to Market New PVC Compound with Recycled Resin." *Recycling Times* (3 July 1990): p. 12.

[89]   *Packaging Digest* (May 1990): p. 14.

[90]   Bennett, R. A. and Lefcheck, D. L. "End Use Markets for Recycled Polyvinyl Chloride. Toledo, OH: University of Toledo College of Engineering, 1990.

[91]   *Modern Plastics, 67* (1) (Jan. 1990): p. 102.

[92]   Raymond, M. "Market Exists for Reprocessed PVC-Processors." *Plastics News* (29 May 1989): p. 7.

[93]   "Two Vinyl Films have Recycled Content." *Plastics News* (26 Nov. 1990): p. 5.

[94]   European Vinyls Corporation. "Face the Facts—Addressing the Issues on PVC in the Environment." Brussels, Belgium, 1989.

[95]   "Collection is the Big Problem Facing PVC." *Plastics World Earth Day Supplement* (22 Apr. 1990): p. 55.

[96]   Environmental Protection Agency, "Methods to Manage and Control Plastic Wastes" EPA 530-SW-89-051. Report to Congress, Feb. 1990.

[97]   *Modern Plastics, 67* (1) (Jan. 1990): p. 106.

[98]   "Cost-Effective Cryogenic Recycling Technology for Plastics Composite Scrap" *Modern Plastics 58* (6) (June 1981): p. 54.

[99]   Reference 98.

[100] Reference 11.

[101] Boehm, V. W., Broyde, B. T., Jr. U. S. Patent 4 038 219.

[102] "Recycling PVC." *Bell Laboratories Record* (Sept. 1988): p. 15.

[103] Reference 97.

[104] Hall, H. J., Vista Chemical. Personal communication.

[105] Reference 84.

[106] Ackermann, K. "Reynolds Aims High with Recycling Unit." *Plastics News* (2 Apr. 1990): p. 6.

[107] Sonoda, K., Mitsubishi Kasei Vinyl. Personal communication.

[108] Bonau, H. "PVC—Recycling of, Technical Applications and Economical Aspects." Recycle '90 Abstracts, 29–31 May 1990, Davos, Switzerland, p. 471.

7 Engineering Thermoplastics

R. H. Burnett, G. A. Baum

Contents

7.1	Introduction	153
7.2	Scope	153
7.3	Materials Recovery and Disposal Options	154
7.4	Sources	155
	7.4.1 Automotive	155
	7.4.2 Consumer Products	157
	7.4.3 Other Commercial and Consumer Industries	158
	7.4.4 Distributors	158
	7.4.5 Materials Recycling Companies	158
7.5	Technology of Recycling	158
	7.5.1 PET	160
	7.5.2 PBT	163
	7.5.3 ABS	163
	7.5.4 Nylons	164
	7.5.5 Polyacetals	164
	7.5.6 PC	164
	7.5.7 Modified Polyphenylene Oxide	165
	7.5.8 Other Polymers	165
7.6	Economics	165
7.7	Global Activities	166
	7.7.1 Europe	166
	7.7.2 Japan	166
7.8	Future Outlook	166
References		167

This chapter attempts to cover the status and outlook of engineering thermoplastics in the waste stream. Although not in the public eye at this writing, the volume of polymers going into durable goods will result in a disposal problem in future years. Preliminary attempts to anticipate and plan for this situation are covered together with a prognosis on possible developments worldwide.

7.1 Introduction

Not many people are thinking about recycling engineering thermoplastics because it is such a departure from the plastic recycling that has existed until this time. In many ways, recycling of engineering thermoplastics constitute the next horizon—the next frontier. In the activity that has occurred to date, there is a parallel with those activities in the mid- to late 1980s. The idea of recycling plastics packaging was just beginning to emerge in 1985, and there were outcries from citizenry and the government that eventually led to contemplation of bans or taxes on plastics packaging. The reaction from industry was to create specialized trade groups to counter these trends and to begin to develop the infrastructure that would allow the plastics in packaging to be removed from the municipal solid waste (MSW) stream through a variety of ways. This led to the establishment of a few bottle-deposit states and, more recently, the development of curbside recycling.

The issue of recycling engineering plastics is, today, at the same point where plastics packaging was in the mid-1980s. Although there is a similarity, there are differences. The similarity is in the degree of public concern being reflected in legislative action, which will have an effect on the industry from the resin producers right through to the manufacturers of the final product. Differences exist in useful lifetimes, collection and separation, and product values. Plastics packaging has a useful lifetime, averaging less than one year, and in many applications this may be only days or weeks. For example, a soft drink bottle is in the supermarket one day, consumed in the home the next day, and out in the trash the third day. In contrast to this, the "durable" products of engineering resins have disposal times of five, seven, or ten years in the future. The collection of postconsumer plastics packaging is an existing problem for the industry. This is not a problem for finished engineering resins products found in automobiles and "white goods" such as refrigerators, stoves, and air conditioning units. The problem with engineering thermoplastics is in the sortation into specific polymer streams in a cost-effective manner. The magnitude of the problem is substantial. A rough estimate is that 6 billion pounds $(2.7 \times 10^6$ t) of engineering thermoplastics went into these durable goods in 1990 alone. Another difference is in the higher value of the engineering materials relative to the commodity plastics. For example, as of this writing, the acetals, nylons, and polycarbonates (PC) are valued at three to four times that of the polyolefins (PO) and polystyrenes (PS).

7.2 Scope

The intent of this chapter is to cover the current recycling practice with respect to engineering thermoplastics and to try to predict future trends in this part of the industry. The term "engineering thermoplastics" is defined as those commercial products which replace traditional engineering materials, such as metals or wood, in various applications where a structural or thermal need exists for a high performance product. The high-performance plastics to be discussed include polybutylene terephthalate (PBT), various types of polyethylene terephthalate (PET) molding resins, PC, the acetals, polyamides (nylon), acrylonitrile-butadiene-styrene (ABS), and other specialty thermoplastics. These materials are used to fabricate parts for durable goods, such as automobiles and appliances. Some examples include automotive body panels, bumper fascias, electrical molded pin connectors, industrial business machine housings, power tools, and various appliance housings.

Cross-linked polyester resins, phenolic resins, epoxy resins, polyurethanes (PUR), and polyureas are not covered in this chapter but are discussed in detail in Chapter 10.

The treatment of engineering thermoplastics in the plastics recycling stream is quite unlike the handling of commodity thermoplastics. The major differences in collection, sortation, and reuse stem from the potential volume of product to be treated, the source of the plastics raw material, and the value added to the finished product after recycling.

The potential volume of recycled engineering thermoplastics is currently only about 10% of the major commodity thermoplastic recyclable products such as polyethylene (PE) or PS [1]. Moreover, relatively small volumes of these polymers are found in the MSW stream; rather, they are derived from industrial scrap.

However, the key difference can be found in the use of the term "engineering thermoplastics," which connotes that these products (unlike the commodity thermoplastics) are used as substitutes for wood and metal in various applications requiring high strength-to-weight ratios, resistance to thermal stress, or both.

Since virgin engineering plastics have higher strength and thermal properties than virgin commodity plastics to begin with, higher values of these properties remain after recycling through their primary lifetime. Therefore, the secondary applications can be of a more functional nature than the flower pots, marina piers, or pallets made from separated commodity plastics or commingled municipal plastics waste. The relatively high original cost of $1.50/lb ($3.30/kg) for engineering thermoplastics also precludes their use in low-profit, nonfunctional applications.

In the subsequent sections of this chapter, the current state of the art for the collection, sortation, and reuse or disposal of the engineering thermoplastics solid waste stream is delineated. The technical, business, and legislative viewpoint of current techniques for collection in those industries where these products are used and why their volume is so small at present is discussed. Methods of sortation, including coding and segregation, and the reuse of these products in their modified forms in new applications, both current and future, is then described.

In each area, the technology, economics, markets, and applications will be defined in an attemnpt to provide an overall picture of this component of the plastics recycling field.

7.3   Materials Recovery and Disposal Options

The conventional recycling and waste disposal techniques used for other plastic materials apply equally well to engineering thermoplastics, with the exceptions noted below. The current options for engineering plastics fall either into the primary (reuse of molding sprues, runners, etc.,) or secondary (conversion into a different and usually lower performance application) categories. Other options, including conversion of the polymer back to low molecular weight oligomer and incineration or landfill, also may apply.

Bio- or photodegradation options as proposed for thin packaging films are not practicable for extruded thick-sheet or injection-molded articles, which are the forms most usually associated with engineering plastics. Within the dark, bone-dry, and oxygen-poor interior of landfills, biodegradation occurs at an extremely slow rate that is measured in decades, and photodegradation is virtually impossible. Moreover, most engineering plastics are used in durable goods applications requiring at least a three-year life span, which is a further reason why neither method is a viable option.

Landfill, which has been extensively practiced in the United States up to the present, is becoming the major impetus towards recycling, since available landfill space, especially in major urban areas of the United States and other international industrialized countries, is being rapidly depleted. Current worldwide disposal methods for plastics are given in Table 7.1 [2].

Table 7.1   Global plastics waste disposal

Region	Landfill/Other %	Recycling/Reuse %	Incineration %
United States	80	3	17
Western Europe	63	8	29
Japan	23	12	65 *

* 25% of incinerated plastics in Japan is converted to electricity
Source: *Plastics News* [2]

7.4   Sources

With the exception of PET soft drink bottles, the majority of engineering plastics products are not found in the municipal waste stream. Products such as nylon-6/6, PBT, PC, and others are usually found in many relatively different and relatively low-volume applications in durable goods, such as nylon fiber scrap, appliances, automotive exterior and interior parts, and electrical circuit boards and pin connectors, to name just a few. Packaging products are virtually nonusers of engineering plastics. This makes collection and sortation of these products more difficult than commodity plastics and limits their recovery to some well-defined sources, such as fiber scrap (nylon and PET), water bottles (PC), automotive large-part dismantling activities, and similar uses.

7.4.1   Automotive

When the subject of engineering thermoplastics is raised within the industry, the automotive end use usually comes first to mind. The total volume of plastics used in the automotive industry was approximately 2 billion pounds ($0.9 \yen 10^6$ t) in 1990, and an increase to at least 3 billion pounds ($1.35 \yen 10^6$ t) is expected by the year 2000 [3]. The cars manufactured in the United States in 1990 had on average almost 300 pounds (136 kg) of plastics in each unit. About 11% of this total or 33 lb/car (15 kg/car) is currently classified as engineering plastics. The different engineering plastics used in various parts of an automobile are listed in Table 7.2 and include PC, nylons, PET, ABS, unsaturated polyesters as sheet-molding compound (SMC), and PUR [4]

At this particular time in the United States, there is no effective way to recover the engineering plastics or any other polymeric materials when the automobile is disposed of after its useful life. The current method to separate the nonmetallic component begins as the wrecked or junked cars are stripped of usable parts, such as the battery, tires, radio, air conditioner, etc., and then crushed and shredded. After removal of the metals component via magnetic separation, the nonmetallic residue or "fluff," also called automotive shredder residue (ASR), is usually landfilled and rarely incinerated. The composition of plastics in ASR after removal of metal and other materials is given in Table 7.3 for 1981 model U. S. automobiles [5].

Considerable work has been done at various industrial and academic institutions (Stevens, Lowell, Polytechnic University, Lehigh, and others) to evaluate the ASR fluff as a plastic raw material stream in its own right. To date, no real uses have been developed. Some potential applications include use as a high-loading filler in a polyolefin matrix or as a concrete additive [6]. The major problem is the low solubility (positive free energy of mixing) of different thermoplastics in each other, which leads to products having low impact strength, flexural modulus, and creep resistance. One example that may be cited is the fabrication of recycled plastics into an end-use item such as plastic lumber, which is structurally weak because of the immiscibility of the base polymer phases. There has not been a significant breakthrough in the development of compatibilizing agents that would help in fabricating commingled plastics with adequate mechanical properties.

Table 7.2   Distribution of engineering plastics in automobiles

Automobile Part	Type of Engineering Plastic Used
Interior Parts	
Trim	ABS, ABS alloys
Instrument Panel	RIM Polyurethane
Instrument Panel Skin	ABS/PVC alloy
Console	ABS
Functional Interior Parts	
Radiator Header Tanks	Nylon-6/6
Brake Reservoirs	Nylon
Distributor Caps	PBT
Fuel Rails	Various high-performance specialty polymers
Exterior Parts	
Bumper Fascia	RIM, RRIM, PUR
	PC/PBT Blends
	Thermoplastic Polyolefins (TPO)
	Thermoplastic Elastomers
Body Panels	Unsaturated Polyesters (SMC, BMC)
	Polyester/PBT Blends
	Nylon and Nylon/Polyphenylene Oxide (PPO™) Blends
	TPO
Exterior Trim	PPO™/PS
	PC and PC/Polyester Blends
Functional Exterior Parts	
Headlamp Bezels	Modified Polyesters
Wheel Covers	Modified Polyesters

Adapted from Reference 4 with permission, Battelle Institute

Table 7.3   Plastics composition of automotive shredder residue (ASR) [5]

Material	1981 Model Wt %
Polyurethanes (foam)	22.6
Unsaturated Polyester (SMC, BMC)	21.9
Polypropylene	19.2
Polyvinyl Chloride	15.5
ABS/SAN	7.3
Nylon	3.7
Acrylics	2.5
Phenolic	2.1
Miscellaneous	5.2
TOTAL	100.0

Improvement in some of the mechanical properties of the base polymers through the addition of recycled PS to a refined stream of postconsumer plastics waste has been reported by researchers at the Center for Plastics Recycling Research. This is discussed in detail in Chapter 9. At least one company is offering plastics lumber and other fabricated products that are claimed to possess adequate mechanical properties on a long-term basis. The material is a glass-filled structural foam core product that also may contain compatibilizing agents [7].

The key point is that the supply of automotive fluff is potentially huge, but considerably more research is needed to determine whether usable products can be prepared from this mixture. There is no easy way to separate the individual components.

One key to success in large-part dismantling of automobiles may be a more universal acceptance of coding of the various plastic parts to aid in identification and sortation. As discussed in Chapter 1 (see also Appendix 2), the packaging industry in cooperation with The Society of the Plastics Industry (SPI) has already developed a molded-in bottle coding system, which is currently either in use or being adopted in several states [8]. A similar system recommended by The Society of Automotive Engineers (SAE) for parts identification in the automotive industry is a molded-in identification on each part [9]. An on-board computer chip with parts identified by plastic type is under consideration for the future. If marking is used more widely, it would help to guard against commingling of immiscible plastics and add value to the recovered material.

The automotive industry has started to develop recycling programs for engineering plastics. However, successful commercial ventures remain to be accomplished, as witnessed in the recently dissolved partnership between General Electric Corporation (GE) and Luria Brothers (Cleveland, Ohio), a scrap metals dealer, formed to recover GE's PC/polyester blends from body panels, bumper fascia, and other exterior parts. The concept was for Luria to dismantle the plastic parts prior to crushing and shredding the car and to ship the plastic back to GE for conversion into a polymer with ABS-type applications in building materials and other nonengineering applications. The products would offer high thermo-mechanical properties compared to ABS and also would be sold at a premium compared to ABS prices.

One reason for the disbanding was the unavailability of supply of adequate quantities of uncontaminated plastic parts on a constant and reproducible basis. It is possible that the GE approach may be feasible in several years when sufficient late-model cars containing large amounts of plastic parts are ready to be scrapped, assuming that a cost-performance advantage for the reprocessed resin can be established relative to virgin competitive products. A second reason was the high labor cost of hand-dismantling of plastics parts from the junked cars.

In any case, we should look at these materials as items of potential value, which, if they can be separated into their component polymer streams, can be effectively reused and have new life. However, the technology does not exist in proven form today to separate these materials in an economically viable fashion.

Some new and novel concepts are being advanced as to how to approach this entire question. One possibility is the idea of disassembling the used automobile rather than shredding or crushing it. It would even seem possible with robotic manufacturing today to merely reverse the polarity of those robots and disassemble the major plastic components in reverse order of how they were put together. These parts then could be reused as such (if in satisfactory condition) or reclaimed using available technology.

7.4.2 Consumer Products

A moderate amount of PC from five-gallon water bottles is recovered under a "buy-back" program from the major manufacturer GE. After about 100 refills and reuses, the bottles are scrapped, ground, and reprocessed by Polymerland, GE's resin distributor. GE envisages this process being repeated by first converting this material to automotive parts and then, as the automotive plastic parts are scrapped, reusing the material again in the building and construction industry.

7.4.3 Other Commercial and Consumer Industries

In the United States, the proportion of engineering plastics recovered or reused from the electrical, communications, or other durable goods industries is essentially zero. This is due to several factors but mainly to the lack of a recovery system and the wide distribution of mixed plastic parts in various components.

The recovery of plastic from so-called "white goods" (i.e., refrigerators, washing machines, etc.,) also is practically nonexistent at present. Some research efforts to recover ABS panels from refrigerator liners has been initiated. It is possible that recovery and reuse of engineering plastics from the above markets can lead to a significant replacement of wood in buildings and construction.

7.4.4 Distributors

GE, through their Polymerland Distributors, has announced plans to implement a scrap buyback program by which various engineering polymers, including ABS and PC, will be repurchased and resold for use in less demanding applications.

Currently, Polymerland offers three ABS grades, a PC, and a PC/PBT alloy. The products are produced from feedstocks sourced from the buy-back program. The supplier says it can offer these resins at a 10–20% cost advantage over the counterpart virgin materials and that the properties are not generally affected.

Markets for these added-value products include automotive, communications, audio, personal care, construction, electrical, small appliances, and packages. The company claims it is developing additional ABS and PC grades as well as nylon- and Noryl™-based resins [10], see section 7.5.7.

Currently, the ultimate success of the program is unclear because of the short period of time since its inception. Other distributors such as General Polymers (Ashland) have or are planning to announce similar programs.

7.4.5 Materials Recycling Companies

Typical of the recycling companies that have become active in engineering thermoplastics is MRC Polymers, Inc. (Chicago, IL). MRC collects recycled plastics primarily from industrial waste and bottle deposit states, and sells various grades of nylon-6/6, PC, PET, commingled polypropylene/low-density polyethylene (PP/LDPE) film, and high-impact polystyrene (HIPS). Prices for these products generally average 20% lower than virgin resins. Other companies, such as Wellman, deal in nylon fiber scrap that are mainly reused in fiber-type applications such as outerware insulation. Supplies of engineering resins other than nylon or PET are at present limited to the 5–10-million-pound (2.25–4.50 $\times$ 10^3 t) range. Individual orders for greater quantities can only be filled at a high price, testifying to the inadequate supply situation.

A count of the approximately 103 existing U. S. and Canadian recyclers indicates that 30% claim to be active in reprocessing engineering plastics, including nylon. Table 7.4 lists these companies [11].

7.5 Technology of Recycling

A major hurdle to the use of recycled engineering thermoplastics in their original high-performance applications is the fact that, in a mixed waste stream, their thermo-mechanical properties are significantly reduced. Most crystalline (engineering) polymers are simply not compatible with

Table 7.4   U. S. and Canadian companies with engineering plastics reprocessing capabilities

Company	Location
Acri-Tech Plastics	Ontario, CA
All States Plastic Company, Inc.	Vancouver, WA
Allied Polymer Products, Inc.	Edmonton, Alberta, CA
Asian Export, Inc.	Newton, MA
Bay Polymer Corp.	Freemont, CA
Brandywine Recyclers	Lebanon, PA
Domtar Packaging	Toronto, Ontario, CA
Genpak Corp.	Glens Falls, NY
Greenline Resins, Ltd.	Woodstock, Ontario, CA
Hafner Industries	New Haven, CT
Interstate Plastic, Inc.	Vancouver, WA
Kenny Plastics Corp.	Flushing, NY
M. A. Polymers	Peachtree City, GA
MRC Polymers	Chicago, IL
National Recovery Corp.	New Haven, CT
New England CRInc.	North Billerica, MA
Plastic Recovery Service	New Berlin, WI
Plastic Recycling Alliance	Wilmington, DE
Plastic Recovery Corp.	Salem, OR
Polymerland Inc.	Parkersburg, WV
Ranier Plastics Inc.	Yakima, WA
Recoverable Resources/R2B2	Bronx, NY
REI Distributors	Somerset, NJ
Shuman Plastics, Inc.	Depew, NY
Southern Industrial Plastic Recyclers	Dunwoody, GA
Star Plastics	Albany, NY
Talco Plastics, Inc.	Whittier, CA
U. S. Recycling Industries	Denver, CO
Union Carbide Plastics Recycling Center	Danbury, CT
Western Gold Thermoplastics	Los Angeles, CA
Woods Brothers Industries, Inc.	Red Oak, TX

Excerpted from Reference 11 by special permission, *Modern Plastics*

most other polymers and tend to form large, weak domains with molecular "fault lines." In addition, the cost of separation of the individual components has prevented recycling of these products.

Considerable work has been done on the addition of a third component, a compatibilizing agent, that would tend to reduce intermolecular domain faults and strengthen physical and chemical bonding between mixed plastics. Much of this work is proprietary as it pertains to engineering plastics. A detailed review of the literature is beyond the scope of this chapter, but a few examples are given.

MRC Polymers, Inc. offers a modified PET product containing a proprietary polymer that accelerates the crystallization rate of recycled PET and, therefore, speeds up the molding process. The product is reported to retain much of the ductility of virgin PET.

A significant improvement in the impact strength of PET/high-density polyethylene (HDPE) blends was achieved using a functionalized styrene-ethylene/butylene-styrene elastomer as a compatibilizing agent impact modifier [12]. Other compatibilizing agents and impact modifiers that were tried, including acrylate rubbers, were evaluated in a subsequent study, but little improvement in mechanical properties was observed [13]. However, another core-shell acrylic impact modifier, when used at a 20 wt % concentration, is claimed to promote very uniform dispersions of elastomer in the resin continuous phase matrix and provides notched Izod impact strengths of 15–18 ft-lb/in (8–9.6 J/cm) of notch [14].

Rohm and Haas offers a reactive impact modifier claimed to couple to polyamides through residual free amine groups. The modifier is added via a vented extrusion compounder, which removes the water of the reaction. At a 20 wt % concentration, the impact strength is increased from 2–3 to 15–20 ft-lb/in (1.0–1.6 to 8.0–10.7 J/cm) of notch [15].

Nylon-6 has been blended with up to 20% of a maleic-modified ethylene-propylene-diene monomer (EPDM) elastomer. Similar experiments have been carried out with nylon-6/6 as well. In the nylon-6 case, significant improvements in impact strength and resistance to delamination after injection molding were observed [16].

A large number of compounding, mixing, and blending systems are being investigated to find more efficient ways to mix virgin and recycled incompatible polymers. The twin-screw, corotating, compounding extruder, for example, is reported to give very good mechanical mixing of incompatible polymers, but it is usually limited to polymers having roughly the same melting point.

Many of the well-known extrusion machinery manufacturers are active in this area and have developed systems for melt-compounding totally or partially immiscible plastics. Advantages such as improved processing or minimizing compounded product inherent viscosity (molecular weight) also are claimed for some machines. A partial listing of some special machinery manufacturers and the equipment features is given in Table 7.5 [17].

Most of the systems have been evaluated with PET or PET/HDPE recycled mixtures and may not necessarily give acceptable results with other types of recycled engineering thermoplastics.

7.5.1 PET

PET is used either solely or as an alloy with other polymers in engineering plastics applications, typically as a molding resin. Estimates of PET used as molding resin grades range from 3 to 5 million pounds (1.4 to 2.3×10^3 t) [18]. DuPont and Hoechst Celanese offer PET molding resin grades under the Rynite™ and IMPET™ names, respectively. In addition, these companies and others offer engineering plastic blends containing various amounts of PET to improve mechanical and aesthetic properties, such as control of warping and improved surface finish. Molding-grade PET requires a high level of purity. Generally, processes used to recover PET soft drink bottles produce polymer of sufficiently high purity and viscosity to qualify for molding applications, (e.g., the recovery process at Rutgers's CPRR generates a PET flake of 99.9% purity. The composition of the flake is given in Table 7.6 [19]. Molding-grade PET in pellets should have good color and an intrinsic viscosity (IV) of 0.60–0.65 dl/g.

Industrial recovery of PET from molded parts is considerably more difficult than recovering PET soft drink bottles. Many of the PET parts are admixed with other polymers, glass or mineral fillers, and other additives. Some degradation leading to lower molecular weight product occurs on remelting and reprocessing the plastic. In most cases, the product is

Table 7.5   Recycling machinery/extrusion systems offered by various companies [17]

Company	Location	Machinery/Extrusion System
Advanced Recycling Technology	Braxel, Belgium	"ET-1" machine offers twin-screw adiabatic compounder with molds for processing reclaim plastic flake
APV Chemical Machinery Inc.	South Plainfield, NJ	Reclaim extruders with self-feeding capacity and ability to handle to 2,700 lb/h (1225 kg/h) of scrap using a pneumatic ram stuffer, 6-inch (152-mm) diameter extruder); useful for film and fiber scrap
American Leistritz	Somerville, NJ	Single- and twin-screw extruders for reprocessing plastic scrap both on a laboratory and commercial scale
Berstorff Corporation	Charlotte, NC	Single- and twin-screw extruder for reclaiming and conversion of plastic waste into various shapes and end-use items
Buss (America)	Elk Grove Village, IL	Reciprocating single-screw extruder for compounding/mixing all types of industrial thermoplastics including reclaimed feedstock
Davis-Standard Division, Crompton & Knowles Corporation	Pawcatuck, CT	Customized extruders, feed screws, drives, and controls for compounding and extrusion of major thermoplastics including reclaimed feedstock
Farrell Corporation	Ansonia, CT	Single-screw extrusion equipment including large-diameter up to 26 inches (660 mm) machines for rates up to 50,000 lb/h (22.7 × 10^3 kg/h)
HPM Corporation	Mt. Gilead, OH	Single-screw extrusion equipment including "double-wave" screw design geared to high efficiency processing
Hartig Plastics Machinery Division, Somerset Technologies, Inc.	New Brunswick, NJ	Single-screw extrusion equipment coupled with a proprietary reclaim system for hard-to-feed, low bulk density materials such as film and fiber scrap
Instamelt Systems	Midland, TX	Unique "screwless" extruder with a smooth rotor housed in a short, eccentrically bored barrel; molten material passes around the circumference of the rotor, exits the barrel into a gear pump, and then through a die; Advantages claimed are short residence time and greater energy savings for both virgin and reclaimed feedstock

(continued)

(Table 7.5 continued)

Company	Location	Machinery/Extrusion System
MPM Polymer Systems Inc.	Clifton, NJ	Custom-designed and standard extruders for compounding both virgin and reclaimed feedstock
NRM Steelastic	Columbus, OH	Complete extrusion systems capable of handling both virgin and reclaimed feedstock for solid and foam sheet, film, profile, etc.
Reifenhauser-Van Dorn Company	Danvers, MA	Single- and twin-screw extrusion systems for most virgin and reclaimed thermoplastic feedstocks; screws are custom-designed for feedstock type and physical form (film, fiber, flake, etc.)
Toshiba Machine Company of America	Elk Grove Village, IL	Corotating twin-screw equipment for compounding and devolatilizing both virgin and reclaimed feedstock; large-size equipment with 2–6.3-inch (50-160-mm) screw diameters featured
Wayne Machine & Die Company	Totowa, NJ	Both single- and twin-screw extrusion systems for virgin and reclaimed feedstock
Welding Engineers, Inc.	Blue Bell, PA	Nonintermeshing twin-screw extruder systems for compounding and degassing virtually any polymer formulation; well-suited for recycled plastics, the extruders allow passage of solid contamination for subsequent melt filtration and feature wide feed and vent openings and ability to control shear; assembled from standard modular components to specific process needs; available for rates from 10 to 40,000 lb/h (4.5 to 18,000 kg/h)
Welex, Inc.	Blue Bell, PA	High-performance single-screw extrusion equipment from 1.25 to 12-inch (31 to 305 mm) diameter; double-vented systems permit feedstock with high levels of volatiles to be processed; complete systems including downstream equipment offered
Werner and Pfleiderer Corporation	Ramsey, NJ	Twin-screw compounding extrusion systems for handling all types of virgin and reclaimed feedstock including scrap film, sheet and flake/pellet; lab units to high volume systems capable of processing over 100,000 lb/h (45,000 kg/h) are available depending on the nature of the feedstock

Table 7.6   Composition of CPRR processed PET flake

Material	Content (Wt %)
PET	99.90
Polyethylene	0.03
Aluminum	0.01
Adhesive, paper, polypropylene, etc.	0.06
	100.00

Adapted from Reference 19 with permission, Plastics Recycling Foundation

unsuitable for reuse in its original application and must be downgraded to a lower performance product.

The major polyester manufacturers are depolymerizing PET scrap in the presence of glycols (glycolysis) or methanol (methanolysis) to low molecular weight polyester diols and dimethyl terephthalate (DMT). The purified products are then used as starting materials for new PET products. Goodyear uses glycolysis to manufacture REPETE™, a new product line containing 10–20% "recyclable" material [20]; Hoechst Celanese uses methanolysis to generate DMT for repolymerization [21]; and Eastman Chemical also uses a methanolysis process for reclaiming X-ray film scrap [22]. The polyester diols produced by glycolysis also are being used for the preparation of rigid polyurethane foams [23], unsaturated polyesters [24], and experimentally in polymer concrete [25].

Depolymerization of both PET and PBT to terephthalic acid (TPA) and the corresponding diol by hydrolysis has been claimed using superheated high-pressure steam at 200–300 °C and 15 atm (1.5×10^3 kPa) [26]. The process has not yet been commercialized. See Chapter 3 for a further discussion of this topic.

7.5.2   PBT

This product, made via the condensation of DMT and 1,4-butanediol, finds a wide variety of uses in various automotive and electrical applications, including distributor caps, computer keyboard caps, and automotive body parts. It is sometimes found in the waste product stream in the pure state but more often is recovered alloyed with other polymers to improve toughness, fire retardance, or other properties. Estimates of PBT available from industrial waste ranged from 1–2 million pounds (0.45–0.9×10^3 t) in 1989 [27]. To date, no commercial activity has occurred to recover or recycle PBT. Laboratory work to depolymerize PBT by methanolysis also has been successful in recovering DMT for reuse in PBT manufacture [28].

7.5.3   ABS

ABS is a mixture of styrene-acrylonitrile copolymer (SAN) with SAN-grafted polybutadiene rubber. Various grades, including blends and grafts with other polymers such as PC, are widely used in the automotive, business machines, telecommunications, and consumer markets.

The transportation market is considered to be the key to recycling ABS because of the large amount of ABS used in this application and the relatively advanced collection features. In 1990, 235 million pounds (105×10^3 t) were used by the transportation industry [29]. ABS polymers can be remelted and reused several times if properly stabilized with various antioxidant and thermal stabilizer packages. One problem that continues to plague the recycling effort is the contamination of the ABS part by other plastics.

During the early 1980s, AT&T recycled large quantities of scrapped ABS telephones through their own in-house developed reprocessing system [30]. Chapter 1, Section 1.6 describes the process used by AT&T. It is not known if this recycling activity continues today.

However, in the United Kingdom, British Telecom is currently recycling 5 million ABS-based telephones per year for precious metals, ferrous and nonferrous metals, as well as ABS [31].

7.5.4 Nylons

The nylons are melt-processible polymers that contain the amide linkage recurring along the polymer chain. They are among the oldest engineering plastics, dating from the mid-1930s. They are also the largest volume of engineering thermoplastics sold at 570 million pounds $(257 \times 10^3$ t) in 1990) [32]. The nylons offer a combination of properties, including high strength at elevated temperatures, ductility, wear and abrasion resistance, low friction, and good chemical resistance, among others. They are widely sold in such diverse applications as transportation (connectors, wire jackets, glass-reinforced radiator header tanks, and other uses), electrical (plugs, connectors), industrial (gears, bearings), consumer (ski boots, racquet sports equipment), appliances (tool housing), and film (food packaging and barrier layers).

Nylons can be remelted and reused several times and, therefore, are a prime candidate for recycling efforts. However, to date most of the activity in nylon collection has been in fiber scrap recovery, where companies such as Wellman have been very active for many years. Since nylon is a hygroscopic polymer, the reground product must be dried to a maximum moisture content of 0.2%. Many applications, such as automotive brake lines, cannot tolerate even minimum quantities of recycle or regrind because of safety factors; therefore, extremely close quality control is a must.

Many nylon end uses require impact modifiers to be added to the formulation to increase toughness. Any deterioration of the formulation caused by the addition of recycled nylon can be extremely detrimental to end-use performance.

Depolymerization of nylon-6 and polyester/nylon mixed scrap using a high-pressure steam process was patented by Allied Chemical Corp. in 1965 [33] and 1967 [34], respectively. The process is described in Chapter 1, Section 1.6. Later patents also issued to Allied teach the recovery of caprolactam monomer from low molecular weight (oligomers) of nylon-6 [35], from process residues [36], and from a liquid process stream of organic and inorganic materials containing 40 to 60% caprolactam [37].

7.5.5 Polyacetals

Acetals are highly crystalline polymers of formaldehyde (homopolymers) or mixtures of trioxane (a formaldehyde trimer) and another oxygenated monomer, such as ethylene oxide. Both types are characterized by excellent load-bearing properties, low coefficient of friction, and good mechanical properties including rigidity. These characteristics allow polyacetals to be used in industrial applications such as timing gears, rollers and bearings, and automotive window-lift mechanisms and cranks, door handles, and other functional parts in the heating and ventilating systems. Major U. S. producers include Hoechst Celanese, DuPont, and BASF.

It is this widespread use of polyacetals together with the relatively low weight of an individual part that has so far prevented any concerted attempt at collection and reuse. Polyacetals are immiscible with most other polymers and, therefore, are not found in alloys or blends. Consequently, no significant market exists at present for recycled polyacetals. It is estimated that no more than 5 million pounds $(2.3 \times 10^3$ t) of polyacetals are currently available for reuse from recovered recycled industrial product. Laboratory experiments on pyrolysis of polyacetals have so far not produced clear-cut routes back to formaldehyde or trioxane monomer [38].

7.5.6 PC

PC is an amorphous engineering thermoplastic prepared by the polycondensation of bisphenol A with carbonyl chloride (phosgene). It has an excellent balance of properties, including toughness, clarity, and high heat-deflection temperatures. Major applications for PCs are found in the automotive, business machine, and appliance markets. Since clarity (in the unfilled state) and toughness are its main attributes, the polymer is widely used in street lighting globes, school windows, riot control and sports equipment, and other similar uses. Major U.S. suppliers include GE Plastics, Dow Chemical Company, and Mobay Corporation (Miles Inc. as of 1 Jan. 1992). PC water bottles and PC/polyester automotive exterior parts are starting to be actively recycled, as previously mentioned. GE has taken a leading role in this area.

7.5.7 Modified Polyphenylene Oxide

Polyphenylene oxide (PPO™) is produced by polymerizing 2,6 dimethylphenol through an oxidative-coupling reaction. It is an amorphous polymer with a softening point about 210 °C. Because the polymer is difficult to process, it is rarely sold or used alone but is found in combination with other materials. It is miscible with PS at all levels and is supplied by GE under the Noryl trade name. A wide variety of grades are offered combining flame retardance, high-heat resistance, excellent processing, and dimensional stability among others. The modified polyphenylene oxides are used in interior and exterior parts of automobiles, such as seat backs, instrument panels, rear spoilers, and wheel covers. Other applications include machine housings, keyboard and printer bases, etc., for the telecommunication and business machine industry, and hair dryers, power tools, and portable mixers for the appliance industry.

Recently, GE demonstrated and reported [39] on the recyclability of Noryl. Used computer housings made of Noryl have been ground, blended with virgin resin, pigmented, and molded into 21×42 inch (53×107 cm) roofing shingles. The shingles are being used in a test program for reroofing of McDonald's® restaurants [40].

7.5.8 Other Polymers

Many other polymers can be categorized as engineering thermoplastics, including polyarylates, polyphenylene sulfides, polyfluorocarbons, polyketones, polysulfones, liquid crystal polymers, etc. For the most part, these products are used in small, part-weight, widely scattered applications in automotive, electrical, and aerospace industries. The difficulty of collection and sortation of these polymers, coupled with their extremely high melt-processing temperatures, has so far precluded any large-scale attempts at postconsumer recycling. Because of their extremely high thermal/hydrolytic stability and their ability to be remelted several times without significant loss of mechanical properties, many of the waste components of molded parts, such as the sprues and runners, can be reused in subsequent molding cycles; consequently, recycling occurs at the primary level.

An example of the excellent thermal stability referred to above is illustrated for a liquid crystal polyester polymer, Vectra A-950™. After five moldings at 300 °C using 100% regrind, 85% of the tensile strength, 91% of the flexural strength, and 100% of the notched Izod impact was retained in the molded part [41]. The property retentions are not unusual compared to similar high-temperature thermoplastics.

7.6 Economics

At this point in time, there are no programs in place in the United States that bring economic incentives to the recycling of engineering resins in durable applications. In fact, it may be said at this particular time that there are no disincentives in place either, as the government and the general public have not yet awakened to the fact that billions of pounds of engineering polymers are headed toward landfills or possibly to incineration. In attempting to calculate the economic incentives for recyclers to enter this field, it is perhaps premature to place any numbers on the game. The major driving force, short of government initiative, would be the economic value of the polymers themselves. At an average price of three to four times that of commodity plastics, there should be value and incentive if the infrastructure can be developed to capture and reclaim these products.

7.7   Global Activities

7.7.1   Europe

The situation outside of the United States is to some degree different, particularly in Europe. On the Continent, automotive recycling and disposal has long been a subject of investigation and development. Several major companies such as Hoechst AG have organized programs to recover large external automotive parts from postconsumer disposal. In one instance, an impact-modified polypropylene bumper is being detached from several 1989 and 1990 Opel models and reclaimed successfully using conventional techniques. Following this model, other resin suppliers, recyclers, automotive companies, and dismantler/shredders are cooperating to search out similar types of applications. The automotive manufacturers themselves, particularly Volvo in Sweden and BMW in Germany, have stated that they will, in the future, design automobiles with the idea of disassembling in mind. Several possibilities are being examined as to how to identify these plastic parts through color-coding, stamped imprints, or the computer chip approach.

In Europe also, incineration has long been favored as an alternative to landfill. In the automotive area, shredding and separation of the ASR followed by incineration is felt to be the only practical means of plastic disposal. Due to the relatively high fuel value of shredder residue (10,000–18,000 Btu/lb) compared to fuel oil (21,000 Btu/lb), the Germans refer to this fraction as "weisskol" (white coal). Recently, the world's three largest chemical companies— BASF AG, Bayer AG, and Hoechst AG—announced a research joint venture for recycling plastics in response to pending German federal legislation [42]. The major purpose of the joint venture will be to explore applications for recycled plastics, develop specifications, and obtain recycle approval by plastics end users. The group will work with mixed plastics and sorted plastic wastes from autos, refrigerators, freezers, electric appliances, construction, and chemical containers.

7.7.2   Japan

In Japan, it does not appear that there are any programs in place to investigate durable goods recycling. The preferred method for disposal of old automobiles is the conventional shredding and landfill disposition of the fluff. One wrinkle in the Far East is that many crushed hulks are shipped as such to other countries in Southeast Asia where material reclamation is facilitated by the availability of very inexpensive labor.

7.8   Future Outlook

Although we are at the very beginning of this phase of plastics recycling, it is the feeling of a number of experts in the field that high performance resins will be a target of opportunity within five to ten years. If the technology for sortation and reclamation of particular polymer types is developed or if a new type of hybrid engineering polymer is developed from automotive fluff through the addition of compatibilizers and fillers, then the infrastructure to recycle the resins will become a reality.

A great deal of research work is going on in universities and in private companies to support the future direction of this type of activity. For example, the work at Rensselaer Polytechnic Institute on selective dissolution and at the University of Pittsburgh on separation using near and supercritical fluids show promise, although the economics of these processes have yet to be proven. The work of the Durables Committee of The Council for Solid Waste Solutions is progressing with several contracts in place relating to life cycle analysis and disposal of ASR. There is no doubt much work remains to be done and will be accomplished in the decade of the nineties as this particular aspect of plastics in the waste stream is addressed and hopefully solved for the benefit of mankind.

References

[1]   *Plastic Newsbriefs*. Society of the Plastics Industry, Vol 9(4) (1989).

[2]   Lashinsky, A. "Japan Isn't Reusing Post-consumer Plastic." *Plastics News*, (6 May 1991): p. 5.

[3]   *Modern Plastics 68* (1) (Jan. 1991): p. 120.

[4]   Manganaj, D. *et al.* "Recycling and Disposal of Waste Plastics from Automotive and Allied Industries." Report by Battelle Institute, Columbus, OH, 1989, pp. 1-22–1-33.

[5]   McClellan, T. R. *Modern Plastics 60* (2) (Feb. 1983): pp. 50–52.

[6]   Curry, M. J. *Secondary Reclamation of Plastics*. Research Report—Phase I. Plastics Institute of America, 1987, pp. 3–5.

[7]   Mack, W. "Turning Plastic Waste into Engineering Products Through Advanced Technology." Presented at Recyclingplas V, Plastics Institute of America, Washington, DC, 1990, p. 81.

[8]   *Chemical Week* (20 Apr. 1988): pp. 14–15.

[9]   "Marking of Plastic Parts." SAE J1344. The Society of Automotive Engineers Inc., Warrendale, PA, revised March 1988.

[10] *Modern Plastics 67* (8) (Aug. 1990): p. 106.

[11] "Waste Solutions."*Modern Plastics,* Supplement (Apr. 1990): pp. 75, 76.

[12] Chen, I. M. and Shiah, C. "Producing Tough PET/HDPE Blends from Recycled Beverage Bottles." *Plastics Engineering* (Oct. 1989): pp. 33–35.

[13] Curry, J. and Kiani, A. "Compounding Recycled PET into Impact-Resistant Plastic." *Plastics Engineering* (Nov. 1990): pp. 38–39.

[14] Smoluk, G. "Key Role for Additives: Upgrade Polymer Recycle." Excerpted by special permission from *Modern Plastics 65* (10) (Oct. 1988): p. 87.

[15] Reference 14.

[16] Fuzessary, S. "Compatibilizing and Impact Modification of Multipolymer System." Paper presented at Recycle '89 Conference, Davos, Switzerland, 10–13 April 1989, pp. 275–281.

[17] *Plastics Technology Manufacturing Handbook and Buyer's Guide*. 1990/1991 Ed. (Mid-July 1990): 193–213.

[18] Hoechst Celanese Corporation, Engineering Plastics Division, Chatham, NJ. Unpublished internal estimates, 1989.

[19] "New Engineering Polymer Source: Recycled PET from Post-Consumer Soft Drink Bottle." Brochure. New Brunswick, NJ: Center for Plastics Recycling Research, Rutgers University, 1987.

[20] Richard, R. E., Boon, W. H., Martin-Schulte, M. L., and Sisson, E. A. "PET Based On Post Consumer Recycle: REPETE Polyester Resin." *Polymer Preprints 2* (32) (June 1991): pp. 144–145; Polymer Technology Conference, Philadelphia, PA.

[21] Barham, V. F. "Closing the Loop for PET Soft Drink Containers." *J. Pkg. Tech.* (Jan./Feb. 1991): pp. 28–29.

[22] *European Chemical News* (20 Nov. 1990): p. 27.

[23] Carlstrom, W. L., Reineck, R. W., and Svoboda, C. R. U. S. Patent 4 223 068, 1980.

[24] DeMaio, A. "Engineering High Performance Thermoset Resins From PET Thermoplastics." 46th Annual Conference, Composites Insitute, The Society of the Plastics Insitute, Inc., 18–21 Feb. 1991, Session 18c, pp. 1–3.

[25] Rebeiz, K., Fowler, D. W., and Paul, D. R. "Recycling Plastics in Polymer Concrete." Plastics Waste Management RETEC, Society of Plastics Engineers, Inc., 17–18 Oct. 1990, Paper 22.

[26] Mandoki, J. W. U. S. Patent 4 605 762, 1986.

[27] Reference 18.

[28] Hoechst Celanese Corporation, Chatham, NJ. Unpublished research, 1980 to date.

[29] Reference 3.

[30] Wehronber, R. H. *Materials Eng.* (Feb. 1982): pp. 44–50.

[31] Day, S. "Post Consumer Recycling of Engineering Thermoplastics in Practice." GE Plastics Recycle Programs. Paper presented at Recycle '90 Conference, Davos, Switzerland, 30 May 1990. Mack Business Services, Moosacherstrasse 14, CH 8804 Au/AH, Switzerland.

[32] Reference 3, p. 115..

[33] Bonfield, J. H., Hecker, R. C., Snider, O. E., and Apostle, B. G. U.S. Patent 3 182 055, 1965.

[34] Lazarus, S. D., Twilley, I. C., and Snider, O. E. U. S. Patent 3 317 519, 1967.

[35] Crescentini, L., Blackman, Jr., W. B., DeCaprio, J. D., Fisher, W. B., Lilley, Jr., R. J., and Wagner, J. W. U. S. Patent 4 311 642, 1982.

[36] Crescentini, L., Fisher, W. B., Mayer, R. E., DeCaprio, J. D., and Nilsen, R. K. U. S. Patent 4 582 642, 1986.

[37] Balint, L. J., Greenburg, J. U. S. Patent 4 764 607, 1988.

[38] Reference 28.

[39] National Plasics Exposition, Chicago, IL, 17–21 June 1991.

[40] Lashinsky, A. "McDonald's Beefs Up Use of Recycled Materials." *Plastics News* (18 June 1991): pp. 1, 32.

[41] Hoechst Celanese Corporation. "Vectra Molding Guidelines and Specifications." VECTRA™ Product Bulletin VC-6, 1989.

[42] Short, H. "Chemical Giants Plan Joint R & D." *Plastics News* (21 May 1990): p. 1.

8   Acrylics

J. B. Schneider

Contents

8.1   Introduction 171

8.2   Industry 172

    8.2.1   Polymerization Procedures 172

    8.2.2   Commercial Production 173

    8.2.3   Applications 174

    8.2.4   Markets 174

8.3   Acrylic Recycling 175

    8.3.1   Sources of Industrial Scrap 176

8.4   Depolymerization Technology 177

    8.4.1   Dry Distillation 177

    8.4.2   Super-heated Steam as HeatTransfer Agent 177

    8.4.3   Molten Metal and Metal Salt Heat Transfer Media 178

    8.4.4   Retort Method 180

    8.4.5   Fluidized Bed 180

    8.4.6   Extrusion Methods 181

8.5   Regrinding Technology 183

    8.5.1   Dry Paint Stripping 183

    8.5.2   Repelletizing and Remolding 183

8.6   Postconsumer Acrylic Recycling 184

8.7   Future Outlook 185

References 185

With the rising overall consumption of acrylic, especially in the transportation market, it may one day be necessary to recover postconsumer acrylic scrap and generate new uses for the recycled material. Currently, preconsumer or industrial acrylic scrap is being collected and recycled. It is either depolymerized to recover methyl methacrylate (MMA) monomer or reground for use in molding and extrusion applications. Another use for the reground acrylic is in the dry stripping of paint from aircraft parts and truck bodies. These examples illustrate how recycled acrylic scrap is presently being handled. If and when the collection of postconsumer acrylic scrap is initiated in the future, it may be similarly processed.

Although the acrylic that is currently being recycled consists primarily of industrial scrap and is often shipped overseas for remolding and extruding, domestic recycling programs are growing and will likely continue to do so. This chapter outlines the current production of acrylic and highlights the major acrylic markets. It then details the history of acrylic recycling technology from the early 1900s to the present. A discussion of some future options in the recycling of acrylic waste is also included.

8.1 Introduction

The use of polymers of acrylate and methacrylate esters has increased over the last decade. Some typical applications of acrylic polymers include automobile taillights, sanitaryware, and impact-resistant signs. There are a number of additives and modifiers which can be introduced into acrylic and methacrylic polymer formulations to obtain a wide range of physical properties from rigid and tough to soft and pliable. Acrylic monomers can also be copolymerized with various other monomers, yielding materials with unique physical properties. This copolymerization technology further broadens the application potential for acrylic polymers.

Most often, the term "acrylic" refers to polymethyl methacrylate (PMMA). The recycling of postconsumer PMMA is not as prevalent as it is for other plastics, such as polyethylene terephthalate (PET) and high-density polyethylene (HDPE). It is not uncommon, however, for fabricators and resin producers to recycle industrial acrylic scrap. This includes acrylic resin and fabricated parts that do not meet customer specifications, parts that either underfill the mold or show evidence of flashing, acrylic scrap that accumulates in fabrication when parts are sawed or sanded, and off-grade resin production. The acrylic scrap may be used in a reground or repelletized form or depolymerized to yield monomeric MMA of high purity. Although the recycling of industrial acrylic is being pursued in this country, much of the acrylic scrap obtained in the United States is exported overseas. It is often the lower purity or partially contaminated acrylic scrap that is exported. This is because some overseas markets are better able to tolerate the lower purity of acrylic while welcoming the reduced raw material costs.

The process of recycling acrylic scrap is somewhat unique in that the polymer may be depolymerized or cracked to yield approximately 100% of the original monomer. This property is characteristic of only a few polymeric materials, including polymers of α-methylstyrene and tetrafluoroethylene. To better illustrate this, Table 8.1 [1] lists several polymers and the corresponding percentage of monomers obtained via thermal decomposition in the absence of oxygen. It is evident from these data that some of the more commonly used plastics, such as polystyrene (PS) and polypropylene (PP), yield only 42% and 2% of monomer, respectively, when subjected to thermal degradation.

Acrylic polymer, based mainly on PMMA, is the most widely used plastic that can be recycled both easily and effectively by a depolymerization process. For this reason, a review of the recyclability of PMMA is worth noting in a chapter of its own. The chapter will briefly review the formation, production, and application of acrylic polymers, including a review of

Table 8.1   Monomer yields from polymers by thermal decomposition

Polymer	Monomer, wt %
Polytetrafluoroethylene	97–100
Polymethyl methacrylate	95–100
Polyα-methylstyrene	95–100
Polymethacrylonitile	85–100
Polystyrene	42
Polyisobutene	32
Polychlorotrifluoroethylene	28
Polyisoprene	12
Polybutadiene	2
Polypropylene	0.2–2
Polymethyl acrylate	<1
Polyethylene	<1

the major acrylic markets. It will then highlight the industrial acrylic recycling technology and address the possibility of recycling postconsumer acrylic in the future.

8.2   Industry

There are three types of PMMA available in the acrylic industry. Acrylic sheet may be prepared either by casting or extrusion methods. Acrylic resin, on the other hand, is sold in a pelletized form which may contain pigments and modifiers for specific applications. A third area in which acrylics are widely used is in paints and coatings. In this discussion of acrylic recycling, however, acrylic paints and coatings will not be addressed.

8.2.1   Polymerization Procedures

Polymers of MMA are generally formed via free-radical polymerization techniques using either azo catalysts or peroxides as initiators. The actual polymerization process used in the commercial manufacture of acrylic depends upon the form of the desired product. Bulk polymerization is used in the production of PMMA molding resins, sheet, and rods. MMA polymers used in adhesives, coatings, and laminates are prepared by solution polymerization. Emulsion polymerization is used in some copolymerization reactions of MMA, while suspension techniques are typically employed when preparing resins for molding and ion-exchange applications [2].

Bulk polymerization is a relatively uncomplicated operation in which the polymer is formed by heating the liquid monomer in the presence of a free-radical initiator, such as benzoyl peroxide. Because the polymerization of MMA is exothermic, however, a problem exists in the transfer of heat as the reaction proceeds. There is a significant increase in viscosity as the molecular weight of the polymer increases that does not allow for the heat generated during polymerization to dissipate. Bulk polymerization is used commercially in the production of some molding resins and cast acrylic sheet.

Solution polymerization entails heating the monomer in the presence of a solvent for the polymer. Benzoyl peroxide and azobisisobutyronitrile are commonly used initiators in solution polymerization processes. The resulting polymer may be used as a solution or isolated by solvent evaporation or precipitation of the polymer with the addition of a nonsolvent. Although the presence of the solvating medium in solution polymerization allows for better control of the exothermic nature of the reaction, some difficulties remain when attempting to isolate the polymer. High polymer molecular weights are also difficult to achieve in solution due primarily to increased chain transfer with the solvent.

In suspension polymerization, a discontinuous phase of monomer droplets are suspended in a medium with the aid of a suspending agent. Once polymerized, the polymer particles are suspended in the medium and may be easily separated by centrifugation. Because MMA is insoluble in water, an aqueous medium is generally used in the preparation of PMMA. Upon drying of the polymer, a fine powder results. In suspension polymerization, the reaction temperature is more easily controlled due to the presence of the aqueous medium about the droplets. It is also possible to control the particle size of the polymer in suspension polymerization. It has been noted that the size of the acrylic particles formed is dependent on the polymerization temperature, the monomer-to-water ratio, and the type and amount of suspending agent used.

Similar to the suspension polymerization process, emulsion polymerizations are generally carried out in an aqueous medium, but in the presence of a surfactant instead of a suspending agent. The suspended polymer product or latex usually contains approximately 35-40% solids but may contain as high as 60-65% polymer. The rates of polymer formation in emulsion polymerizations are more rapid than in suspension or bulk processes, and higher polymer molecular weights can be achieved. The reaction temperature in emulsion polymerization can be easily controlled due to the presence of an excess of the aqueous medium [3].

8.2.2 Commercial Production

The casting of acrylic sheet may be carried out using a continuous caster or a batch-type cell casting operation. In the continuous casting procedure, the MMA monomer is added to a syrupping tank where it is partially polymerized by heating to a predetermined viscosity. The reaction is then quenched and the MMA syrup is transferred to a promoting reactor. Here, the initiators, UV stabilizers, impact modifiers, pigments, and other additives are introduced. The mixture is then degassed in order to eliminate the presence of bubble imperfections in the final cast product. The degassed mixture is poured between two moving stainless steel belts which first enter water zones. The water zones serve to heat the belts and to reinitiate the polymerization of the MMA monomer/polymer mixture. The belts then enter several ovens for the final polymerization. The acrylic sheet is rolled off of the stainless steel belts and cut to desired sheet sizes. In continuous casting operations, acrylic sheet of a very uniform thickness is obtained.

Cell casting of acrylic involves pouring the promoted MMA syrup between large glass plates which are then heated to promote curing. The glass plates are bound using spring clips which maintain tension as the acrylic contracts during polymerization. Acrylic prepared by cell casting has improved optical properties and less surface irregularities than does continuous cast acrylic. Cell cast acrylic sheet may be cast to a thickness as great as 4.5 inches (114.3 mm) and as thin as 0.030 inch (0.76 mm). The thickness of continuous cast acrylic, on the other hand, ranges from 0.125 to 0.375 inch (3.18 to 9.53 mm).

Another way of preparing acrylic sheet is by extruding pelletized PMMA. Acrylic resins are formed either by bulk or suspension polymerization processes. Bulk-produced acrylic may be followed by extrusion into pellets or directly into sheet. The acrylic sheet may be as thick as 0.250 inch (6.35 mm). Although extruded acrylic sheet is similar in quality to cell and continuously cast acrylic, it has a lower polymer molecular weight. This results in the reduction of some physical properties including heat distortion temperature, chemical resistance, and stress limitations.

In addition to acrylic sheet, acrylic resins are sold for molding, especially in injection molding and extrusion applications. The pellet formulations vary in molecular weight, depending on the application, and may contain various impact modifiers and pigments.

8.2.3 Applications

Acrylic polymers offer some unique properties, including optical clarity, toughness, chemical resistance, and weatherability. Transparent acrylic sheet has an optical clarity comparable to that of optical glass. Thus, one of the original and continuing uses of acrylic sheet is in aircraft windows and canopies. It is also used for window glazing in safety applications such as storm windows and patio doors. Indoor and outdoor lighting panels are often made of acrylic, not only because of its optical clarity but for its resistance to yellowing as well. Transparent and translucent skylights may also be fabricated using acrylic sheet.

The weatherability of acrylic has allowed for extended outdoor use in commercial signs. Impact modified acrylic sheet is used in some sign applications to reduce breakage due to severe weather conditions and vandalism. A resistance to yellowing, even over long periods of time, extends the use of both acrylic sheet and molding compounds to indoor and outdoor lighting applications. In addition, acrylic sheet is useful in enclosure applications and tinted sunscreens for malls, swimming pools, and restaurants. The use of acrylic sheet is increasing in the sanitaryware market due to the smoothness of the acrylic surface and resistance to household chemicals.

Acrylic molding resins also exhibit excellent weatherability and optical clarity and are, thus, widely used in the automotive industry. The primary use has been in taillights, but acrylic molding compounds are also used in instrument panel covers, side markers, and decorative trim. Acrylic resins are used in touch-tone phone buttons, display cases, and appliance knobs and panels. There is a rising potential for acrylics in the audio and visual markets, as well as data storage, due to excellent optical properties and the ease of fabrication. Another area in which acrylic resins are beginning to be used is in medical applications [4]. For example, acrylic sample vials are less fragile than glass and have good clarity, allowing the transmission of ultraviolet light for sample analysis.

8.2.4 Markets

The total consumption of acrylic in the United States in 1990, as shown in Table 8.2, was 751 million pounds (338×10^3 t), an increase of 12 million pounds (5.4×10^3 t) over the 1989 consumption [5]. Approximately 39.4% of the total acrylic consumption was in the form of cast sheet, while 28.4% was in molding and extrusion compounds. The remaining 32.2% of acrylic consumed in 1990 consisted primarily of acrylic coatings, impact modified acrylic and other special grades, and a smaller amount in emulsion polymers, transesterification resins, and other

Table 8.2   Acrylic pattern of consumption–1990 [5]

Market	Million Pounds	(Thousand Tonnes)
Cast Sheet [a]	296	(133)
Molding and extrusion compounds	213	(96)
Special grades [b]	85	(38)
Coatings	102	(46)
Other [c]	55	(25)
TOTAL	751	(338)

[a] Includes imports [b] Impact modified, etc. [c] Emulsion polymers, transesterification resins, etc.

Note: Used with permission, *Modern Plastics*

minor markets. Table 8.3 shows a breakdown of the U. S. acrylic cast sheet and molding/ extrusion markets. The consumption of acrylic in the consumer, industrial, and sign markets increased in 1990. The bulk of cast acrylic sheet and molding/extrusion compounds, however, was consumed by the building market.

Table 8.3   U. S. acrylic sheet and molding/extrusion markets–1990 [5]

Market	Cast Sheet Million Pounds	(Thousand Tonnes)	Molding/Extrusion Million Pounds	(Thousand Tonnes)
Building	115	(52)	110	(50)
Consumer	5	(2)	6	(3)
Industrial	45	(20)	45	(20)
Signs	10	(5)	10	(5)
Transportation	45	(20)	42	(18)
TOTAL	220	(99)	213	(96)

Note: Used with permission, *Modern Plastics*

The major markets for acrylics in the United States are listed in Table 8.4 [6]. The largest single market for acrylics in 1990 was in window glazing and skylight applications, consuming 140 million pounds (63×10^3 t). The lighting fixture and plumbing markets consumed 45 and 43 million pounds (20.3 and 19.4×10^3 t), respectively, in 1990, while the amount of acrylic in transportation applications in North America totaled 51 million pounds (23×10^3 t) . The panels and siding market consumption of acrylic was 28 million pounds (12.6×10^3 t) and consumption in appliance applications was about 10 million pounds (4.5×10^3 t) in 1990.

Table 8.4   Major U. S. acrylic markets–1990 [6]

Market	Million Pounds	(Thousand Tonnes)
Applliances	10	(5)
Building		
Glazing and skylights	140	(63)
Lighting fixtures	45	(20)
Panels and siding	28	(13)
Plumbing	43	(19)
Transportation		
Cars, vans, light trucks	39	(17)
Other vehicles	12	(6)
TOTAL	317	(143)

8.3   Acrylic Recycling

Most of the acrylic that is currently being recycled is preconsumer industrial scrap. Industrial scrap is obtained primarily from off-specification production material and fabrication scrap. It is possible, however, that a small amount of postconsumer acrylic is being recovered from automobile and construction scrap but the recycling of industrial scrap is much more prevalent. Automobile scrap, known as automotive shredder residue (ASR) contains mixed-plastics and nonplastic material. Separating the mixed plastics from the other materials is a technical problem facing the industry today. A possible method of recovering PMMA from ASR is presented in Section 8.6. If acrylic parts are removed from automotive and construction scrap before disposal or further treatment, these materials present a potential source for postconsumer acrylic scrap. Industrial scrap, on the other hand, is generally "clean" acrylic scrap, which eliminates the need for a difficult and costly sorting step. A review of industrial acrylic recycling is given here and may be used to demonstrate how acrylic scrap could be handled in the future when, and if, a postconsumer collection system is developed.

8.3.1 Sources of Industrial Scrap

Industrial acrylic scrap is readily available because, like many industrial processes, the acrylic industry generates waste in the form of off-specification material and fabrication scrap. The scrap acrylic generated by acrylic producers is relatively pure in that it is not in a mixed-plastics form, although it may contain small amounts of dirt and other impurities. Recycling industrial acrylic scrap not only aids in the reduction of solid waste in landfills but is also cost-effective for acrylic producers. Instead of disposing of the waste generated in acrylic production facilities, it can be used to partially replace virgin starting materials and, thus, reduce raw material costs. For example, it has been estimated that the cost of MMA monomer obtained from the depolymerization of recycled PMMA is approximately 20¢/lb (44¢/kg) [7]. This estimate is based on the internal recovery cost to the producer and is not a market price. Virgin MMA, on the other hand, was priced in March of 1991 at 71¢/lb ($1.57/kg) [8]. The reduction in raw material costs when even partially replacing virgin MMA with recycled monomer is significant. It also is important to note that there is virtually no difference in purity between recycled and virgin MMA monomer.

Industrial scrap may consist of cast or molded material that does not meet customer specifications. This may be due to the presence of voids in the acrylic sheet or to the mischarging of pigments, fillers, or other additives to the syrup. Recyclable acrylic sheet may be material that has prematurely polymerized, in which case the unused monomer vaporizes, also leaving voids in the sheet. Voids may also be caused by the incomplete degassing of the intermediate syrup. A variation in one or more of the physical properties of the acrylic product, including optical clarity, toughness, or thermal stability, may cause the cast or molded material to be off-specification. In addition, the presence of water in the monomer can cause the acrylic sheet to be rejected as off-specification material. On occasion, promoted syrups are rejected due to formulation changes and can be recycled. These partially polymerized syrups may either be cracked to yield monomer, or the polymerization may be completed and the resulting acrylic sheet treated as off-specification material. In extrusion and injection molding processes, scrap acrylic is generated when the color or additive formulations of the resin to be extruded are changed. The sprues, runners, flashing, and purging materials obtained during the molding of acrylic products are recyclable. Acrylic scrap is also generated during the fabrication of sheet. This includes sawdust, chips, and trimmings of sheet and thermoformed products. Sawdust is also generated when acrylic parts are sanded or polished. Similar scrap results from quality control operations in a manufacturing plant as acrylic samples are gathered for testing. Acrylic scrap from each of these sources can be collected quite readily for recycling.

There are several potential uses for industrial acrylic scrap. It may be ground into a powder and added to virgin acrylic in certain applications (Section 8.5). It may also be ground and sold to the aircraft or automotive industries for dry paint stripping applications (Section 8.5.1). In addition to grinding, it is possible to depolymerize or crack recycled acrylic to yield MMA monomer which is then used in the synthesis of new acrylic polymers (Section 8.4).

8.4   Depolymerization Technology

The thermal degradation of PMMA occurs via a free-radical chain reaction at approximately 300–400 °C. The free-radicals then initiate the unzipping or depropagation of the polymer chain as shown in Equation 8.1 [9].

$$
\text{\textasciitilde\textasciitilde CH}_2\!-\!\underset{\underset{\text{COOCH}_3}{|}}{\overset{\overset{\text{CH}_3}{|}}{C}}\!-\!\text{CH}_2\!-\!\underset{\underset{\text{COOCH}_3}{|}}{\overset{\overset{\text{CH}_3}{|}}{C}}\!-\!\text{CH}_2\!-\!\underset{\underset{\text{COOCH}_3}{|}}{\overset{\overset{\text{CH}_3}{|}}{C}}\text{\textasciitilde\textasciitilde} \;\rightarrow\; \text{\textasciitilde\textasciitilde CH}_2\!-\!\underset{\underset{\text{COOCH}_3}{|}}{\overset{\overset{\text{CH}_3}{|}}{C}}\cdot \;+\; \text{CH}_2\!=\!\underset{\underset{\text{COOCH}_3}{|}}{\overset{\overset{\text{CH}_3}{|}}{C}} \tag{8.1}
$$

The depolymerization of PMMA can be found in the literature as early as 1935 in a DuPont patent, which involved the dry distillation of the PMMA and the collection of monomeric MMA [10]. Other patents describe the use of super-heated steam [11], molten metals [12], and molten metal salts [13] to effect the depolymerization. During the last decade, the most common commercial method for recovery of MMA monomer from polymerized acrylic involved the use of a molten metal bath. Recently, however, this has been phased out in the United States due to governmental pollution and safety regulations.

8.4.1   Dry Distillation

As described in a patent issued to DuPont [14], the dry distillation method for the thermal degradation of PMMA involved placing acrylic scrap in a conventional distillation flask. The scrap was heated over an open flame at atmospheric pressure to a temperature above the depolymerization point of the polymer. The monomeric vapors were either condensed and collected or reacted directly with other monomers to form new acrylic polymers. When the condensate was collected, it was further fractionally distilled yielding monomer of additional purity. The dry distillation was also carried out at pressures greater than atmospheric, allowing for the heat-cracking of the polymer in the distillation flask prior to the vaporization and condensation steps. The dry distillation process was also carried out at reduced pressures. It was necessary, however, at any pressure to exceed the depolymerization temperature in order for monomer liberation to occur. The patent also introduced the idea of flash distilling the acrylic scrap, a process by which the scrap was made to contact a highly heated surface and was immediately vaporized.

The DuPont patent claims that the dry distillation process may be used to recover monomer from many types of acrylic scrap including off-specification material. This method has since been characterized, however, by low heat efficiency. The heating of the dry acrylic scrap using a flame or electrical heater was not uniform in the distillation flask. The acrylic chips lining the flask were heated much more quickly and to higher temperatures than was the material near the center of the flask. As a result of the higher temperatures, a decomposition residue formed on the walls of the distillation vessel. The residue was difficult to remove and interfered with the potential for a continuous process.

8.4.2   Super-heated Steam as Heat Transfer Agent

The depolymerization of acrylic scrap using super-heated steam as the heat transfer agent involved the use of nitrogen, or any inert gas, as a carrier. The nitrogen flowed in the direction opposite the steam, resulting in a wind-sifting action within the steam-heated cleavage column, shown in Figure 8.1 [15]. With PMMA of mixed particle size, the initial depolymerization of the larger acrylic particles occurred in the bottom, higher temperature region of the cleavage column, and the depolymerization of the finer particles in the upper region of the column. The

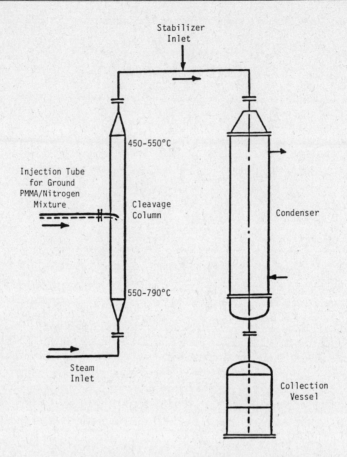

Figure 8.1 Apparatus for the thermal depolymerization of PMMA using super-heated steam as the heat transfer agent [15]

temperature range in the bottom region of the cleavage column, where the super-heated steam was introduced, was 550–790 °C. In the upper region, the temperatures were approximately 400–550 °C. Prior to depolymerization, the PMMA was ground to a particle size of less than 0.24 inch (6 mm). The small particle size allowed the scrap to be carried into the column by the inert gas. The monomer vapors were stabilized using an inhibitor to limit the potential for repolymerization of the recovered monomer. The vapors were then condensed. The condensate was steam-distilled to yield MMA monomer of 99.4–99.7% purity. It is important to note that in this process, according to the patent, the formation of organic residues, a side reaction that can occur in the depolymerization of acrylics and other plastics, was not evident.

8.4.3 Molten Metal and Metal Salt Heat Transfer Media

Depolymerization methods using molten salts such as barium sulfate and other inert materials, including sand, proved to be rather inefficient as far as monomer recovery primarily due to the insufficient transfer of heat. Molten metals or molten metal salts, however, provided a more efficient heat transfer medium for the cracking of PMMA. The metals used in such a process included lead, bismuth, cadmium, tin, rubidium, selenium, tellurium, thallium, and zinc, with lead being the metal most commonly used. Molten sodium was unsatisfactory as a liquid heat transfer medium because of the emission of a considerable amount of sodium vapors. The metals were heated to approximately 400–500 °C. This temperature range was high enough to depolymerize PMMA while limiting the generation of metallic vapors from the liquid heat transfer medium. Alloys of the metals mentioned above were also used as heat transfer media, including alloys with other metals of the group or with aluminum, antimony, and magnesium. In addition, silver or cadmium halides were employed. In a 1958 patent [16], as shown in Figure 8.2, a perforated cage containing acrylic chips was partially immersed in the molten metal bath. Upon contact with the molten metal, depolymerization occurred, releasing monomeric vapors which were condensed and collected. This procedure, however, was limited to a batch process because after depolymerization the cage was removed, refilled with scrap, and re-immersed into the bath. There was no means for continually introducing the scrap into the cage. The monomer purity claimed in this process reached as high as 98% MMA.

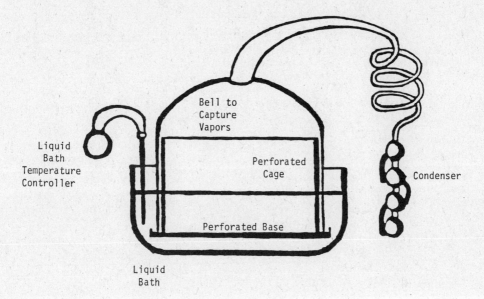

Figure 8.2   A perforated cage containing PMMA chips was partially immersed in a molten metal bath in this thermal depolymerization process

In more recent processes, the acrylic scrap was brought in direct contact with the surface of the liquid heat transfer medium [17]. The direct contact process was virtually continuous because the acrylic scrap was fed onto the surface of the liquid heat transfer medium through a hopper, as shown in Figure 8.3. In this process, the decomposition residue which accumulated during depolymerization floated on the surface of the molten metal and was removed with the medium when the residue reached a predetermined height. The heat transfer medium, in this case, consisted of one or more metallic salts which did not solvate or react with the acrylic scrap. The metallic salts employed were potassium nitrate, sodium nitrate, potassium chloride, sodium chloride, lithium chloride, lithium nitrate, or any mixture of these.

Molten metals, including those mentioned above, were also used as liquid heat transfer media. The melting point of the medium was generally below 450 °C. The bulk of the experimental work in the patent involved a mixture of lithium chloride and other alkali metal chlorides having melting temperatures below 450 °C. The specific gravity of the molten salt or metal was also important. A specific gravity of 2.0–3.0 allowed for good surface contact with the scrap. For example, when using 42 wt % lithium chloride with 58 wt % potassium chloride at 400–450 °C as the heat transfer medium, the MMA monomer obtained had a purity of 98% and was collected at 2.0 lb/min (0.92 kg/min). The feed rate of PMMA (5–10 mm^3 mesh) was 2.2 lb/min (1 kg/min).

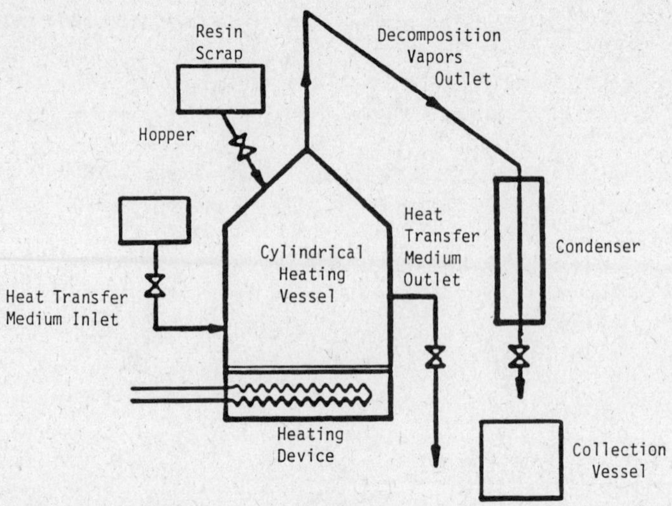

Figure 8.3   Apparatus for continuous thermal depolymerization of PMMA by gradual feeding of scrap through a hopper resulting in direct contact with a liquid heat transfer medium [17]

8.4.4 Retort Method

Although the most common way to depolymerize acrylic scrap has been using a molten metal bath, problems do exist with this method. The build-up of solids and decomposition residues in the medium requires frequent cleaning. As a result, acrylic recyclers have evaluated other depolymerization processes.

One such experimental process uses a retort in place of a molten heat transfer medium [18]. The reactor was designed with a series of parallel "U"-tubes, shown in Figure 8.4, which were heated using natural gas, oil, or similar fuels. The U-tubes extended into a bed of acrylic scrap. As the scrap came into contact with the heated tubes, it was immediately pyrolyzed as was the case with molten heat transfer medium. The depolymerization vapors were condensed using a direct-contact condenser. The condensate contained approximately 90–95% MMA monomer, which was distilled to 99–100% purity.

In this process, however, there still exists the problem of residual build-up on the tubes and the interior wall of the reactor. The residue consisted primarily of organic decomposition products and inorganic fillers, pigments, and other additives. As a result, the depolymerization had to be stopped and the reactor furnace flushed with air and steam, which converted the residue to a granular ash. The ash was then either vacuum-drawn or blown out of the reactor. The retort method proposed the installation of two reactor furnaces so that one could be cleaned as the other remained in operation, allowing for the continuous depolymerization of acrylic scrap.

8.4.5 Fluidized Bed

A fluidized bed for the depolymerization of scrap plastics and rubber tires is being used commercially in Germany [19, 20]. This method involves the fluidization of a solid such as sand using the upward flow of an inert gas, which mobilizes the solid. Examples of the plastics used in the fluidized bed depolymerization process include polyethylene (PE), PP, PET, and polyvinyl chloride (PVC). Recently, a fluidized bed reactor was successfully used for the depolymerization of acrylic scrap [21]. The unit consisted of a fluidized bed of alumina with nitrogen as the fluidizing medium. It was electrically heated to approximately 510 °C. The

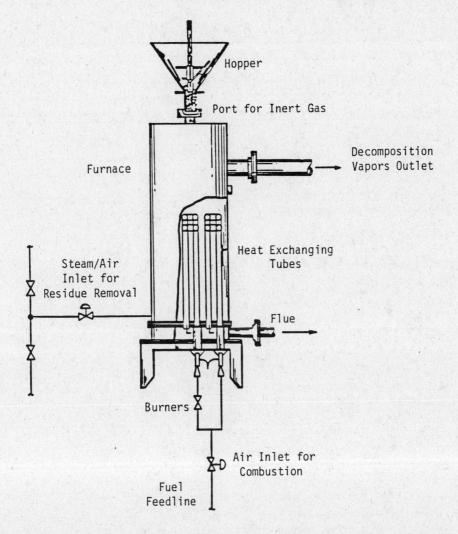

Figure 8.4   Retort with parallel "U" tubes extending into a bed of PMMA scrap for direct contact thermal depolymerization [18]

acrylic scrap was fed directly onto the surface of the fluidized alumina and was depolymerized upon contact. The vapors were collected using a conventional condenser. The accumulation of organic and inorganic residue was apparent but did not interfere with the depolymerizing capability of the process. As the unwanted particulate formed, it was distributed throughout the alumina and circulated within the unit as part of the heat transfer medium. Periodically, the alumina or sand in the process was replaced as the amount of residues present in the fluidized bed exceeded an acceptable limit. A related method of depolymerizing PMMA employed hot sand in a rotary calciner [22]. In this process, the reactor consisted of a rotating glass tube in an electrically heated furnace. The tube was swept with nitrogen and heated to 400 °C. The scrap PMMA was mixed with approximately 67 wt % of sand and introduced into the sample chamber. With this amount of sand, the walls of the chamber remained virtually residue-free, and the PMMA did not form a putty-like ball as it did at lower concentrations of sand. There were, however, heat-transfer limitations with this method, which led to a need for further investigation of the use of extruders for acrylic depolymerization.

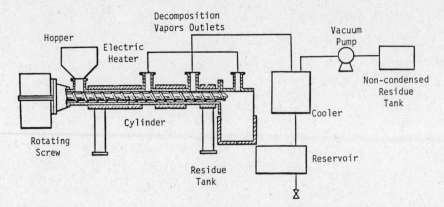

Figure 8.5   Externally heated single-screw extruder apparatus for thermal depolymerization of PMMA [23]

8.4.6 Extrusion Methods

Methods using extruders in the depolymerization of PMMA have involved the use of either a single or a twin-screw vented extruder. In using an extruder for the depolymerization of PMMA, as shown in Figure 8.5, the polymeric acrylic scrap was fed into a hopper and loaded into the cylinder containing a single-screw [23]. The scrap was heated via the rotating motion of the screw, the kneading action that occurred between the surface of the screw and the inside surface of the cylinder walls, and the external heating of the cylinder. The external heating was achieved by using electric heaters. As a result of the heating, the acrylic scrap entered a melt phase that served to seal the hopper side of the cylinder, preventing the escape of the vapors of depolymerization. Modified cylinders had the capability of maintaining a constant back-flow of nitrogen in the cylinder while loading the scrap acrylic. The acrylic then entered a melting zone of about 250 °C followed by a thermal decomposition zone of 500–600 °C. The vapors escaped the cylinder through vents and were condensed and collected in a reservoir. It was possible in such a set up to use a multistage distillation column to condense the vapors of depolymerization. This allowed for simultaneous collection and purification of the MMA monomer.

The formation of inorganic residues from fillers, pigments, and other additives, as well as organic residues that formed during depolymerization, had been a problem in many acrylic scrap cracking systems, including the liquid heat transfer media and retort methods. In this extruder system, however, the screw was extended beyond the cylinder and into a residue tank in which any solid byproducts of polymerization were collected. The thread of the screw was designed to prevent the adhesion of the residue on the inner surface of the cylinder. A light flow of air was introduced into the thermal decomposition zone to further prevent the adhesion of carbon residues to the chamber. If residue deposits did occur, the loading of acrylic scrap into the cylinder was stopped and air forced through the chamber to remove the deposits. This oxidized the carbon residue, which was then released as carbon dioxide. In the depolymerization of nonfilled PMMA using a single-screw extruder, the yield of liquid products was 99.6%, which contained 95 wt % MMA monomer [24].

The use of a twin-screw extruder for the depolymerization or cracking of acrylic scrap has also been described [25]. The extruder was equipped with external cast-bronze electric heaters. Similar to the single-screw extruder, the acrylic scrap was converted to a melt phase in a plasticating zone that was held at a lower temperature than the depolymerization section of the extruder. As in the single-screw extruder, the vapors exited the barrel through vents and were condensed using a direct-contact liquid condenser. Any polymer that had not depolymerized exited the extruder with the aid of a single-screw through a die. The crude monomer obtained was of similar purity to that obtained using a liquid heat transfer medium (Section 8.4.3). The condensate was then distilled to obtain MMA monomer of even higher purity.

8.5   Regrinding Technology

Acrylic scrap collected from production and fabrication processes, in addition to being depolymerized, can be ground to mesh sizes specific for certain applications. For example, off-specification extruded acrylic sheet can be reground and mixed in small amounts into new extruded sheet formulations. This scrap recovery technique is less involved than depolymerization processes and remains economical for acrylic producers because of the increasing cost of MMA monomer. A cell-cast acrylic sheet also can be reground, but it cannot be used in new sheet because of the demand in cell-casting for very high optical clarity and excellent surface quality. The cell-cast regrind must be recycled in a molding product application where the physical and optical property specifications are not as stringent.

8.5.1   Dry Paint Stripping

Acrylic scrap that has been reground may be used in dry paint stripping methods. Conventionally, solvents such as methylene chloride have been used to remove paint and other organic coatings from the metallic surfaces of airplanes and truck bodies. The use of sand and other inorganic particles as blasting media to remove paints and coatings is too abrasive and tends to damage the metallic surface. Acrylic scrap as the blasting medim, however, is a soft material and leaves the primer layer undamaged [26].

A patent issued to DuPont in 1988 [27] describes the use of ground PMMA containing approximately 65% aluminum trihydrate, aluminum oxide, or barium sulfate for paint stripping. This combination did very little damage to the metallic surface. The filled PMMA particles, ground to a particle size of 20–50 mesh, were hurled against the painted surface with the use of a propellant such as water, air, or another gas. The filled PMMA removed the paint more rapidly than unfilled PMMA and other organic blasting media, including ground unsaturated polyester. A more recent patent [28] recommended curing unsaturated polyester with approximately 10–20% MMA, grinding to 20–60 mesh, and using the combination as a blasting medium for paint stripping. In this instance as well, the paint stripping was rapid and resulted in virtually no damage to the metallic surface.

Removing paints and coatings using the dry stripping method meets military specifications for blasting media [29]. These specifications require a loss of 7% or less of the blasting medium after four cycles of use. Therefore, not only is the blasting medium made of recycled materials but it also is recyclable as a paint stripping material.

Regrinding acrylic scrap and using it as a blasting medium in applications where conventional inorganic blasting media may cause damage to the coated surface has proved successful. Paints and coatings can be removed not only from metallic surfaces in airplane and truck body applications but from textured dash boards and plastic sheet molding compounds as well [30]. Interestingly, this technique is gentle enough to be used on circuit boards and other nonmetallic surfaces [31]. This method is more efficient than the use of solvents or conventional blasting media to remove paints and coatings and is environmentally favorable.

8.5.2   Repelletizing and Remolding

American Commodities, Inc. (Southfield, Michigan) has recently begun collecting off-specification acrylic taillights from the automotive industry and regrinding them. The recycled taillights consist of parts that did not fill the mold, parts containing flash, or parts that were cracked, scratched, or otherwise damaged during automobile assembly. The company has been regrinding the rejected taillights and recompounding the acrylic scrap into red and amber pellets. The pellets are sold and molded into reflectors for trailers, road construction, and similar applications [32].

8.6 Postconsumer Acrylic Recycling

The major automotive companies in the United States initiated the exploration of recycling shredder residues from automobiles as a result of the energy crisis in the 1970s. Some of the early work in this area was directed toward the recovery of acrylic from ASR by The Ford Motor Company. This section summarizes the work done by Ford and others concerning the recycling of acrylic from ASR. This is an example of postconsumer acrylic recycling and shows how postconsumer acrylic scrap may be handled in the future if a collection system is eventually established.

The scrap from junked automobiles consists of metallic and nonmetallic fragments. The metallic portion of the automobile scrap is separated magnetically, leaving what is referred to as the "light fluff" or ASR, which consists of mixed plastics and other materials. The largest portion of the residue consists of dirt, tar and other impurities. As would be expected, the ASR also contains glass, rubber, fabric, and a small amount of metal. The remaining plastics fraction of the ASR contains PMMA, polyurethanes, polyolefins, and other thermosets and thermoplastics. According to the patent example, the amount of PMMA (30.2%) present in the nonmetallic fraction of the ASR was sufficient to be considered for recovery.

The patented process [33] is shown in Figure 8.6 for the recovery of MMA from ASR. In the patent, the ASR was extracted with alkanes, including pentane, hexane, heptane, octane, or mixtures of these. The extraction yielded an insoluble residue that was filtered through a screen of 16 mesh. The larger particulate that did not filter through the screen contained the PMMA and was extracted again using acetone as the extraction solvent. The soluble fraction contained the PMMA, which was separated from the insoluble residue. This fraction was then evaporated to dryness, pyrolyzed, and the overhead vapors condensed. The recovered condensate, analyzed by gas chromatography, consisted of approximately 95.5 wt % of MMA monomer. If the solid obtained from the soluble fraction was compression molded prior to pyrolytic heating, the condensate after pyrolysis contained 98.0% by weight of MMA monomer.

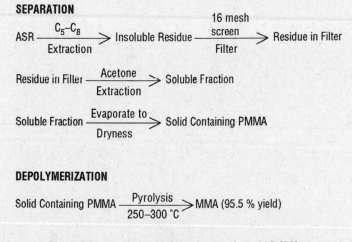

SEPARATION

ASR $\xrightarrow[\text{Extraction}]{C_5-C_8}$ Insoluble Residue $\xrightarrow[\text{Filter}]{\substack{\text{16 mesh}\\\text{screen}}}$ Residue in Filter

Residue in Filter $\xrightarrow[\text{Extraction}]{\text{Acetone}}$ Soluble Fraction

Soluble Fraction $\xrightarrow[\text{Dryness}]{\text{Evaporate to}}$ Solid Containing PMMA

DEPOLYMERIZATION

Solid Containing PMMA $\xrightarrow[\text{250–300 °C}]{\text{Pyrolysis}}$ MMA (95.5 % yield)

Figure 8.6   Recovery of MMA monomer from ASR [33]

An alternative to the extraction/pyrolysis method yielding MMA monomer was to use the mixed plastics-containing ASR as an aggregate in polymer concrete applications [34]. A virgin monomer (or monomer obtained by depolymerization), in this case MMA, was used as a binder for varying amounts of aggregate from 40 to 70 wt %. The molded products exhibited good surface quality and acceptable overall appearance. As the amount of aggregate increased, however, the compressive strength of the polymer concrete decreased. The compressive and tensile splitting strengths of the polymer concrete containing the ASR aggregate were lower, in general, than those of virgin PMMA but were acceptable for some applications. The molded products of the recycled auto scrap polymer concrete systems had strengths of a medium range and relatively low density. This polymer concrete could be used in applications where the density of conventional concretes is excessive (e.g., in construction projects where weight considerations can be prohibitive).

It also has been shown that the ASR can be milled and compression molded into panels used in construction applications, rivaling masonite and particle board [35]. This process does not require the addition of virgin binding resins. Although the compression molded panels showed good rigidity, they were of moderate strength and impact resistance. It then became important to study the use of various additives in order to improve the compression molded properties which would, in turn, improve the panel application potential [36, 37]. These additives were virgin materials and included linear low density polyethylene, HDPE, PP, PS, and acrylic. When some of the additives were used, the resulting impact strengths of the molded products were comparable to virgin PS and acrylic parts.

8.7 Future Outlook

As the consumption of acrylic products continues to grow, so will the rate of production and the need for additional uses of recycled industrial acrylic scrap. This will include the development of new and improved depolymerization techniques as well as the expansion of regrinding and remolding technologies. Eventually, the need for postconsumer acrylic collection and recycling may arise. This would require the development of novel separation and sorting techniques. In addition, depolymerization, extrusion, and molding processes would necessarily expand to include recycled acrylic obtained from appliances, construction, and transportation applications.

References

[1] Grassie, N., Scott, G. *Polymer Degradation and Stabilisation*. Cambridge: Cambridge University Press, 1985.
[2] Kroschwitz, J. I. Exec. ed. *Concise Encyclopedia of Polymer Science and Engineering*. New York: Wiley Interscience, 1990.
[3] Sandler, S. R. and Karo, W. *Polymer Syntheses Volume 1*. San Diego: Academic Press Incorporated, 1974.
[4] Mengel, G. R. *Modern Plastics Mid-October Encyclopedia Issue* (1991): p. 20.
[5] *Modern Plastics, 68* (1) (Jan. 1991): p. 114.
[6] Reference 5, pp. 119, 120..
[7] Minghetti, E., Aristech Chemical Corp. Private communication. July 1991.
[8] *Chemical Marketing Reporter* (25 March 1991).
[9] Reference 1, p. 25.
[10] Strain, D. E. U. S. Patent 2 030 901, 1935.
[11] Mannsfeld, S., Paulsen, K., Buchhoiz, B. and Buchhoiz, E. U. S. Patent 3 494 958, 1970.
[12] Segui, E. D. and Alarcon, B. C. U. S. Patent 2 858 255, 1958.
[13] Tatsumi, T., Yoshihara, H. and Uesaka, G. U. S. Patent 3 886 202, 1975.
[14] Reference 10.
[15] Reference 11.
[16] Reference 12.

[17]   Reference 13.

[18]   Aristech Chemical Corp. Unpublished research. 1987.

[19]   Kaminsky, W. in "Resource Recovery and Conservation, 5." Amsterdam: Elsevier Scientific Publishing Company, 1980.

[20]   Staffin, H. K. and Staffin, R. U. S. Patent 4 161 389, 1979.

[21]   Thompson, II, J. E., Aristech Chemical Corp. Private communication. Oct. 1990.

[22]   Aristech Chemical Corp. Unpublished research. 1981.

[23]   Tokushige, H., Kosaki, A. and Sakai, T. U. S. Patent 3 959 357, 1976.

[24]   Reference 23.

[25]   Aristech Chemical Corp. Unpublished research. 1987.

[26]   Ackermann, K. *Plastics News* (23 July 1990): p. 5.

[27]   Hochberg, J., Young, R. B., U. S. Patent 4 759 774, 1988.

[28]   Risley, L. F. U.S. Patent 4 947 591, 1990.

[29]   "Plastic Media, for Removal of Organic Coatings." Military Specification, MIL-P-85891 (AS), 6 May 1988.

[30]   Reference 26.

[31]   Reference 28.

[32]   PM & E/PC Recycling Equipment Report, 6 September 1990.

[33]   Mahoney, L. R. U. S. Patent 3 965 149, 1976.

[34]   Crawford, W. J., Manson, J. A. *Polym. Prepr. (Am. Chem. Soc., Div. Polym. Chem.), 24*(2) (1983): p. 432.

[35]   Deanin, R. D., Yniguez, A. R. *Polym. Mat. Sci. Eng., 50* (1984): p. 143.

[36]   Spaak, A. *Conservation and Recycling, 8*(3/4) (1985): p. 19.

[37]   Deanin, R., Busby, D. M., DeAngelis, G. J., Kharod, A. M., Margosiak, J. S., Porter, B. G. *Polym. Mat. Sci. Eng., 53* (1985): p. 826.

9 Commingled Plastics

K. E. Van Ness, T. J. Nosker

Contents

9.1   Introduction 189

9.2   Materials and Sources 191

    9.2.1 Plastic Containers 191

    9.2.2 Collection and Sortation 191

9.3   Manufacturing Processes 193

    9.3.1 Klobbie-based Intrusion Processes 193

    9.3.2 Continuous Extrusion 194

    9.3.3 Reverzer Process 195

    9.3.4 Compression Molding 195

    9.3.5 Scientific and Economic Evaluations 196

9.4   Structure-Property Relations for Commingled Materials 196

    9.4.1 Processing and Materials 196

    9.4.2 Compositional Analysis of NJCT 197

    9.4.3 Macroscopic Features of "Lumber" Profiles 198

    9.4.4 Mechanical Testing Procedures 199

    9.4.5 Stress-Strain Relations for NJCT 199

    9.4.6 Discussion of Mechanical Tests for NJCT 200

9.5   NJCT in Combination with Recycled Polystyrene (RPS) 200

    9.5.1 Materials and Mechanical Testing Procedures 200

    9.5.2 Stress-Strain Relations for RPS/NJCT Blends 200

    9.5.3 Microstructure of RPS/NJCT Blends: Positron
          Lifetime Studies 202

    9.5.4 Scanning Electron Microscopy 206

    9.5.5 Implications of the Structure-Property Relations for
          RPS/NJCT Blends 208

9.6   New Commingled Processes 210

    9.6.1 Mechanical/Physical Commingled Processes 210

    9.6.2 CPRR/WEI Refined Commingled Technology 211

      9.6.3 Chemical Process: Selective Dissolution 219

      9.6.4 A Physical/Chemical Approach 224

9.7   Global Activities 224

      9.7.1 Japan 224

      9.7.2 Europe and Canada 224

Acknowledgments 225

References 225

Current technologies concerned with the recycling of the mixed (commingled) plastic waste component of the municipal solid waste (MSW) stream are described. Traditional manufacturing processes designed to mold the contaminated, commingled waste plastics into bulky "lumber" profiles are outlined. Mechanical properties of profiles made from mixed rigid containers collected at curbside are presented. Compressive properties of blends of postindustrial (regrind) polystyrene (PS) and postconsumer plastic containers are discussed. The microstructure/morphology of the blends are shown through scanning electron microscopy and positron annihilation techniques. Newer, more innovative technologies which reduce the level of contamination of the commingled feedstock are described. One new type of technology uses physical/mechanical processes to produce a refined polyolefin feedstock which is compatible with conventional processing systems, either alone or in combination with additives: for example, fillers, compatibilizers, or recycled thermoplastics. Another type of technology uses a chemical process to separate the mixed plastics by homopolymer, which may then be distributed through conventional channels for processing.

9.1 Introduction

In order to consider the current possibilities for the recycling of postconsumer plastic wastes, we first identify which of these are both available and collectible from the MSW stream. Of the major nonfiber plastics consumed in the U.S., we see from Table 9.1 [1] that plastic packaging, generally comprised of products with a short useful lifetime, is the end-use category with the greatest contribution by weight. At about 15 billion pounds (6.8×10^6 t) per year, plastic packaging is about 4% by weight, and well under 10% by volume, of the approximate 400 billion pounds (181×10^6 t) per year attributed to the total MSW stream [2].

Table 9.1   Major nonfiber plastics uses–1990

Use	Million Pounds	(Thousand Tonnes)
Packaging	14,821	(6723)
Building	11,885	(5391)
Transportation	2335	(1059)
Appliances	1246	(565)
Electrical/Electronic	2312	(1049)
Furniture	1183	(537)
Toys	760	(345)
Housewares	1476	(670)

Source: *Modern Plastics* (Jan. 1991)

While most of the plastic packaging ultimately enters the waste stream, only those items that are readily collectible can be recycled effectively. As Table 9.2 [3] shows, plastic containers, the major target of community recycling collection programs, currently constitute about 7 billion pounds (3.2×10^6 t) or approximately one-half of the total plastic packaging wastes. Most of these containers are readily collectible, processible, and salable, and, in addition, can serve as feedstocks for a large number of reclamation businesses. The approximately 6 billion

Table 9.2   Plastics in packaging (by end use)–1990

End Use	Million Pounds	(Thousand Tonnes)	%
Containers*	7241	(3285)	48.9
Film	5608	(2544)	37.8
Coatings	1186	(538)	8.0
Closures	786	(357)	5.3
TOTAL**	14,821	(6724)	100.0

* Approximately half are bottles
** Does not include adhesives
Source: *Modern Plastics,* (Jan. 1991)

pounds (2.7×10^6 t) of plastic films used in packaging are also technically reclaimable and reusable, but are now relatively difficult to collect and reuse, since the required infrastructures are not yet in place. A closer scrutiny of the types of plastic containers and their constituent polymers is further useful in identifying a viable feedstock for the production of products made from recycled plastic wastes.

Of the different types of plastics used in rigid packaging in the U.S. (Table 9.3) [4], high-density polyethylene (HDPE) and polyethylene terephthalate (PET), molded into milk jugs and beverage bottles, respectively, are the most easily identified. Since these materials have significant properties as well, they are prime targets for recycling. About 360 million pounds (2×10^3 t)) of postconsumer bottles, or greater than 9% or all such bottles, were recycled in 1990 [5]. However, nonfood product applications utilizing 1 billion pounds (450×10^3 t) per year of recycled HDPE and PET bottles have been identified [6]. Therefore, the major current problem in plastics recycling is getting more waste plastic containers collected, sorted, and placed into the existing reuse infrastructure. Fortunately, many states and communities are initiating programs to collect these materials, and the recycling rate for plastics is expected to climb rapidly over the next few years.

Table 9.3   Plastics used in rigid packaging*–1990

Material	Million Pounds	(Thousand Tonnes)	%
High-Density Polyethylene (HDPE)	3556	(1613)	38.8
Low-Density Polyethylene (LDPE)	1154	(523)	12.6
Polyethylene Terephthalate (PET)	1225	(556)	13.4
Polypropylene (PP)	913	(414)	10.0
Polystyrene (PS)	1517	(611)	16.5
Polyvinyl chloride (PVC)	447	(203)	4.9
Other	353	(160)	3.8
TOTAL	9165	(4157)	100.0

* Closures, coatings, containers (no film)
Source: *Modern Plastics,* (Jan. 1991)

Nevertheless, regardless of the increased recycling of a large portion of plastic containers as individual generic cleaned resins, barely a dent will have been made in the solid-waste problem overall, or even in the plastic portion of the MSW. As of the date of this publication, the only plastics, both valuable and readily collectible, which are being used for resin recovery in any appreciable amounts from the MSW include PET soft drink containers and unpigmented HDPE milk and water jugs. Fortunately, mixed or "commingled" plastic wastes in the form of both films and plastic items (containers, toys, etc.) that are left after the valuable and easily identifiable plastics have been removed can be processed into "lumber profiles" and into many other useful products, generally bulky in character, thereby greatly increasing the potential for the recycling of plastics. The use of whole mixed plastic waste (as received from the plastic component of the MSW stream) as a feedstock to fabricate salable molded articles is known as commingled plastics processing. There is presently only a bare beginning of such an industry in the United States. Increasing the volume of this type of reclamation will be a real challenge for plastics recycling. In this chapter, we consider in turn the sources of these materials, the various technologies for producing commingled products, their structure-property relations, and the most promising of newer, more innovative commingled processes.

9.2   Materials and Sources

9.2.1   Plastic Containers

Commingled plastics processes are capable of utilizing many feedstocks, including postconsumer waste plastics, a variety of plastic plant scraps, and various percentages of fillers. The recycling of rigid plastic containers, the most widely available and readily collectible of postconsumer waste plastics, is the focus of attention here.

As mentioned earlier, some rigid containers are easy to identify and, hence, easy to separate from the MSW stream. Examples of these are plastic milk bottles made from unpigmented HDPE and plastic carbonated beverage bottles made from PET. Containers such as these are said to be standardized by the plastics industry; that is, a particular polymer has been identified with a particular packaged product (e.g., HDPE with milk and PET with soft drink). However, in general, plastics containers are not standardized for a particular packaged product; containers for household cleaners, cooking oils, foods, motor oils, and the like may vary in shape, color, and polymeric constitution from manufacturer to manufacturer (see Table 9.4)[1] . Since these containers are not easily identifiable, to separate them by resin would be difficult and expensive. In many states, to facilitate sortation, there has been legislation to require the SPI identification symbols to be shown on certain types of containers. Many molders have voluntarily incorporated these symbols on their molds. Regardless, the resulting products made from the recycled resins would be of limited value due to the varied pigmentations and additives used in the original processing of the containers. Therefore, the mixed plastic containers left after the removal of HDPE milk jugs and PET soft drink bottles are typically either sent directly to a landfill or used by a commingled plastics processor to produce salable articles. For convenience, these remaining mixed plastics will be referred to hereinafter as "tailings," a term adopted from the mining industry. In the next section, we discuss briefly the collection and sortation of this important commingled feedstock. For more detail on collection and sortation, please see Chapter 2.

9.2.2   Collection and Sortation

At the present time, most recycling programs that collect plastics request only unpigmented HDPE milk and water jugs and PET soft drink bottles from among the plastic containers.

---

[1] Changes made by industry as to the type of polymer used to package a given product further complicate the separation process; for example, while Table 9.4, which reflects the situation in Nov. 1989, shows a Palmolive™ bottle as made from PVC, several months later the bottle is made from PET.

Table 9.4   Plastic materials used for common household products

Material	Polymer
Glue Bottle	LDPE
Surf™ Detergent	HDPE
Tic-Tac™ Case	PS
Palmolive™ Bottle	PVC
Oil Bottle	HDPE
Soft Drink Bottle Base cup	HDPE
Ocean Spray™ Juice Bottle	PET
Tape Dispenser	PS
Baby Oil Bottle	PVC
Windshield Wiper Fluid	HDPE
Milk Bottle	HDPE
Mouthwash Bottle	PET
Egg Carton	PS
Honey Bear	LDPE
Vegetable Oil Bottle	PET
Pert Plus™ Shampoo Bottle	HDPE
Cookie Tray	PS
Single Service Juice	LDPE
Soft Drink Bottle	PET
L'eggs™ Container	PS
Medipren™ Bottle	HDPE
Sitff Stuff™ Spray Bottle	LDPE

Source: Teresa M. Simmons, "Plastics Recycling Awareness Program," sponsored by the National Science Foundation—Center for Plastics Recycling Research, Nov. 1989

Practically, these comprise 80% of the mixed plastics placed at curbside for recycling, leaving 20% as a potential commingled feedstock. Since it has been demonstrated that the volume of recyclables is maximized by collecting mixed recyclables at curbside [7], recycling programs generally begin with the collection of these, followed by their transport to a materials recovery facility (MRF), where newspapers, tin-plated steel cans, aluminum and other nonferrous metals, and glass are separated from the plastics. The plastics in turn are separated into three categories: (1) unpigmented HDPE; (2) clear and green PET; and (3) the remaining mixed plastics or tailings. The HDPE and PET are sold directly to reclaimers; the tailings, while generally sent to landfills, provide a feedstock for the fabrication of commingled plastics products.

The exact polymeric composition of the tailings, which consist principally of polyolefins, will naturally vary from one municipality to another and from year to year as collection and sortation practices change. The CPRR estimated that tailings received from central New Jersey during the mid- to late 1980s consisted predominately of HDPE (about 80%) and small amounts of other thermoplastics (see Table 9.5 in Section 9.4.2 and Table 9.18 in Section 9.6.3).

Additional commingled waste plastics are readily formed by the blending of plant scraps, recycled thermoplastics, or fillers either with each other or with mixed plastic waste such as tailings. In the next section, several of the traditional manufacturing processes capable of processing mixed plastics wastes are described.

9.3 Manufacturing Processes

There are several types of manufacturing processes that have been developed specifically for processing commingled plastics. These processes may be roughly categorized into four basic types: intrusion processes based on Klobbie's design, continuous extrusion, the "Reverzer" process, and compression molding. Each of these processes is capable of producing products from a variety of macroscopically inhomogeneous mixtures of waste plastics, all containing some degree of contamination. Because of the heterogenous nature of these mixtures, commingled processes are limited to producing products of large cross section, where small internal imperfections may be of little consequence for the mechanical properties. Properties of the products are measured by testing several of the large samples, thereby averaging the effects due to the inclusions upon the bulk material. Experiments conducted in such a way on a variety of samples produced using different combinations of commingled plastics as feedstocks have been found to yield properties that are heavily dependent upon the feedstock composition, as is shown in Section 9.4.

9.3.1 Klobbie-based Intrusion Processes

In the 1970s, Eduard Klobbie of the Netherlands began developing a system for processing "unsorted thermoplastic synthetic resin waste material into an article having the working and processing properties of wood [8]." His system consisted of an extruder, several long, linear molds of large cross section mounted on a rotating turret, and a tank of cooling water into which the turret is partially submerged. Since the Klobbie has been presented and illustrated previously [9], our discussion here of its operation is necessarily brief. More detail can be obtained from references [10–12].

The extruder first works and softens the thermoplastic mixture, and then forces this material into one of the molds without using a screen-pack or an extrusion nozzle. After this mold has been filled, the turret rotates one position in order to fill the next mold. Eventually, each mold slowly passes underneath the coolant level in the tank, where the plastic is cooled, solidified, and shrinks away from the mold, pulling coolant into the gap between the product and mold surface. After the turret is further rotated to raise the cooled molds from the liquid, the finished parts are removed from the molds. This process is a cross between conventional injection molding and extrusion, and therefore may be termed an "intrusion" process. Klobbie never patented his system in the U.S., but he did patent the use of foaming agents in such equipment [13]. This design is extremely forgiving as to the variety of feedstocks which may be processed, as long as sufficient structural integrity is built into the molding system. Currently, Lankhorst Recycling Ltd. (Sneek, Netherlands) owns the rights for the Klobbie process in the Netherlands and Belgium; Lankhorst has sold the rights for the rest of the world to Superwood International Ltd. (Dublin, Ireland). In August 1991, Superwood International went into receivership [14] (see Section 9.7.2 for a discussion of Superwood's activities).

In the 1980s, several companies were producing equipment, which (in the opinion of the authors) is based on Klobbie's design, but with some variations. These include Advanced Recycling Technology (ART) (Brakel, Belgium); Hammer's Plastic Recycling (Iowa Falls, IA); and Superwood (Ireland). Of the three, only ART is currently selling this type of equipment to interested parties, with no royalties.

ART's machine, the ET/1 (Extruder Technology 1), shown in Figure 9.1, is an adiabatic extruder, capable of processing most types of densified, mixed, thermoplastic materials.[2] The ET/1 was designed to accept a maximum of 30–40% contamination in the form of unsoftened/

---

[2] On 13 Nov. 1990, a federal court in The Hague ruled that BWR Ltd. (North Holland) used the ET/1 system in a way that violated the patent rights held for the Klobbie process by Lankhorst Recycling, Ltd. Lankhorst and BWR reportedly have settled the matter. ART is appealing the decision [19].

Figure 9.1 The ET/1 adiabatic extruder from Advanced Recycling, Ltd. (Belgium)

unmelted polymers and other nonpolymeric materials such as paper, glass, dirt, metal, etc. [15, 16]. The remaining 60+% of material consists of polyolefins, which soften/melt during processing, thereby encapsulating the contaminants. While the ET/1 can accept either industrial or postconsumer scrap, most systems in operation today use the latter as their main feedstock [17]. This system can mold objects of either constant or tapered cross section, up to 6×6 inches (15.2×15.2 cm) in area and 12 feet (3.7 m) in length, at a rate of 450 lb/h (204 kg/h)on a continuous 24-hour, 3-shift basis [18]. The ET/1 is capable of making objects where the length is large compared to the cross-sectional area (i.e., posts, poles, stakes, planks, slats, etc.). Mechanical testing and microstructural analysis of lumber produced by the ET/1 are discussed in Section 9.4.

Hammer's Plastic Recycling Corporation (Iowa Falls, IA) probably the largest producer of commingled plastic lumber [20], utilizes a process covered by several U.S. patents [21–24]. The Hammer process is characterized by, among other things, closed molds, a heated nozzle, and a screen pack, which serve to increase the molding pressure over that of the original Klobbie design, thus requiring no blowing agents. The plant in Iowa typically processes both postconsumer and postindustrial plastics in the following typical percentages: 65% low-density polyethylene (LDPE), 20% HDPE, 5% polypropylene (PP), 5% PET, and 5% miscellaneous [25]. Two types of products are produced: (1) thick-wall moldings such as pallets, pig drinkers, and bench ends; and (2) linear lumber-like profiles.

9.3.2 Continuous Extrusion

A variation of the previously mentioned technique that may be used to produce linear profiles from mixed plastic waste is to continuously extrude molten polymeric material of large cross section into cooled dies in a manner similar to continuous pipe manufacture. There are special considerations required when adapting this technology to large profiles made from commingled plastic waste. For example, provisions must be made to cool the extruded material for a fairly long time because of the large cross sections produced. The limiting factor in this case is that polymers generally have low thermal conductivity, which causes large profiles to have considerable temperature gradients between the interior and the outer skin during cooling. Another important consideration is the assurance of consistency of raw material in terms of melt index and other rheological properties. The lack of consistency in raw materials could result in surging and in a lack of dimensional stability for the product. Assurance of some consistency for post consumer commingled plastics may be increased by washing and float separating.

This type of process is practiced by Tri-Max of New York, and they have patents pending. Their product is described as "a glass-reinforced highly filled composition with a structural foam core" [26]. The design of the product is such that the glass-fiber reinforced outer skin gives good strength in flexure as is desirable for beams, while the foamed core results in a density similar to a medium hardwood such as southern yellow pine. The plastics component of the feedstock is derived from postconsumer commingled plastics waste.

9.3.3 Reverzer Process

The earliest U.S. patented process for fabricating products of large cross section from commingled plastics was by the Japanese company Mitsubishi Petrochemical. The process, referred to as the "Reverzer" process, is described in some detail in references [27] and [28]. Since this technology has been presented before along with numerous figures [29], herein we describe only briefly a few of the operational principles.

In this process, commingled waste plastics are softened in a hopper and then mixed in a screw to develop a well-mixed uniform batch of fluidized plastic. Three systems exist for transforming the output of the Reverzer into useful products: flow molding, extrusion, and compression molding. In flow molding, molten plastic is fed under very low pressure into thin molds made of sheet metal. The extrusion system allows the machine to function as an extruder that operates at low pressure, generally filling long linear molds. Adaptation to compression molding makes use of a special device to develop high pressures necessary for filling large molds with sizable surface area. The Reverzer process is capable of producing many different-shaped items from commingled plastic wastes and is not limited to linear shapes as the previous two processes are. This process is also extremely versatile in that it is capable of accepting a large variety of mixtures of contaminated commingled scrap as feedstock.

Mitsubishi apparently did not achieve marketing success with this equipment, however. Manufacture of the Reverzer was terminated after only eight units were produced. Efforts are well underway to develop and eventually produce a similar type of machine in the United States by the American Plastics Recycling Group (Ionia, MI).

9.3.4 Compression Molding

The most successful technology for the compression molding of commingled plastics was developed in Germany by Erich Weichenreider, and is called Recycloplast [30]. This process mixes batches consisting of 50–70% thermoplastics with other materials, melting via friction the portions of the mix characterized by low melting points. An automatically adjusted scraper then removes the melted material from the plasticator and presses it via a heated extruder die into premeasured, roll-shaped loaves. The loaves are then conveyed to a press charging device, which fills a sequence of compression molds alternately. Products are cooled in the molds to a temperature of 40 °C and ejected onto a conveyor, which carries the product to a storage area. Flashing from the mold process may be transferred to a granulator for in-house recycling. One of the major differences between this process and those mentioned previously is that plant size is necessarily large, and the capital investment large by comparison. Alternatively, this process has high throughput and is capable of producing finished thick-walled products such as pallets, grates, benches, and composting boxes. For more information on the Recycloplast technology, see Section 9.7.2.

9.3.5 Scientific and Economic Evaluations

Traditional commingled processes such as those just described are dedicated primarily to the fabrication of bulky products, which in many cases are candidates as substitutes for products made from wood, to be used in applications where physical properties are widely assumed not to be critical. Regardless of the degree of truth in this assumption, in order for commingled products such as these to compete more realistically with wood over a variety of even more critical applications, it is reasonable and desirable that they have mechanical properties at least approaching those of wood. While good mechanical properties are important, purely economic considerations may prove to be somewhat less so if there are compelling advantages, such as superior resistance to environmental stresses, associated with the use of commingled products over, for example, pressure-treated lumber.

In fact, economic data for these traditional methods are readily available. The range of cost for processing plastic lumber is estimated to be 19–25¢/lb (42–55¢/kg) [31]. This translates into a cost-per-unit length about twice that of pressure-treated lumber. On the other hand, there is the issue of environmental impact; specifically, the effect of both pressure-treated lumber and commingled plastic products upon aquatic ecosystems, which is currently being studied by the CPRR in collaboration with the New Jersey College of Medicine and Dentistry.

As for mechanical properties, it is incredible that up until 1989 there existed no published, reliable data on commingled products made from waste plastics. This reflects, in part, a hesitation shown by the scientific community, academic and industrial, both to mechanically test commingled products and to study blends made from recycled plastics. It is not that scientists were unaware of the commercial significance of the mechanical blending of mixed plastics waste [32] but that they chose to study blends made from virgin rather than recycled resins. Ostensibly, by understanding more fully the processing of well-characterized polymer pairs and their resultant properties and morphologies, such information would facilitate the commercial blending of mixed plastic wastes. Unfortunately, yet understandably, as a result of some very interesting and thorough studies [33, 34], most polymer scientists remain convinced of the fruitlessness of trying to make useful products from mixtures of incompatible plastics [35] such as are used commonly by the packaging industry.

In contradiction to this belief, recent studies show that good and consistent properties for commingled products made from postconsumer plastic wastes have been measured. The next section presents data on these properties and discusses their significance for the plastics recycling industry.

9.4 Structure-Property Relations for Commingled Materials

The first published data on the mechanical properties of commingled products manufactured from mixed plastic wastes was reported by a group of researchers from the CPRR [36]. The results of the mechanical tests and the complementary microstructural/morphological studies, which were performed at Washington and Lee University [37], are presented in some detail below.

9.4.1 Processing and Materials

The CPRR used an ET/1 intrusion molding machine (see Section 9.3.1) from ART of Belgium. The materials used as a commingled feedstock in the production of samples by the ET/1 at the CPRR were obtained from the postconsumer plastic waste streams of several New Jersey communities which are recycling plastic containers. More specifically, the feedstock consisted entirely of tailings (see Section 9.2.2), hereinafter referred to as New Jersey Curbside Tailings (NJCT).

9.4.2 Compositional Analysis of NJCT

The composition by homopolymer of NJCT was estimated by the CPRR and is shown in Table 9.5 [38]. The percentage of polyolefins is approximately 88% by weight, consisting of about 80% HDPE, 4% LDPE, and 4% PP. The remaining 12% is attributable to polyvinyl chloride (PVC), PS, PET (nonsoft drink bottles), ferrous and nonferrous metals, foodstuff, dirt, paper, adhesive, and plastic labels.

Table 9.5   Polymeric composition of NJCT

Polymer	Percent
HDPE (non-milk jug)	80
LDPE	4
PP	4
PS	3
PVC	4
PET (non-soft drink bottle)	4
Other	1
TOTAL	100

Source: CPRR estimates

Elemental analysis by neutron activation was performed at the University of Virginia for several batches of NJCT. Table 9.6 shows the results of these studies for two batches of tailings

Table 9.6   Neutron activation studies of NJCT*

Element	Surface Samples A	B	1st Depth Samples A	B	2nd Depth Samples A	B	3rd Depth Samples A	B
Aluminum	900 ±170	860 ±150	960 ±180	590 ±120	790 ±150	690 ±130	1150 ±220	520 ±100
Antimony	60 ±15	50 ±20	70 ±20	70 ±20	85 ±20	50 ±20	85 ±20	50 ±40
Bromine	8.5 ±2	7.5 ±2	11 ±2	7.5 ±2	17 ±3	8 ±2	14.5 ±3	3 ±2
Calcium	1500 ±400	<1200	1200 ±400	<1200	<1200	<1400	<1400	<1200
Chlorine	19700 ±2700	26800 ±3700	23000 ±3200	23000 ±3500	26200 ±3600	24200 ±3400	21600 ±3000	21200 ±3000
Copper	90 ±20	250 ±50	100 ±20	85 ±20	75 ±20	190 ±40	1000 ±200	700 ±200
Magnesium	2100 ±600	2000 ±600	2600 ±800	1200 +400	2300 ±700	1700 ±500	1800 ±600	1200 ±400
Silicon	<500	<500	<500	<500	<500	<500	<500	<500
Sodium	1500 ±100	1500 ±100	1300 ±100	1500 ±100	1500 ±100	1700 ±100	1650 ±100	1300 ±100
Titanium	200 ±50	180 ±50	150 ±40	170 ±45	150 ±40	230 ±60	130 ±30	150 ±40

* All data given in ppm; columns represent samples from different batches of NJCT according to depth
Note: Antimony and titanium ppm were calculated with flux; "<####" indicates the minimum detectable ppm

as a function of depth from the outer surface of the sample [39]. The most prevalent element, chlorine, is due to the presence of PVC. With the exception of aluminum, which may be present in the form of chips which were not removed from the tailings, the samples were fairly homogeneous with respect to the distribution of these trace elements.

In summary, it was found that the commingled materials produced from recycled, postconsumer, rigid plastic containers consisted primarily of HDPE, with traces of metallic elements mixed uniformly (excepting Al) throughout.

9.4.3   Macroscopic Features of "Lumber" Profiles

A cross-sectional profile of a piece of "lumber" consisting of 100% NJCT is shown in Figure 9.2. The pieces are solid around the perimeter of the cross section, while the area around the center (core) contains numerous pores or voids, varying in size. These voids are believed to be caused by a combination of factors. Wherever the polyethylene (PE) phase crystallizes, there will be significant shrinkage. Because of the bulky nature of the profiles and because polymers do not conduct heat well, the periphery of the extruded product solidifies first, shrinking and pulling away from the mold. The remaining core of molten material will in turn cool slowly, crystallize, and shrink. However, due to the solidified outer skin, the external dimensions of the piece will stay approximately the same, thereby resulting in the formation of internal voids or pores. Voids are also thought to be caused by water vapor and other gases that were not vented during processing. A comparison of the density of the outer region with the overall density of a profile gives an estimate of the volume fraction due to voids equal to 10%.

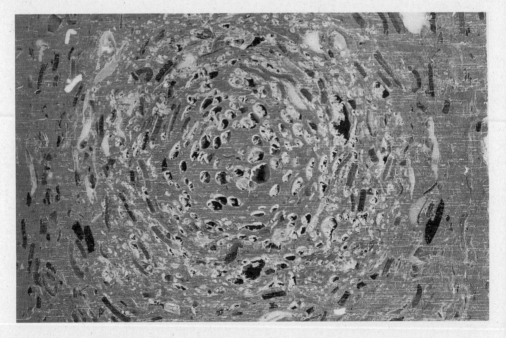

Figure 9.2   Typical cross section of plastic "lumber" showing cylindrical symmetry and various imperfections such as inclusions and voids

A second macroscopic observation is that all cross sections exhibit cylindrical symmetry as evidenced by the elliptical nature of the second phase inclusions, the long axes of these inclusions being oriented roughly along circular arcs centered about the centroid of the cross section.

9.4.4 Mechanical Testing Procedures

Results of initial tests conducted on samples showed variability depending upon the position of the sample taken along the length of the profile. Figure 9.3 displays the average variation of specific gravity as a function of position along the bar [40]. Therefore, all mechanical tests were performed on lumber profile samples $2.5 \times 2.5 \times 5$ inch ($6.4 \times 6.4 \times 12.8$ cm) cut from the mid-section of the profile. All tests were performed as closely as possible to American Society of Testing and Materials (ASTM) standards. "Dogbone" shapes were not used since researchers have reported large inconsistencies in measurements of mechanical properties on commingled postconsumer plastic scrap using dogbone samples. Voids, metal particles, and nonmelted transparent polymer chunks were noted as phases that may serve as flaws during deformation [41]. These flaws are large compared to the fixed thickness required of ASTM dogbone samples.

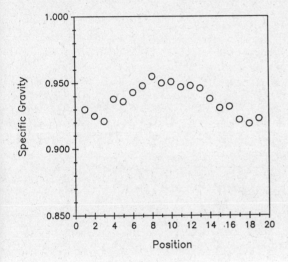

Figure 9.3   Average variation of specific gravity as a function of position along the bar for 100% NJCT

9.4.5 Stress/Strain Relations for NJCT

The results of the compressive mechanical tests performed on the samples produced from the ET/1 intrusion molding machine are presented in Table 9.7 [42].

Table 9.7   Compressive properties of NJCT samples

Sample Designation	Specific Gravity	Compressive Modulus	Yield Stress (2% Offset)	Compressive Strength (10%)	Day Produced
005	0.944	114×10^3 psi (789 MPa)	2578 psi (17.8 MPa)	3049 psi (21.0 MPa)	Day 1
006	0.925	114×10^3 psi (784 MPa)	2704 psi (18.6 MPa)	3159 psi (21.8 MPa)	Day 1 + 3
009	0.919	115×10^3 psi (794 MPa)	2729 psi (18.8 MPa)	3207 psi (22.1 MPa)	Day 1 + 21
020	0.937	115×10^3 psi (794 MPa)	2818 psi (19.4 MPa)	3253 psi (22.4 MPa)	Day 1 + 75

Comparing the samples of NJCT, the maximum variation in specific gravity was about 3%, while the values of compressive modulus were identical, within experimental error. The spread in values of yield stress and compressive strength was larger, but still reasonable, at 9% and 7%, respectively. A typical compressive stress/strain curve for NJCT is presented graphically in Figure 9.4.

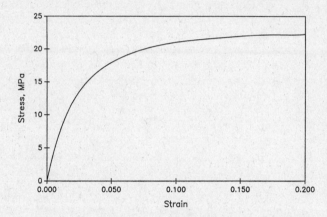

Figure 9.4 A typical compressive stress-strain curve for 100% NJCT
Source: "Recycle PS, Add Value to Commingled Products," Nosker, Renfree, Morrow, *Plastics Engineering* (Feb. 1990)

The properties given in Table 9.7 show an unexpected measure of consistency, particularly since the feedstock materials for these samples were obtained entirely from the mixed, postconsumer recycling stream at different times, the only similarity between batches being the removal of the PET soft drink and unpigmented HDPE milk bottles.

9.4.6 Discussion of Mechanical Tests for NJCT

While the compressive properties of NJCT shown in Table 9.7 are consistent, they are also low in comparison with similar properties for building materials such as wood. To compete more successfully with such materials, the CPRR directed further studies toward developing a commingled technology that would enhance the NJCT's relatively low mechanical properties. Therefore, tailings were mixed with a more rigid polymer (i.e., PS). The results and implications of this work, described in the following section, are taken from previously published works [43–45].

9.5   NJCT in Combination with Recycled Polystyrene (RPS)

9.5.1 Materials and Mechanical Testing Procedures

The PS used in this study was densified reground PS obtained from Mobil Chemical Company's expandable polystyrene (EPS) production facility. Compositions by weight percent of 0/100, 10/90, 20/80, 30/70, 35/65, 40/60, and 50/50 recycled polystyrene (RPS)/NJCT were tested. Compositions greater than 50% RPS were not tested because such samples did not shrink enough upon cooling, making it impossible to remove them from the molds. All samples were processed as before by an ET/1 intrusion molding machine (see Sections 9.3.1 and 9.4.1). The sample size used for the compressive testing was 1 in² (6.4 cm²) in area and 5 in (12.8 cm) in length and was sectioned roughly from the mid-section of a profile which was 1 in² × 7.9 ft (6.4 cm² × 2.41 m).

9.5.2 Stress-Strain Relations for RPS/NJCT Blends

The values of compressive properties reported for 100% NJCT listed in Table 9.8 represent averages obtained from all of the batches of NJCT which were tested (see Table 9.7). The

Table 9.8 Mechanical properties for blends of RPS and NJCT

Composition PS/NJCT	Compressive Modulus 10³ psi	(MPa)	Yield Stress psi	(MPa)	Ultimate Stress psi	(MPa)
0/100	115	(794)	2701	(18.63)	3171	(21.87)
10/90	144	(996)	3102	(21.39)	3221	(22.22)
20/80	163	(1127)	3861	(26.63)	3861	(26.63)
30/70	198	(1363)	4353	(30.02)	4353	(30.02)
35/65	239	(1649)	4953	(34.16)	4953	(34.16)
40/60	222	(1534)	4753	(32.78)	4753	(32.78)
50/50	220	(1518)	5322	(36.71)	5322	(36.71)

Source: "Recycle PS, Add Value to Commingled Plastics," Nosker, Renfree, Morrow, *Plastics Engineering* (Feb. 1990)

average compressive modulus of all 100% NJCT samples tested was 115×10^3 psi (794 MPa), with only a few percent variation in the other compressive properties. Inclusion of an amount of RPS as small as 10% by weight increased the compressive modulus by 27%, from 115×10^3 psi (794 MPa) to 144×10^3 psi (996 MPa). The addition of 20% RPS by weight increased the compressive modulus by 42%, the yield stress by 43%, and the ultimate stress by 22% (Table 9.8).

At the level of 35% RPS, the value of the compressive modulus increased over that of 100% NJCT by 108%. Above 35% RPS by weight, the modulus decreases slightly, appearing to level off around 220×10^3 psi (1515 MPa). Figure 9.5 shows the relation between the compressive modulus and the percent PS by weight of the tested samples. Similarly, Figures 9.6 and 9.7 show the yield stresses and ultimate stresses for these blends, respectively.

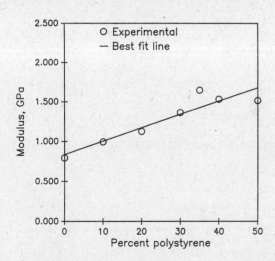

Figure 9.5 Compressive modulus as a function of percent RPS for RPS/NJCT blends
Source: "Recycle PS, Add Value to Commingled Products," Nosker, Renfree, Morrow, *Plastics Engineering* (Feb. 1990)

To ascertain whether or not the consistent and good properties of these blends were limited to compressive tests, the CPRR subjected the RPS/NJCT profiles to three-point bending in order to measure the flexural properties. Results show flexural moduli follow the same trend as the compressive, only at higher magnitude [46]. For example, at 100% NJCT, the flexural modulus is 171×10^3 psi (1178 MPa), while at 35/65 RPS/NJCT it is 315×10^3 psi (2169 MPa), an increase of 84%. To show the relation between the measured properties and the associated

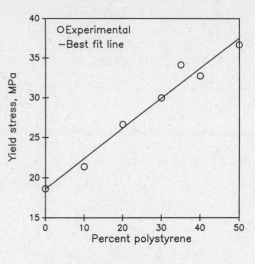

Figure 9.6   Compressive yield stress as a function of percent RPS for RPS/NJCT blends
Source: "Recycle PS, Add Value to Commingled Products," Nosker, Renfree, Morrow, *Plastics Engineering* (Feb. 1990)

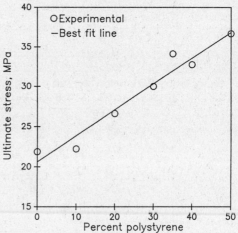

Figure 9.7   Compressive ultimate stresss as a function of percent RPS for RPS/NJCT blends
Source: "Recycle PS, Add Value to Commingled Products," Nosker, Renfree, Morrow, *Plastics Engineerin,*g (Feb. 1990)

microstructures/morphologics for the commingled lumber profiles, two principal techniques were used: positron lifetime spectroscopy and scanning electron microscopy.

9.5.3   Microstructure of RPS/NJCT Blends: Positron Lifetime Studies

Positron annihilation techniques have been used for some time to study the microstructure of molecular substances [47, 48]. Positrons are emitted from a radioactive source and then, after a time (lifetime) on the order of 0.1–5.0 ns, undergo mutual annihilation with electrons in the sample. Measured lifetimes depend upon the electron density of the material, thereby yielding microstructural information. In this study of lumber profiles made from RPS/NJCT blends, measured positron lifetimes were used to estimate: (1) the degree of polymeric homogeneity throughout the extent of the tested profiles; (2) the relative percent crystallinity of PE phase as a function of percent PS; and (3) the size and number of microvoids, a measure of free volume on a molecular level, within the samples [49–51]. These applications are discussed in turn below.

Figure 9.8 shows the measured longest lifetimes (which reflect annihilations occurring exclusively within the polymeric microstructure) of the central, inner region, or "core," in direct comparison with those of the outer region, or "skin," as a function of the percent RPS;

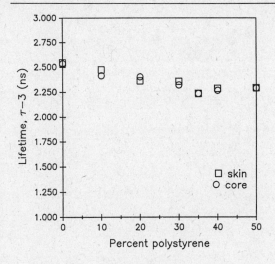

Figure 9.8   Longest lifetimes for outer "skins" and inner "cores" as a function of percent RPS for RPS/NJCT blends
Source: *Phys. Stat. Sol. (a)*, *124*, 67 (1991)

Figure 9.9 shows a comparison of the corresponding intensities for these samples. These results indicate that, on average, the polymeric microstructure of the porous cores is virtually identical to that of the more dense skin, implying that there is little or no migration or segregation of the polymeric components throughout the extent of the samples. Measurements which were taken from different profiles showed little variation in observed lifetimes and intensities, certainly within the limits of experimental error.

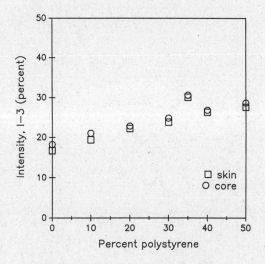

Figure 9.9 · Intensities of longest lifetimes for outer "skins" and inner "cores" as a function of percent RPS for RPS/NJCT blends
Source: *Phys. Stat. Sol. (a)*, *124*, 67 (1991)

For polymers that crystallize, an estimate of the percent crystallinity is obtained by comparing the number of annihilations that occur in ordered regions with the number occurring in amorphous regions. For binary, incompatible blends, where one component can crystallize and the other cannot, it is possible to subtract out the contributions due to the latter, thereby obtaining an estimate of percent crystallinity for the former. Doing this for the RPS/NJCT blends and assuming for simplicity that the NJCT is 100% PE, the percent crystallinity for the PE phase was found, with the notable exception of the 35/65 RPS/NJCT mixture, to remain fairly constant as a function of percent PS (Figure 9.10). At 35% RPS, the anomalous, decreased ability of the PE phase to crystallize was confirmed by wide-angle X-ray diffraction (Figure 9.11) and differential scanning calorimetry (Figure 9.12), both studies in striking agreement with the positron data.

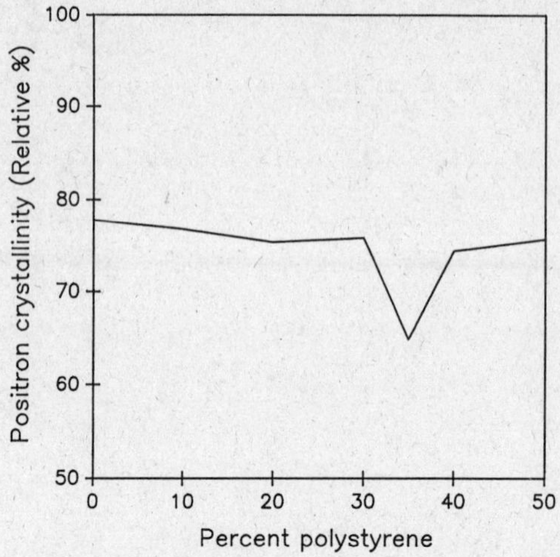

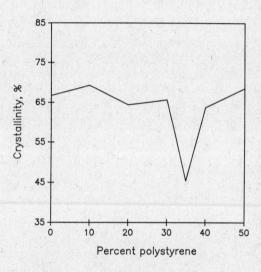

Figure 9.10   Positron crystallinity of the PE phase as a function of percent RPS for RPS/NJCT blends
Source: *Phys. Stat. Sol. (a)*, *124*, 67 (1991)

Figure 9.11   X-ray crystallinity of the PE phase as a function of percent RPS for RPS/NJCT blends
Source: *Phys. Stat. Sol. (a)*, *124*, 67 (1991)

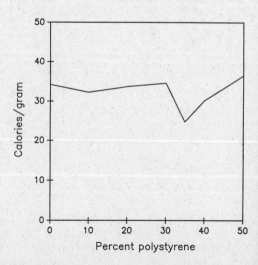

Figure 9.12   DSC results in calories/g for the PE phase as a function of percent RPS for RPS/NJCT blends

Following a quantum mechanical model [52, 53], positron spectroscopy was used to estimate the number and size of microvoids (0.6–1.0 nm in diameter) as a function of percent RPS (Figure 9.13, Table 9.9). The line in Figure 9.13 is calculated from a linear weighted sum of contributions due to the pure components and, as such, represents the relation expected for components that do not interact appreciably with each other. On the other hand, the experimentally determined radii shown lying below the line in Figure 9.13 show that these blends are more tightly packed on a molecular level, indicating a good degree of effective adhesion between the phases. The most densely packed composition, 35% RPS, corresponds with both the anomalous decrease in the percent crystallinity of the PE phase and the increase in the compressive modulus at this composition. The phenomena at 35% RPS are not understood at this time.

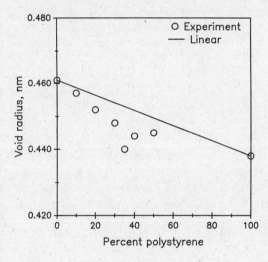

Figure 9.13   Microvoid radius from positron annihilation as a function of percent RPS for RPS/NJCT blends

Source: *Phys. Stat. Sol. (a)*, *124*, 67 (1991)

Table 9.9   Size and number/volume of microvoids as a function of percent RPS for RPS/NJCT blended materials

Composition PS/NJCT	R (nm)	ΔN (rel. units)
0/100	0.461	0.00
10/90	0.457	1.62
20/80	0.452	3.29
30/70	0.448	4.25
35/65	0.440	8.02
40/60	0.444	5.75
50/50	0.445	6.53
0/100	0.438	11.20

Note: R denotes average size of radii, N the number/unit volume

In order to study further the sizes and distributions of the constituent phases, several of the samples were fractured at both room temperature and liquid nitrogen temperatures, and the resulting surfaces were observed with an ISI-40 scanning electron microscope (SEM) [54, 55].

9.5.4   Scanning Electron Microscopy

Capitalizing upon the macroscopic features of the profiles (see Section 9.4.3), it is convenient to superpose a cylindrical coordinate system over the cross section of the profile to describe spatially the principal morphological features of the commingled products. As Figure 9.14 shows, the z-axis is superimposed upon the long, central axis of the product, parallel to the direction of melt flow down the mold cavity. The r or radial direction is perpendicular to z and is directed outwardly from the centroid of the cross section. Finally, the θ or circumferential direction lies in the plane of the section and is at right angles to the radial direction.

In the discussion which follows, we describe fracture surfaces which are perpendicular to the z, θ, and r directions (Figure 9.14), lying in the θr, rz, and zθ planes, respectively. These surfaces are derived from test bars, 5/16 × 5/16 × 3/4 inch (0.79 × 0.79 × 1.9 cm) in size, sectioned from the outer, more dense regions of the molded products (Figure 9.15). The test bars were notched to better control the impact fractures, which were performed at both room temperature and at the temperature of liquid nitrogen. Samples were then mounted, gold-coated, and examined using an ISI-40 SEM.

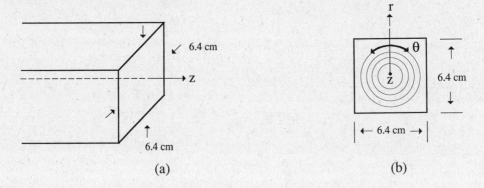

(a)   (b)

Figure 9.14   Superposed cylindrical coordinate system used to analyze the microstructure of commingled products

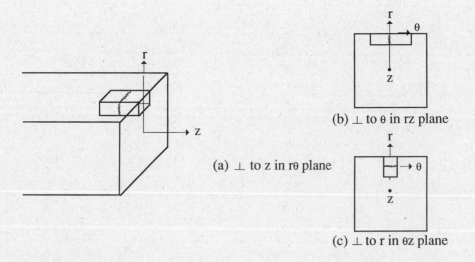

(a) ⊥ to z in rθ plane

(b) ⊥ to θ in rz plane

(c) ⊥ to r in θz plane

Figure 9.15   Diagram showing the orientation of test bars and fracture surfaces for examination under the scanning electron microscope

From 10 to 30 wt % RPS, perpendicular to z, a ductile matrix of PE is seen to surround islands of circular/elliptical particles of RPS (Figure 9.16).   The number of dispersed particles increases with concentration.   Also, some of the larger particles appear to become more elliptical with increasing RPS.   Typically, the dispersions vary in diameter from 1 to 5 μm.

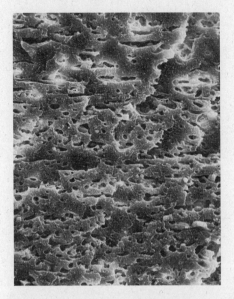

Figure 9.16   Scanning electron micrograph for sample of 30/70 RPS/NJCT, fractured at the temperature of liquid nitrogen. The fracture surface is perpendicular to the z-axis. Toluene was used to extract the dispersed PS phase, which appears as dark, circular and elliptical holes, representing cross sections of drawn fibers.
Reprinted by permission from *Nature*, Vol. 350, p. 563. Copyright 1991 MacMillan Magazines Ltd.

At 35 wt % RPS, still perpendicular to z, the morphology is very uniform, with both polymers forming continuous phases.   Moreover, the phases are entangled or interlocked, forming a three-dimensional interpenetrating network (Figure 9.17).   This morphology appears to be related to the optimum value for the compressive modulus.

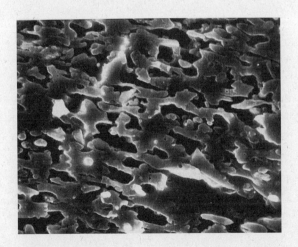

Figure 9.17   Scanning electron micrograph for sample of 35/65 RPS/NJCT fractured at the temperature of liquid nitrogen. The fracture surface is perpendicular to the z-axis. Toluene was used to extract the PS, shown here as the darker phase. A co-continuous, entangled morphology is indicated.
Reprinted by permission from *Nature*, Vol. 350, p. 563. Copyright 1991 MacMillan Magazines Ltd.

At 40% RPS, perpendicular to z in the $r\theta$ plane, the phases are still entangled, but the PE and RPS components are not nearly as uniformly distributed as in the case of the 35/65 formulation. In particular, the PE component is noticeably thin, almost ribbon-like in many areas, while the RPS domains appear larger in extent (Figure 9.18).

Finally, at 50% RPS, the PS phase is the matrix and HDPE the dispersed phase, consisting of circular/elliptical fibers and plates (Figure 9.19).

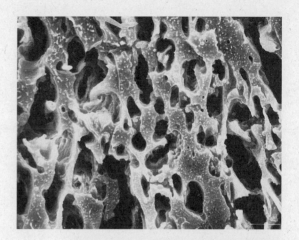

Figure 9.18 Scanning electron micrograph for sample of 40/60 RPS/NJCT fractured at the temperature of liquid nitrogen. The fracture surface is perpendicular to the z-axis. Toluene was used to extract the PS, shown here as the darker phase. The morphology is less uniform than 30/70 and 35/65 RPS/NJCT.
Reprinted by permission from *Nature*, Vol. 350, p. 563. Copyright 1991 MacMillan Magazines Ltd.

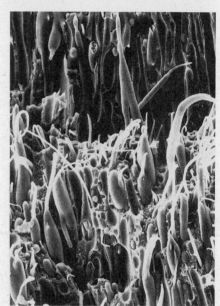

Figure 9.19   Scanning electron micrograph for sample of 50/50 RPS/NJCT fractured at room temperature. The view is at 45° to the z-axis. No solvent extraction was applied. PS is the matrix which contains numerous rods/plates of NJCT.
Source: Original figure appeared in a paper presented at the 49th ANTEC of The SPE, Montreal, Canada (May 1991)

Fracture surfaces perpendicular to θ in the rz plane show structures increasingly lamellar with increasing concentration of RPS, indicating that the second phase particles are in the form of rods/plates, which have long axes parallel to the z-axis and are interspersed regularly in the matrix (Figure 9.20). Fracture surfaces perpendicular to r in the θz plane emphasize both lamellar structure and cylindrical symmetry, generally showing a smooth surface of the more brittle RPS for compositions of RPS equal to or greater than 10% (Figure 9.21).

9.5.5 Implications of the Structure-Property Relations for RPS/NJCT Blends

Blends of virgin polyethylene (VPE) and virgin polystyrene (VPS) have been studied extensively in the laboratory [56–60]. It is stated in the literature that blends of PS and PE invariably yield poor mechanical properties, often incorporating the worst properties of the homopolymers [61]—low elongation at break (PS) and low modulus, yield stress, and tensile stress (PE).

Research at the CPRR, on the other hand, shows something quite different for lumber profiles made from the blending of NJCT with a specific type of RPS. Both mechanical testing

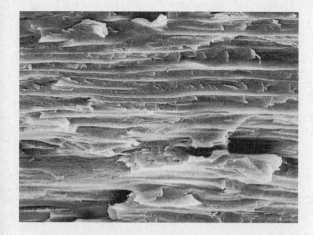

Figure 9.20   Scanning electron micro-craph for sample of 35/65 RPS/NJCT fractured at the temperature of liquid nitrogen. The fracture surface is perpendicular to the θ-direction. No solvent extraction was applied. Alternating sheets/plates of RPS and NJCT are shown.
Reprinted by permission from *Nature*, Vol. 350, p. 563.   Copyright 1991 MacMillan Magazines Ltd.

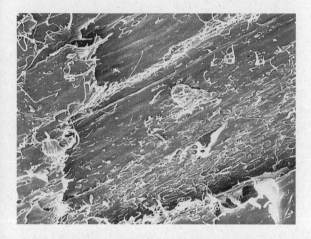

Figure 9.21   Scanning electron micro-craph for sample of 35/65 RPS/NJCT fractured at room temperature. The fracture surface is perpendicular to the r-direction. No solvent extraction was applied. The surface is a smooth sheet of PS.
Reprinted by permission from *Nature*, Vol. 350, p. 563.   Copyright 1991 MacMillan Magazines Ltd.

and morphological analysis indicate that samples containing from 0–40% RPS act and look like fiber-reinforced composites. Measured compressive moduli are equal to or greater than values calculated from the linearly weighted sum of contributions due to the pure components for compositions up to 40% RPS [62–64]. Lending strong support to this theory are scanning electron micrographs which identify numerous PS fibers running parallel to the long axes of profiles with these compositions [65].

The anomalous strengthening at 35% RPS seems to be associated with a region of dual phase co-continuity, wherein the phases are interlocked with each other. This type of morphology, rarely cited in the literature, is poorly understood [66].

Attempts by the CPRR to duplicate these findings by blending NJCT with densified, postconsumer RPS obtained from different sources have not been successful to date [67]. That is, in comparison with blends made using Mobil's densified regrind PS, lumber profiles processed in the ET/1 and using other types of RPS are characterized mechanically by relatively low strength and morphologically by inhomogeneous, coarse dispersions, in agreement with the studies of virgin PS/PE blends. Moreover, mechanical tests indicate that the densified, postconsumer RPS has been seriously degraded [68].

Nonetheless, while it is not as yet understood why particular RPS/NJCT blends have superior properties/morphologies, the results of these studies suggest strongly that composite materials with good mechanical properties can be fabricated from recycled blends made from properly selected and processed materials, even from incompatible polymer pairs. Additional studies will be required to determine the effect of processing and material parameters upon the desired end product. In the next section, the most recent studies which may lead to new commingled processes are described.

9.6   New Commingled Processes

During the past year, numerous publications have appeared, describing studies and processes that are candidates forming the foundation for the continued research and development of commingled plastics recycling in the 1990s. Generally, these processes can be categorized as either mechanical/physical or chemical in nature, depending upon the method of separation or refinement (if applicable) of the mixed plastics waste to be used by the processor. These processes are described below.

9.6.1   Mechanical/Physical Commingled Processes

Researchers have recently reported on the addition of modifiers to blends of PET and HDPE in order to reduce brittleness and increase impact strength. One study by Union Chemical Laboratories used virgin polymers which were selected to have physical properties as close as possible to those used in the manufacture of the PET bottle [69], while a study by Exxon Chemical Co. used soft drink bottles obtained from the MSW stream [70].

A study by Quantum Chemical Corporation [71, 72] reports that recycled, extrusion-grade HDPE, contrary to what might be expected, shows no loss in flexural modulus, low temperature brittleness, tensile strength at yield, or tensile strength at break; and that "postconsumer (washed and pelletized) HDPE resins exhibit adequate processibility to be used in a variety of extrusion processes." On the negative side, recycled resins in comparison with virgin controls showed considerable losses in elongation at break, moderate losses in tensile impact, unfavorable odors, and unacceptable levels of contamination which resulted in "blow holes" in thin-walled bottles. Efforts are under way to reduce contamination and odor levels utilizing washing, extrusion, and stabilization methods. (More details of this study are given in Chapter 4.)

A group of researchers at Rensselaer Polytechnic Institute (RPI) reports that compatibilization and intensive mixing via the use of an intermeshing twin-screw extruder improves some of the properties of recycled commingled pellets made from postconsumer plastics wastes [73]. They tested model blends of both virgin and postconsumer plastics waste. Based upon the approximate percentages of homopolymers used in the manufacture of rigid plastic containers and plastics used in packaging (exclusive of PET and PVC), the model blends consisted of 70% HDPE, 15% PS, 11% PP, and 4% LDPE (branched). These numbers represent more closely the percentages for plastics used to manufacture plastic containers as was shown in Table 9.3 and should not be confused with the composition of municipal mixed plastic wastes that are currently being collected (see Section 9.2.2).

The study at RPI reports a significant loss of 30% in the tensile modulus for the waste blend relative to the virgin blend, both of which were intensively mixed, suggesting that the waste blend was degraded during the mixing process. Addition of the most effective compatibilzer, a PS-poly(ethylene-butylene)-PS block copolymer, increased the tensile strain to failure by over five-fold but further decreased the tensile modulus by 14%. Future action planned by this group involves impact testing and morphological analyses using optical microscopy. The goal of the project is to produce high-valued, thin-walled products from pellets derived from the postconsumer plastics waste stream.

Since the fall of 1989, the CPRR at Rutgers has been designing a system which uses mechanical/physical processes to refine a multicomponent plastic waste stream and to blend the refined matrix with other thermoplastics. The overall objective of this refined commingled technology is to determine the range of heterogenous mixtures (commingled blends, alloys and compounds) that possess a range of optimized properties and are compatible with conventional distribution channels (and processing methods). The resultant family of postconsumer recycled polymer feedstocks can, in turn, be used to produce high value-added and long life cycle products at many plastics manufacturing locations, both domestic and foreign. Successful implementation of this technology would expand the feedstock base for both generic resin and commingled reclamation. A pilot study, conducted jointly by the CPRR and Welding Engineers, Inc. (WEI), is described in some detail below. Unless otherwise noted, this material is condensed from references [74] and [75].

9.6.2 CPRR/WEI Refined Commingled Technology

For this research, the resin recovery plant at Rutgers was utilized to obtain a cleaned, polyolefin fraction of NJCT (also see Section 9.4), hereinafter referred to as refined NJCT (RNJCT). A schematic of the process is shown in Figure 9.22. Once processed through the resin recovery plant, the dry flake RNJCT contain typically >95 wt % PE, <5% PP, and <1% of other materials. These materials were subsequently dry-blended and melt processed using a WEI nonintermeshing twin-screw extruder (NITSE). The RNJCT were blended separately with a postconsumer green PET from curbside collection and a postconsumer black PS from Amoco Chemical Co. in the formulation ratios shown in Table 9.10. In addition, for comparative purposes, a virgin HDPE injection grade was used. The resins were molded at 177 °C and 271 °C, temperatures above and below the melting point of PET, into standard ASTM D-638 test bars for evaluation of flexural, tensile, and notched Izod impact properties.

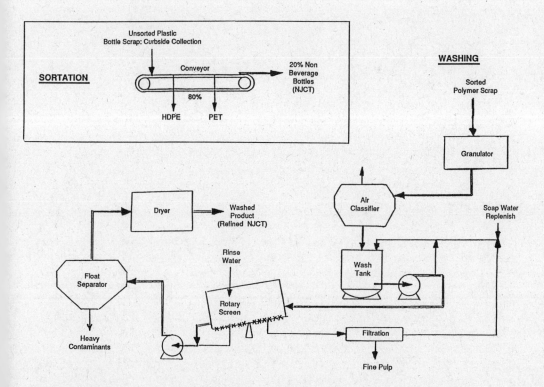

Figure 9.22   Resin recovery process used by CPRR to obtain a cleaned, polyolefin fraction of NJCT
Source: Original figure appeared in a paper presented at the 49th ANTEC of The SPE, Montreal, Canada (May 1991)

Table 9.10   Sample formulations for CPRR/WEI study

Sample Number	Formulation		
	Refined NJCT (%)	Postconsumer Black PS (%)	Postconsumer Green PET (%)
1	100	0	0
2	90	10	0
3	70	30	0
4	95	0	5
5	75	0	25

Source: Original table appeared in a paper presented at the 49th ANTEC of The SPE, Montreal, Canada (May 1991)

The WEI facility used a 51-mm diameter nonintermeshing twin-screw extruder (NITSE), shown in Figure 9.23, to melt, compound, degas, filter, and pelletize the polymers. This process is shown schematically in Figure 9.24. Recent studies indicate that there are certain advantages of the NITSE over the corotating intermeshing twin-screw extruder (i.e., polymer melts are uniformly mixed with minimum degradation) [76]. Features of the NITSE are shown schematically in Figure 9.25.

Figure 9.23   Photo of Welding Engineers Inc. counter rotating nonintermeshing twin-screw extruder
Reprinted with permission from Welding Engineers Inc.

Contaminants not removed during the flotation step were trapped via use of a WEI-brand Kleen Screen™ melt filter. This melt filter is an in-line screen designed so that a small portion of the flow through one screen can backwash the other within seconds. Solids trapped by the Kleen Screen included unmelted polymer flakes, steel caps, and aluminum cap rings. This step is thought to further improve the properties of the end product, since it has been shown that higher levels of contamination result in loss of some physical properties [77].

Vacuum venting of the gases served to minimize void content and odor and smoke levels in the product. The release of volatiles is known to be facilitated by renewal of the surface of a melt, which was enhanced by cross channel flow induced by staggered screw flights in the vent section. For a further discussion of mixing and devolatilization in NITSEs, please see [78] and [79], respectively.

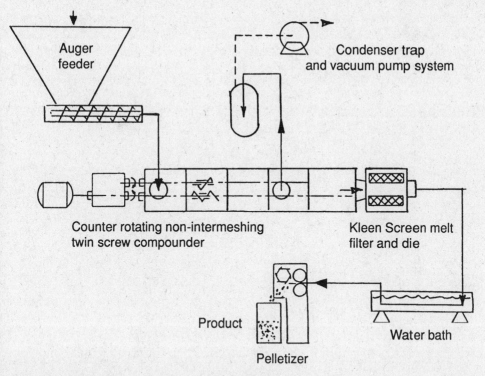

Figure 9.24   Welding Engineers facility: process schematic
Source: Original figure appeared in a paper presented at the 49th ANTEC of The SPE, Montreal, Canada (May 1991)

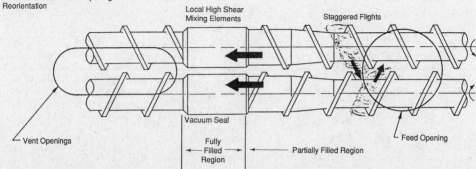

Figure 9.25   Features of the Welding Engineers Inc. counter rotating nonintermeshing twin-screw extruder
Source: Original figure appeared in a paper presented at the 49th ANTEC of The SPE, Montreal, Canada (May 1991)

Table 9.11 shows the measured physical properties of sample formulations and those of a virgin HDPE. Figures 9.26 through 9.29 are bar graphs of the same data presented in a different format for ease of comparison. Tables 9.12 and 9.13 show these data standardized as percent differences over virgin HDPE and RNJCT, respectively. In the bar graphs, labels have been placed atop each bar to indicate the temperatures at which the samples were molded. It is believed that a high molding temperature produces a high level of crystallinity and has the effect of annealing the test bars. This explanation is substantiated for the virgin HDPE by higher tensile and flexural properties and a decrease in impact energy absorption. In many cases, the floatable RNJCT, molded at the higher temperatures, did not exhibit an increase in physical property values, reflecting some degradation. It is important to note that in all cases the measured physical properties of the RNJCT were as high as or higher than those of the virgin HDPE. In fact, the impact properties of the RNJCT showed a 78% improvement over the virgin material molded under similar conditions.

Table 9.11   Physical properties of refined commingled plastics processed through CPRR and WEI facilities [a]

Sample Number	Mold Temp °C	Flexural Modulus 10³ psi (MPa)		Flexural Strength psi (MPa)		Tensile Strength psi (MPa)		Izod Impact ft-lb/in (J/m)	
VHDPE[b]	177	226	(1557)	3016	(20.8)	2813	(19.4)	1.48	(79.6)
VHDPE	271	247	(1701)	3147	(21.7)	2915	(20.1)	1.32	(70.5)
1	177	234	(1612)	3248	(22.4)	3118	(21.5)	2.65	(141.5)
1	271	247	(1701)	3161	(21.8)	2973	(20.5)	2.10	(112.1)
2	177	289	(1991)	3669	(25.3)	3234	(22.3)	1.24	(66.2)
3	177	323	(2225)	4423	(30.5)	3234	(22.3)	0.54	(28.8)
4	177	263	(1812)	3466	(23.9)	3002	(20.7)	1.57	(83.8)
4	271	261	(1798)	3509	(24.2)	3176	(21.9)	1.91	(102.0)
5	177	326	(2246)	4814	(33.2)	3495	(24.1)	0.85	(45.5)
5	271	326	(2246)	4220	(29.1)	3074	(21.2)	0.97	(51.8)

[a] Values reported are the average of five with ranges in the order of 2%; [b] VHDPE = Virgin HDPE
Source: Original table appeared in a paper presented at the 49th ANTEC of The SPE, Montreal, Canada (May 1991)

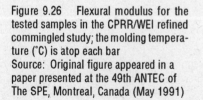

Figure 9.26   Flexural modulus for the tested samples in the CPRR/WEI refined commingled study; the molding temperature (°C) is atop each bar
Source: Original figure appeared in a paper presented at the 49th ANTEC of The SPE, Montreal, Canada (May 1991)

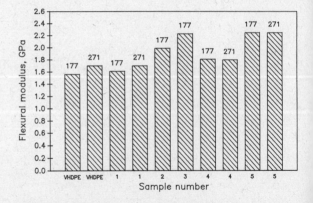

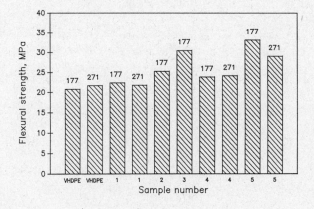

Figure 9.27   Flexural strength for the tested samples in the CPRR/WEI refined commingled study; the molding temperature (°C) is atop each bar
Source: Original figure appeared in a paper presented at the 49th ANTEC of The SPE, Montreal, Canada, May 1991

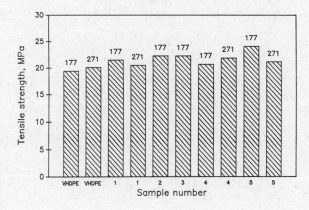

Figure 9.28   Tensile strength for the tested samples in the CPRR/WEI refined commingled study; the molding temperature (°C) is atop each bar
Source: Original figure appeared in a paper presented at the 49th ANTEC of The SPE, Montreal, Canada (May 1991)

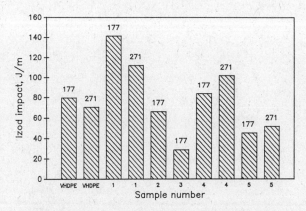

Figure 9.29   Izod notched impact for the tested samples in the CPRR/WEI refined commingled study; the molding temperature (°C) is atop each bar
Source: Original figure appeared in a paper presented at the 49th ANTEC of The SPE, Montreal, Canada (May 1991)

The addition of 10% recycled PS to RNJCT resulted in significant improvements in flexural properties, while impact properties dropped to approximately the levels of virgin HDPE. Addition of 30% PS resulted in a further trade-off of flexural for impact properties.

While similar trends were observed for additions of PET to RNJCT, here it is important to consider the molding temperature. During the extrusion process, after pelletizing, it was observed that some PET strands were oriented in the pellet. It is assumed that these fibers remained solid at the lower molding temperature, orienting in a random fashion, while at the higher temperature, they melted and flowed. Scanning electron microscopy was used to check this assumption.

Table 9.12   A comparison of the physical properties of RNJCT formulations to virgin HDPE

Sample Number	Mold Temp °C	Flexural Modulus % Diff.	Flexural Strength % Diff.	Tensile Strength % Diff.	Izod Impact % Diff.
NJCT (1)	177	4	8	11	78
NJCT (1)	271	0	0	2	59
2	177	28	22	15	-17
3	177	43	47	15	-64
4	177	16	15	7	5
4	271	6	11	9	45
5	177	44	60	24	-43
5	271	33	34	5	-27

Table 9.13   Percent changes in physical properties of RNJCT with various recycled plastic additions

Addition (Sample #)	Mold Temp °C	Flexural Modulus % Diff.	Flexural Strength % Diff.	Tensile Strength % Diff.	Izod Impact % Diff.
10% PS (2)	177	24	13	4	-53
30% PS (3)	177	38	36	4	-80
5% PET (4)	177	12	7	-4	-41
5% PET (4)	271	6	11	7	-9
25% PET (5)	177	39	48	12	-68
25% PET (5)	271	32	34	3	-54

Micrographs were taken of surfaces fractured by impact at both room temperature and the temperature of liquid nitrogen. Samples were fractured along surfaces both perpendicular and parallel to the direction of melt flow. For brevity, only micrographs of the PET/RNJCT blends are presented here.

For samples fractured at the temperature of liquid nitrogen, Figures 9.30 through 9.33, a leaching technique [80] was used to extract the PET phase, which appears as the darker phase. Samples of 25% PET molded at 177 °C are shown in Figures 9.30 and 9.31.   The view

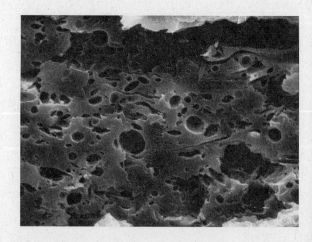

Figure 9.30   Scanning electron micrograph for sample of 25/75 RPS/NJCT, molded at 177 °C and fractured at the temperature of liquid nitrogen. The fracture surface is perpendicular to the direction of melt flow. Solvent extraction was used to remove the PET, which is the darker phase. A nonuniform, coarse distribution of PET particles is shown.

perpendicular to the melt flow shows a nonhomogeneous distribution of cross sections of dispersed particles of PET (Figure 9.30). As anticipated, a parallel view (Figure 9.31) shows that the PET is not drawn into fibers.

In contrast, samples of 25% PET molded at 271 °C show a homogeneous distribution of fibers, less than 3 μm in diameter and greater than 20 μm in length (Figures 9.32–9.33). The

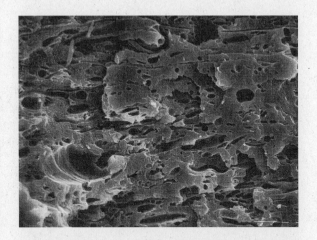

Figure 9.31 Scanning electron micrograph for sample of 25/75 RPS/NJCT, molded at 177 °C and fractured at the temperature of liquid nitrogen. The fracture surface is parallel to the direction of melt flow. Solvent extraction was applied. A nonuniform, coarse distribution of PET particles is shown. There are no drawn fibers.

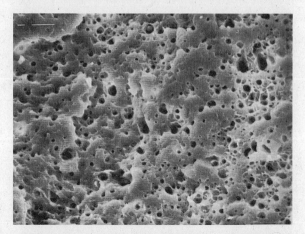

Figure 9.32 Scanning electron micrograph of a sample of 25/75 PET/RNJCT, molded at 271 °C and fractured at the temperature of liquid nitrogen. The fracture surface is perpendicular to the direction of melt flow. Solvent extraction was applied. Numerous cross sections of uniformly distributed PET fibers from 1 to 3 μm in diameter are shown.
Source: Original figure appeared in a paper presented at the 49th ANTEC of The SPE, Montreal, Canada (May 1991)

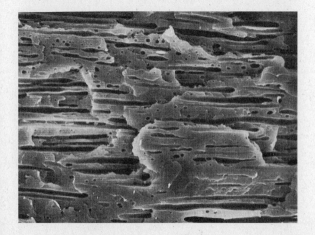

Figure 9.33 Scanning electron micrograph of a sample of 25/75 PET/RNJCT, molded at 271 °C and fractured at the temperature of liquid nitrogen. The fracture surface is parallel to the direction of melt flow. Solvent extraction was applied. Lengthwise cross sections of uniformly distributed PET fibers are visible.
Source: Original figure appeared in a paper presented at the 49th ANTEC of The SPE, Montreal, Canada (May 1991)

lack of improvement of tensile and flexural properties for this sample over the nonfibrous sample is due to the lack of any appreciable bonding between the PET and RNJCT matrix. This absence of bonding is shown in Figure 9.34, a view perpendicular to melt flow for a 25% PET sample fractured at room temperature. Here no solvent extraction was applied, and the PET fibers appear as the lighter of the phases.

Figure 9.34   Scanning electron micrograph of a sample of 25/75 PET/RNJCT, molded at 271 °C and fractured at room temperature. The fracture surface is perpendicular to the direction of flow. No solvent extraction was applied. Fibers of PET (lighter phase) are not bonded to the RNJCT matrix.
Source: Original figure appeared in a paper presented at the 49th ANTEC of The SPE, Montreal, Canada (May 1991)

Blends of recycled PS and RNJCT were characterized by a uniform distribution of PS fibers, oriented parallel to the direction of melt flow. Diameters of fibers were from 1 to 5 μm, and the lengths of these ranged from 10 to 20 μm.

The CPRR/WEI combined system of washing, float separation, degassing, and melt filtration produced recycled composite materials with good mechanical properties, suggesting that there was a minimum of polymer degradation. It is also possible that the measured properties were enhanced by the presence of wood fibers and other mixed polymer additives. Compounding refined commingled materials with compatibilizing agents such as silanes, maleated polyolefins, and organometallic titanates and zirconates, something not tried in the pilot study, could result in additional improvement of the product.

A preliminary economic analysis [81] for 20- and 40-million-pound-per-year (9 and 18×10^3 t) processing plants for the base compounds of extrusion grade polymer materials are encouraging (see Tables 9.14 and 9.15). Total costs range from 7.8–12.8¢/lb (17–28¢/kg) for a 20-million-pound-per-year (9×10^3 t) operation to 6–11¢/lb (13–24¢/kg) for a 40-million-pound-per-year (18×10^3 t) operation, depending upon the cost of the feedstock. The reported estimate to compound/pelletize is based upon operational costs.

Table 9.14   Prelimiinary economics of sorting/reclaiming/compounding system (CPRR): 20-million-pound-per-year (9×10^3 t) capacity plant

Category/Segment	Cost, ¢/lb (¢/kg)							
Feed Stock Cost	-2.5	(-5.5)	0		+1.0	(+2.2)	+2.5	(+5.5)
Sort/Granulate Cost	+1.5	(+3.3)	+1.5	(+3.3)	+1.5	(3.3)	+1.5	(+3.3)
Reclaim Cost	+4.0	(+8.8)	+4.0	(+8.8)	+4.0	(+8.8)	+4.0	(+8.8)
Compound/Pelletize Cost	+4.8	(+10.6)	+4.8	(+10.6)	+4.8	(+10.6)	+4.8	(+10.6)
TOTAL	+7.8	(+17.2)	+10.3	(+22.7)	+11.3	(+24.9)	+12.8	(+28.2)
Selling Price (50% M/U*)	+11.7	(+25.7)	+15.4	(+33.9)	+16.9	(+37.2)	+19.2	(+42.2)
Selling Price (100% M/U)	+15.6	(+34.3)	+20.6	(+45.3)	+22.6	(+49.7)	+25.6	(+56.3)

* M/U = Mark Up

Table 9.15   Prelimiinary economics of sorting/reclaiming/compounding system (CPRR): 40-million-pound-per-year (18 × 10³ t) capacity plant

Category/Segment	Cost, ¢/lb (¢/kg)							
Feed Stock Cost	-2.5	(-5.5)	0		+1.0	(+2.2)	+2.5	(+5.5)
Sort/Granulate Cost	+1.8	(+4.0)	+1.8	(+4.0)	+1.8	(+4.0)	+1.8	(+4.0)
Reclaim Cost	+3.2	(+7.0)	+3.2	(+7.0)	+3.2	(+7.0)	+3.2	(+7.0)
Compound/Pelletize Cost	+3.5	(+7.7)	3.5	(+7.7)	+3.5	(+7.7)	+3.5	(+7.7)
TOTAL	+6.0	(+13.2)	+8.5	(+18.7)	+9.5	(+20.9)	+11.0	(+24.2)
Selling Price (50% M/U*)	+9.0	(+19.8)	+12.8	(+28.2)	+14.3	(+31.5)	+16.5	(+36.3)
Selling Price (100% M/U)	+12.0	(+26.4)	+17.0	(+37.4)	+19.0	(+41.8)	+22.0	(+48.4)

* M/U = Mark Up

9.6.3   Chemical Process: Selective Dissolution

At Rensselaer Polytechnic Institute a chemical process for recycling commingled plastics is underway. This process, called selective dissolution, does not produce a commingled plastic product, but rather separates the mixed plastics waste stream by homopolymer. The following discussion consists of excerpts from sources [82, 83] that describe an overview of the process itself, the experimental procedure, experimental results, and an economic analysis of a representative process.

The selective dissolution method takes advantage of the incompatibility between most pairs of polymers in order to separate the components of a mixed plastics waste stream. Two technologies, selective dissolution and low solids flash devolatilization, are basic to this process, which has evolved naturally from previous work on the impact modification of polymers [84, 85]. Briefly, this is where two incompatible polymers are dissolved in a common solvent to form a homogeneous mixture, followed by phase separation due to rapid removal of the solvent by flash devolatilization. During devolatilization, the polymer/polymer/solvent mixture plunges deeply into a region of two phases, resulting in phase separation by spinodal decomposition and generating a microdispersion. If one of the phase volumes is small, a uniformly sized dispersion results.

Figure 9.35 shows the basic flow diagram of the reclamation process, starting with the separation treatment of the shredded mixed plastic stream by selective dissolution, followed by filtration, blending, and low solids devolatilization and compositional quenching. The separation process involves using a controlled sequence of solvents and solvation temperatures. The dissolution process is rapid at low polymer concentrations (10% by weight or less), which in turn makes filtration feasible due to low viscosity. Filtration removes insoluble contaminates such as metals, glass, cellulose, and some pigments. After filtration, the polymer solution may be blended with additives such as stabilizers and impact modifiers to add value to the product.

The apparatus used for this reclamation process is shown schematically in Figure 9.36. The selective dissolution apparatus performs the first stage of the reclamation. The shredded commingled plastic wastes, typically 5.5 lb (2.5 kg), are placed in the dissolution column with screen packs at both ends. A gear pump is used to circulate solvent, typically two liters, through the heat exchanger and dissolution column. Selective dissolution is performed in the sequential batch mode. A unique temperature for a given solvent is used to selectively extract a single polymer from the commingled stream. The polymer thus obtained is isolated via flash devolatilization. The recovered polymer is dried in an oven at reduced pressure, ground in a mill, and then redried. Meanwhile, the recovered solvent is returned to the dissolution reservoir to extract the next polymer at a higher temperature. It is this series of controlled temperature-solvent extractions that separates the individual polymers from a commingled waste stream.

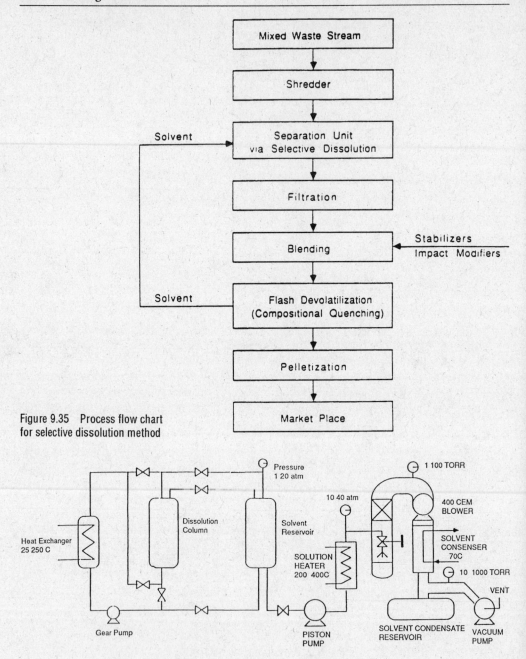

Figure 9.35   Process flow chart for selective dissolution method

Figure 9.36   Reclamation apparatus for selective dissolution method

The second phase, low solids flash devolatilization, is a relatively simple process and is readily scalable. The polymer concentrations are typically 5–10% by weight. The feed solution is metered into a heat exchanger by a positive displacement pump. The pressure is maintained sufficiently high to ensure that no boiling occurs. This pressure is regulated by a back-pressure or flash valve and is constrained at a minimum value, typically 10–40 atm (1011–4053 kPa). The temperature upstream of the flash valve (200–300 °C) is one of the independent variables that governs the devolatilization step. The flash chamber pressure at 5–10 torr (0.66–1.32 kPa) is the other independent variable. Together they determine the polymer concentration after the flash, typically 60–95%, and the after flash temperature, typically 0–100 °C. To a first approximation, the flash achieves vapor-liquid equilibrium.

The first trial solvent used was tetrahydrofuran (THF), which obtained a four-way split when applied to a commingled polymer sample composed of virgin polymers commonly used in packaging: PVC, PS, PP, LDPE, HDPE, and PET. Table 9.16 illustrates the dissolution temperatures using THF as the dissolution solvent [86]. For this synthetic waste stream, the separation efficiencies are very good, usually >99%.

Table 9.16   Selective dissolution using tetrahyrofuran

First cut	:	Room Temperaure
PS		> 99%
PVC		> 99%
Second cut	:	70 °C
LDPE		>99%
Third cut	:	160 °C
HDPE		> 99%
PP		> 99%
Fourth cut	:	190 °C
PET		> 99%

Note: Experiments were conducted with virgin polymers; extraction percents were measured using a gravametric technique

The use of xylene as the solvent yields a six-way split. The dissolution temperatures using xylene are shown in Table 9.17 for a synthetic waste stream consisting of virgin polymers and in Table 9.18 for a sample of NJCT supplied by the CPRR [87]. Except for the HDPE, the percentages of the various polymers in the sample were too low to allow for the measurement of physical properties. Such measurements are being carried out for the HDPE.

Table 9.17   Selective dissolution using xylene

First cut	:	Room Temperaure
PS		>99%
Second cut	:	75 °C
LDPE		>99%
Third cut	:	105 °C
HDPE		>99%
Fourth cut	:	118 °C
PP		>99%
Fifth cut	:	138 °C
PVC		>99%
Sixth cut	:	> 142 °C
PET		>99%

Note: Experiments were conducted with virgin polymers; extraction percents were measured using a gravametric technique.

Separation experiments using xylene also have been performed on a small scale with wire and cable insulation [88], which consist primarily of PVC and LDPE, and on multilayered polymer samples. Physical property measurements were to be conducted on the separated polymers.

The preliminary economic analysis for this process [89] given in Tables 9.19 and 9.20 assumes that THF is the dissolution solvent, even though it is most likely not the solvent of

222   Commingled Plastics

Table 9.18   Selective dissolution using xylene: NJCT

Polymer	% Present	Temperature °C
PS	3.3	ROOM
LDPE	3.5	75
HDPE	75.3	105
PP	8.1	120
PVC	3.4	138
PET & Insolubles	6.4	—

Table 9.19   Manufacturing cost summary for selective dissolution method assuming commingled waste cost of 10¢/lb (22¢/kg)

Job Title	Polymer Reclamation Project		Cost Index Type / Cost Index Value	CE Index 346
Fixed Capital	$5,532,848			
Working Capital	$1,106,570			
TOTAL Capital Investment	$6,639,418			
		$/Year	$/lb	($/kg)
Manufacturing Expenses				
Direct				
Raw Materials		$6,245,110	0.1178	(0.2598)
Solvents		$151,652	0.0029	(0.0064)
Operating Labor		$768,600	0.0146	(0.0322)
Supervisory & Clerical Labor		$153,720	0.0029	(0.0064)
Utilities: Electric Power		$1,504,947	0.0284	(0.0626)
Maintenance & Repair		$640,885	0.0121	(0.0267)
Operating Supplies		$128,177	0.0024	(0.0053)
Laboratory Charges		$153,720	0.0029	(0.0064)
Patents & Royalties		$437,105	0.0082	(0.0182)
TOTAL Direct Expenses		$10,183,916	0.1922	(0.4240)
Indirect				
Overhead		$1,094,243	0.0207	(0.0456)
Local Taxes		$110,657	0.0021	(0.0046)
Insurance		$55,328	0.0010	(0.0022)
TOTAL Indirect Expenses		$1,260,228	0.0238	(0.0525)
TOTAL Manufacturing Expenses		$11,444,144	0.2160	(0.4765)
Depreciation		$553,285	0.0104	(0.0229)
General Expenses				
Administrative Costs		$273,561	0.0052	(0.0115)
Distribution & Selling Costs		$728,653	0.0137	(0.0304)
Research & Development		$364,401	0.0069	(0.0152)
Total General Expenses		$1,366,615	0.0258	(0.0571)
TOTAL EXPENSES		$13,364,044	0.2522	(0.5560)

Table 9.20   Manufacturing cost summary for selective dissolution method assuming zero cost for commingled waste feedstock

Job Title	Polymer Reclamation Project	Cost Index Type	CE Index
		Cost Index Value	346
Fixed Capital	$5,532,848		
Working Capital	$1,106,570		
TOTAL Capital Investment	$6,639,418		

	$/Year	$/lb	$/kg
Manufacturing Expenses			
Direct			
Raw Materials	$0	0.0000	0.0000
Solvents	$151,652	0.0029	0.0064
Operating Labor	$768,600	0.0148	0.0326
Supervisory & Clerical Labor	$153,720	0.0030	0.0066
Utilities: Electric Power	$743,322	0.0143	0.0315
Maintenance & Repair	$640,885	0.0123	0.0271
Operating Supplies	$128,177	0.0025	0.0055
Laboratory Charges	$153,720	0.0030	0.0066
Patents & Royalties	$437,105	0.0084	0.0185
TOTAL Direct Expenses	$3,177,181	0.0612	0.1349
Indirect			
Overhead	$1,094,243	0.0210	0.0463
Local Taxes	$110,657	0.0021	0.0046
Insurance	$55,328	0.0011	0.0024
TOTAL Indirect Expenses	$1,260,228	0.0242	0.0533
TOTAL Manufacturing Expenses	$4,437,409	0.0854	0.1882
Depreciation	$553,285	0.0106	0.0234
General Expenses			
Administrative Costs	$273,561	0.0053	0.0117
Distribution & Selling Costs	$728,653	0.0140	0.0309
Research & Development	$364,401	0.0070	0.0154
TOTALI General Expenses	$1,366,615	0.0263	0.0580
TOTAL EXPENSES	$6,357,309	0.1223	0.2696

choice due to its high cost and relatively high volatility. Because high pressure is required for the PET extraction, capital equipment costs are high. Also, environmental concerns require that all THF be recovered. Such an estimate then serves as the worst case scenario.

It is estimated that at least a 50-million-pound-per-year (22.5×10^3 t) plant is required to operate cost effectively. Based upon the use of only electrical power at 4.5¢/kwh, a cost of 10¢/lb (22¢/kg) for the commingled waste, an additional raw material cost for impact modifiers, and with no heat integration, a final product cost of 25.2¢/lb (55.6¢/kg) is achieved. The incorporation of heat integration and a zero raw material charge gives a processing cost of 12.2 ¢/lb (26.9¢/kg). Tables 9.19 and 9.20 detail the estimated cost per pound assuming a cost for the commingled feedstock of 10¢/lb (22¢/kg) and 0¢/lb (0¢/kg), respectively. The raw materials cost in Table 9.19 includes rubber used as an impact modifier. A more recent analysis, done by BFGoodrich Co., is in reasonable agreement (that is, 15% higher) with the RPI estimates [90].

9.6.4 A Physical/Chemical Approach

A unique approach to separating mixed plastics wastes was reported recently by Beckman of the University of Pittsburgh [91]. The method is based on the densities of the materials being separated but also upon the density of the medium, which can be varied over a wide range and controlled to a sensitivity of $\pm .001$ g/cm^3. The new process uses the unique properties of a fluid near its critical point to allow fine separations at mild temperatures and pressures. In the study, CO_2 and mixtures of CO_2 and SF_6 are being used. The initial work, done in a small-scale batch separation unit, showed that the major thermoplastics found in postconsumer waste streams could be separated cleanly from each other (i.e., PP, HDPE, PS, PET, and PVC). By effecting small incremental changes of pressure, pure CO_2 efficiently separated LDPE, HDPE, and PP. Separation of green PET, clear PET, and PVC also has been demonstrated, and separation of light- and dark-colored HDPE is possible. The different densities exhibited by PET in the neck and body of PET bottles can be separated by CO_2/SF_6 mixtures. The possibility of separating various components of wire and cable scrap also exists.

Work on this system is continuing. An automated semi-batch operation unit is under construction, and several configurations for a continuous unit have been proposed [92].

9.7 Global Activities

9.7.1 Japan

A 1985 publication [93] lists 25 municipalities that were attempting to process postconsumer plastic wastes into salable articles. However, a more recent communication [94] indicates that only one of these municipalities, Kusatsu City, remains somewhat active, producing flower pots. Since they have been unable to find a market, the pots are distributed without any charge to the residents of Kusatsu City.

There is some recycling of postindustrial plastic wastes [95]. The raw materials used to mold products include mainly LDPE, HDPE, and sometimes PP. PS can be mixed less than 10%, but PVC is prohibited to be mixed. Sometimes PVC agricultural waste film is pulverized, washed and dried, and then used for secondary PVC products. About 100 companies recycle roughly 220 million pounds (100×10^3 t) per year.

9.7.2 Europe and Canada

Superwood International Ltd. has established plants in England, Ireland, and Toronto (Alabama in the United States as well) which use a system named "Superflow" to fabricate solid plastic products to be used in applications as substitutes for wood [96]. Based upon modifications of the Klobbie (see Section 9.3.1), Superwood's Jumboflow machine is capable of an output of 2.5 times that for the original Klobbie. In addition to being larger than the Klobbie, the Jumboflow has both a relatively longer mixing zone and a slower extruder speed. Each machine is capable of running 24 hours per day, seven days per week, and has a capacity of 2.2 million pounds (1×10^3 t) annually. It is estimated that about 22 million pounds (10×10^3 t) of plastics waste were processed by Superwood during 1990. Feedstocks for Superwood products consist primarily of thermoplastics, HDPE, LDPE, and PP, obtained from both curbside collection programs and postcommercial operations.

In October, 1989, Superwood bought Plastics Recycling Ltd. from Braithwaite PLC in the U.K. This company processes supermarket plastics waste into large boards or sheets of plastic, called "Stokbord" [97].

In August 1991, Superwood International went into receivership due to loss of revenues as a result of a fire in a plant in Ireland. Two Canadian companies, Superwood Ontario Ltd. and Superwood Western Ltd., formed a joint venture and purchased the rights to the Klobbie patents from Lankhorst Recycling Ltd. to build and sell the machines in North America [98].

Wormser Kunststoff Recycling GmbH (WKR), a privately held German firm located in Worms, Germany, processes postconsumer plastics waste into products which are used in applications as substitutes for wood [99]. This system, named the "Worms technology" and based upon the modification of standard recycling equipment, uses a roller extruder that reportedly is capable of handling up to 15% of heterogenous nonpolymeric material such as sand, glass, stones, and metal parts.

Advanced Recycling Technologies, Ltd. (ART) of Belgium utilizes the ET/1 extruder to process mixed plastics waste into objects of bulky cross section (see Sections 9.3.1 and 9.4). In 1987, ET/1 machines were operating in twelve plants in Europe and Russia, with many more pending in other locations [100]. Currently, there are estimated to be over 40 machines installed worldwide [101].

The ET/1 extruder is used as part of a complete waste management/recycling program owned and operated by Stadtereinigung Nord (SN) in Flensburg, Germany. SN has added finishing steps to its ET/1 product line, marketing some as kits. In addition to the traditional lumber profiles and products, they are fabricating value-added items such as sound-absorbing walls for expressways [102].

Recycloplast-based recycling systems (see Section 9.3.4), such as the Ramaplan facility of Munich, Germany, have witnessed extensive product development since 1986, resulting in finished materials of improved strength and appearance over the original Recycloplast products [103]. This technology is designed specifically for the processing of mixed plastics waste and not for the recovery of homopolymer resins. Currently, Ramaplan charges in part for disposal of postconsumer plastics and in full for postindustrial scrap [104].

Acknowledgments

With regard to the studies done at the CPRR and at Washington and Lee University, the authors would like to thank the following organizations for their generous support: the Plastics Recycling Foundation; the Council for Solid Waste Solutions; the New Jersey Commission for Science and Technology; the National Science Foundation; the Center for Innovative Technology, Herndon, Virginia; the University of Virginia, Department of Nuclear Engineering and Engineering Physics; Rutgers, The State University of New Jersey; and Washington and Lee University. Also, we would like to thank the following individuals for their effort and support in our research: Darrell R. Morrow, James J. Donaghy, and Richard W. Renfree.

References

[1] *Modern Plastics, 68* (1) (Jan. 1991): pp. 119–120.
[2] Nosker, T. J., Rankin, S., Morrow, D. R. "Reclamation Techniques for Post-Consumer Plastics Packaging Wastes." Published in the PHARMTEC conference proceedings, Philadelphia, Sept. 1989.
[3] *Modern Plastics, 68* (1) (Jan. 1991): p. 120.
[4] *Modern Plastics, 68* (1) (Jan. 1991): p. 114–116.
[5] The Council of Solid Waste Solutions. Private communication. October 1991.
[6] Bennett, R. "New Product Appications, Evaluation, Markets for Recycled Products and Expansion of National Data Base." Speakers' Handouts, Thirteenth Meeting Industrial Advisory Council/Seminar VI, Center for Plastics Recycling Research, 25 Sept. 1991.
[7] *Technology Transfer Manual—Plastics Collection and Sortation: Including Plastics in a Multi-Material Recycling Program for Non-Rural Single Family Homes.* Piscataway, NJ: Center for Plastics Recycling Research, Rutgers University, Nov. 1988.
[8] Klobbie, E. U.S. patent 4 187 352, 1974.

[9]   Leidner, J. *Plastics Waste: Recovery of Economic Value*. New York: Marcel Dekker, Inc., 1981.
[10] Reference 8.
[11] Reference 9.
[12] "The Klobbie." REHSIF Bulletin, No. 400. Rehsif S. A., Switzerland.
[13] Reference 8.
[14] Bregar, W. "Ireland's Superwood in Receivership." *Plastics News* (2 Sept. 1991): pp. 1, 19.
[15] Brewer, G. *Resource Recycling, VI* (3) (1987): p. 16–19.
[16] Maczko, J. "The Economics of the ET-1 Operation." Presented at Seminar V: Plastics Collection and Commingled Plastics; Critical Elements in Post-consumer Plastics Recycling at Rutgers, The State University of New Jersey, Center for Plastics Recycling Research, 23 March 1990.
[17] Reference 15.
[18] Reference 16.
[19] Lauzon, M. "Dutch Firm Wins First Lawsuit in Plastic Lumber Patent Fight." *Plastics News* (10 Dec. 1990): p. 2.
[20] *Resource Recycling, IX*(9) (1990): p. 90.
[21] Hammer, F. and Harper, B. U.S. patent 4 626 189, 1986.
[22] Hammer, F. and Harper, B. U.S. patent 4 738 808, 1988.
[23] Hammer, F. and Harper, B. U.S. patent 4 797 237, 1989.
[24] Hammer, F. and Harper, B. U.S. patent 4 824 627, 1989.
[25] Hammer, F. "Production and Marketing of Products From Mixed Plastic Waste." Presented at SPE RETEC, Charlotte, NC, 30–31 Oct. 1989.
[26] Mack, W. "Turning Plastic Waste Into Engineered Products Through Advanced Technology." Presented at Recycling Plas V, 23–24 May 1990.
[27] "Recycling Equipment for Mixed Scrap." *Polymer Age, 4*(12) (1974).
[28] "The Reverzer." REHSIF Bulletin, No. 300. Rehsif S. A., Switzerland.
[29] Reference 9.
[30] Reference 15.
[31] Reference 16.
[32] Traugott, T. D., Barlow, J. W., and Paul D. R. *J. of Appl. Polym. Sci., 28*(1983): p. 2947–2959.
[33] Fayt, R., Jerome, R., and Teyssie, Ph., *J. Polym. Sci.: Polym. Phys. Ed., 19* (1981): p. 1269.
[34] Lindsay, C. R., Paul, D. R., and Barlow, J. W. *J. Appl. Polym. Sci., 26* (1981): p. 1.
[35] Elias, Hans-Georg. *Macromolecules 2: Synthesis, Materials, and Technology*. 2d ed. New York: Plenum Press, 1984.
[36] Renfree, R. W., Nosker, T. J., Rankin, S., Kasternakis, T. M., and Phillips, E. M. "Physical Characteristics and Properties of Profile Extrusions Produced from Post Consumer Commingled Plastics Waste." 1989 ANTEC Proceeding, pp.1809–1812.
[37] Van Ness, K. E. and Donaghy, J. J. "The Analytical Study of Commingled Products." Presented at Seminar V: Plastics Collection and Commingled Plastics; Critical Elements in Postconsumer Plastics Recycling at Rutgers, The State University of New Jersey, Center for Plastics Recycling Reseach, 23 March 1990.
[38] Applebaum, M., Morrow, D. R., Nosker, T. J., and Renfree, R. W. "The Processing and Properties of Post Consumer, Non-beverage Bottlescrap, Utilizing Non-intermeshing Twin Screw Extruder Technology." Presented at SPE RETEC," 17–18 Oct. 1990.
[39] Williamson, T. G., and Morrison M. C., Department of Nuclear Engineering and Engineering Physics, University of Virginia. Private communication. July 1989.
[40] Reference 36.

[41] Ramer, R. A., Byun, S. G., and Beatty, C. L. "Properties of Extruded Sheet Recycled High Density Polyethylene (HDPE) Milk Bottles and Commercially Available Post Materials." Presented at the SPE RETEC, Charlotte, NC, 30–31 Oct. 1989.

[42] Reference 36.

[43] Nosker, T. J., Renfree, R. W., and Morrow, D. R. "Improvements in the Properties of Commingled Plastics By the Selective Mixing of Plastics Waste." Proceedings, SPE RETEC, Charlotte, NC, Oct. 1989.

[44] Nosker, T ., Renfree, R. W., and Morrow, D. R. *Plastics Engineering, XLVI*(2) (1990): p. 33.

[45] Morrow, D. R., Nosker, T. J., Van Ness, K. E., and Renfree, R. W. "Structure/ Properties/Processing—Improvements in Commingled Plastics Products." Presented at Second Annual Plastics Recycling Fair, Society of Plastics Engineers, Burbank, California, 26 Apr. 1990.

[46] Nosker, T. J., Van Ness, K. E., Renfree, R. W., Morrow, D. R., and Donaghy, J. J. "Properties and Morphologies of Recycled Polystyrene/Curbside Tailings Materials." Presented at SPE 49th ANTEC, Montreal, 5–9 May 1991.

[47] Stevens, J. R. in *Methods of Experimental Physics 16B*. Fava, R. A., Ed. New York: Academic Press, 1980.

[48] Tao, S. J. *Appl. Phys., 3* (1) (1974): p. 1–7.

[49] Reference 37.

[50] Reference 45.

[51] Donaghy, J. J. and Van Ness, K. E., Thompson, M. E., Cassada, D. C., Nosker, T. J., Morrow, D. R., and Renfree, R. W. "Positron Lifetimes in Blends of Recycled Polystyrene and Polyethylene." *Phys. Stat. Sol. (a)124*, 67 (1991).

[52] Goldanskii, A. V., Onishuk, V. A., and Shantarovich, V. P. *Phys. Stat. Sol. (a), 102* (1987): p. 559.

[53] Thompson, M. E. "Application of a Quantum Mechanical Model in Determining Microvoid Radii in Polymer Blends." Honors Thesis, Washington and Lee University, 1990.

[54] Reference 37.

[55] Reference 46.

[56] Van Oene, H. *J. of Colloid and Interface Science, 40* 1972): p. 448.

[57] Han, C. D., Villamizar, C. A., and Kim, Y. W. *J. Appl. Polym. Sci., 21* (1977): p. 353.

[58] Lindsay, C. R., Paul, D. R., and Barlow, J. W. *J. Appl. Polym. Sci., 26* (1981): pp. 1–8.

[59] Fayt, R., Hadjiandreou, P. and Teyssie, Ph. *J. Polym. Sci.: Polym. Chem. Ed., 23* (1985): p. 337

[60] Schwartz, M. C., Barlow, J. W. and Paul, D. R. *J. Appl. Phys., 35* (1988): p. 2053.

[61] Fayt, R., Jerome, R., and Teyssie, Ph. *J. of Polym.Sci., Polym. Lett. Ed., 19* (1981): pp. 79–84.

[62] Reference 43.

[63] Reference 44.

[64] Reference 46.

[65] Reference 46.

[66] Jordhamo, G. M., Manson, J. A., and Sperling, L. H. *Polym. Eng. and Sci., 26* (1986): p. 517.

[67] The Center for Plastics Recycling Research. Private communication. March 1991.

[68] Reference 67.

[69] Chen, I. M. and Shiah, C. M. *Plastics Engineering, XLV*(10) (1989): p. 33.

[70] Wissler, G. E. "Commingled Plastics Based on Recycled Soft Drink Bottles." Presented at the ANTEC '90, Dallas, TX, 7–11 May 1990.

[71] Gibbs, M. L. "An Evaluation of Postconsumer Recycled High Density Polyethylene in Various Molding Applications." Presented at the ANTEC '90, Dallas, TX, 7–11 May 1990.

[72] Gibbs, M. L. "Postconsumer Recycled HDPE: Suitable for Blowmolding?" *Plastics Engineering, XLVI*(7) (1990): p. 57.

[73] Swint, S. S., Webber, K. D., and Chung, C. I. "Value-added Recycling of Post-consumer Commingled Plastics Waste." Presented at the RETEC titled "Plastics Waste Management: Recycling and its Alternatives," 17–18 Oct. 1990.

[74] Reference 38.

[75] Applebaum, M. D., Van Ness, K. E., Nosker, T. J., Renfree, R. W., and Morrow, D. R. "The Properties and Morphologies of Fiber Reinforced Refined Post-consumer Plastics Compounded on a Non-Intermeshing Twin Screw Extruder." Presented at the SPE 49th ANTEC, Montreal, 5–9 May 1991.

[76] Applebaum, M. D. and Bigio, D. I. "A Feasibility Study for the Continuous Mixing of Simulants for Energetic Compositions on a Non-Intermeshing Twin Screw Extruder." Presented at JANNAF conference, 13–15 Nov. 1990.

[77] Gibbs, M. L. Private communication. 1990.

[78] Kowalski, R. C. "Extruder Halogenation of Butyl Rubber." *Chem. Eng. Prog.* (May 1989).

[79] Lindt, J. T. "Devolatilization in Twin Screw Extruders." A.I.Ch.E. Annual Meeting, N.Y., 1987.

[80] Traugott, T. D., Barlow, J. W., and Paul, D. R. *J. of Appl. Polym. Sci., 28* (1983): pp. 2947–2959.

[81] Morrow, D. R., Center for Plastics Recycling Research. Private communication. 1989.

[82] Nauman, E. B. U.S. Patent 4 594 371, 1986.

[83] Lynch, J. C. and Nauman, E. B. "Recycling Commingled Plastics via Selective Dissolution." Proceedings of the 10th International Coextrusion Conference, 1989, pp. 99–110.

[84] Nauman, E. B., Ariyapadi, M., Balsara, N. P., Grocela, T., Furno, J., Lui, S., and Mallikarjun, R. "Compositional Quenching: A Process for Forming Polymer-in-Polymer Microdispersions and Interpenetrating Networks." *Chem. Eng. Commun., 66* (1988): pp. 29–55.

[85] Nauman, E. B., Wang, S. T., Balsara, N. P. "A Novel Approach to Polymeric Microdispersions." *Polymer, 27* (1986): pp. 1637–1640.

[86] Reference 83.

[87] Nauman, E. B., Rensselaer Polytechnic Institute. Private communication. 1990.

[88] Reference 87.

[89] Reference 83.

[90] Reference 87.

[91] Super, M. S., Enick, R. M., Beckman, E. S. "Separation of Thermoplastics Using Near- and Supercritical CO_2 as a Precursor to Recycling." Conference Proceedings, 49th ANTEC, Society of Plastics Engineers, Inc., Montreal, Canada, 5–9 May 1991, p. 1130.

[92] Beckman, E. J. and Enick, R. M. "Separation of Thermoplastics by Density Using CO2 in the Vicinity of the Critical Point as a Precursor to Recycling." Speakers' Handouts, Thirteenth Meeting Industry Advisory Council/Seminar VI, The Center for Plastics Recycling Research, Rutgers University, Sept. 1991.

[93] "Plastic Waste: Resource Recovery and Recycling in Japan." Tokyo: Plastic Waste Management Institute, 1985.

[94] Nakane, K. Private communication. July 1990.

[95] Reference 93.

[96]   Bunyan, R. J. "Woodlike Products from Waste Plastics." Presented at Recycling Plas
        IV, The Plastics Institute of America, Washington, D.C., 24–25 May 1989.
[97]   Reference 96.
[98]   Reference 14.
[99]   Grace, R. "Europe-Wide Recycling Plan Sought." *Plastics News* (29 Nov. 1989): pp.
        1, 12.
[100]   Reference 15.
101   Maczko, J., MidAtlantic Plastic Systems, Inc. Private communication. 1991.
[102]   Brewer, G. "The Plastics Recycling Loop: Case Studies from the U.S. and Abroad."
        Presented at SPE RETEC, Charlotte, NC, 30–31 Oct. 1989.
[103]   Reference 102.
[104]   Reference 102.

10 Thermosets

W. J. Farrissey

Contents

10.1	Introduction	233
10.2	Sources and Volumes	233
	10.2.1 PURs	234
	10.2.1.1 FlexibleFoams	234
	10.2.1.2 RigidFoams	235
	10.2.1.3 RIM and Cast Elastomers	235
	10.2.2 Phenolics	235
	10.2.2.1 Phenolic Molding Resins	236
	10.2.3 Unsaturated Polyesters	236
	10.2.4 Epoxy Resins	237
	10.2.5 Melamine and Urea Resins	237
10.3	Recycling Technologies	238
	10.3.1 Size Reduction	238
	10.3.2 Material Recycling	238
	10.3.2.1 PUR Flexible Foams	238
	10.3.2.2 PUR RIM	239
	10.3.2.3 Phenolics	242
	10.3.2.4 Epoxy Resins	243
	10.3.2.5 Unsaturated Polyester SMC	245
	10.3.3 Chemical Recycling: Hydrolysis/Glycolysis/Pyrolysis	247
	10.3.3.1 Stream Hydrolysis	249
	10.3.3.2 Glycolysis	250
	10.3.3.3 Pyrolysis	253
	10.3.4 Energy Recovery	255
10.4	Applications	255
10.5	Global activities	258
References		259

Thermoset resins, at 9 billion pounds (4.05×10^6 t) in annual sales, present a significant opportunity for recycle and resource recovery. Addition of reground process scrap back into uncured resin formulations has been examined for polyurethanes (PURs), phenolic resins, unsaturated polyester sheet molding compound (SMC), and epoxy resins. Other recycle and recovery technologies examined include thermal processing for PURs (compression molding and extrusion), hydrolysis/glycolysis/pyrolysis for PURs and SMC, and energy recovery from PURs and mixed plastic waste from auto scrap recycling activities. Applications for recycled material include the original applications in regrind recycle, rebond carpet underlay for flexible PUR foam scrap, booms for oil spill containment from rigid polyisocyanurate (PIR) foam scrap, hydroponic gardening substrate from rebonded flexible foam scrap, and a blasting medium for selective paint removal.

10.1 Introduction

Recyclers of thermoplastic material have had to combat the widespread public belief that plastics are not recyclable. Encouragingly, considerable progress has been made in correcting that misperception. On the other hand, those interested in the recycling of thermoset materials have the added burden that not even their plastics coworkers are aware that thermosets can be recycled. However, we must recycle if we are to broaden the base of thermosets, and continue to offer to design engineers the broadest possible choice in materials selection to achieve a specific performance goal [1–8]. It will be the objective of this chapter to provide as broad a spectrum as possible of recycling technologies for various thermoset resins. Perhaps a recycle technology explored in one area will be adaptable to another. Hopefully, we can look forward to solutions to our recycle challenge before other, perhaps less effective, solutions are externally imposed [9].

10.2 Sources and Volumes

Over 9 billion pounds (4.05×10^6 t) of thermoset resins were sold in 1990, representing about 15% of the 61 billion pounds (27.4×10^6 t) of plastic materials sold in the U.S. As shown in Table 10.1, the major components of the mix were PURs and phenolics at about one-third each of the total, with unsaturated polyester and melamine urea resins splitting most of the

Table 10.1 Global thermoset plastic demand

Resin	U.S. Million Pounds	(Thousand Tonnes)	World Million Pounds	(Thousand Tonnes)
Urethanes	3265	(1484)	10,297	(4670)
Epoxies	464	(211)	1433	(650)
Phenolics	2827	(1285)	5686	(2680)
Unsaturated Polyester	1227	(558)	3394	(1540)
Urea, Melamine	1439	(654)	2997	(1360)
TOTAL	9222	(4192)	23,807	(10,900)
TOTAL POLYMERS	61,480	(27,945)	140,326	(63,640)

Notes: U.S. volumes are for 1990 [12]
World volumes are for 1988 [11]

remainder, and epoxy resins garnering the smallest share of the major thermosets. Also shown in Table 10.1, for comparison, are volume estimates for these resins on a global basis. It should be emphasized that the volume figures given here are for resins only and do not include the fillers, reinforcements, processing aids, stabilizers, etc., which can be a significant portion of the weight of the finished part. In unsaturated polyester recipes, for example, the resin itself may be only 30% by weight of the total formulation. The magnitude of the recycle/disposal problem for these resins, therefore, can be considerably larger than the resin figures would indicate. A recent report by Franklin Associates [10] describes the method of disposal of the major plastics materials. Interestingly, in contrast to the major thermoplastics, which are disposed of typically through Municipal Solid Waste (MSW) facilities, the majority of thermoset resin products are disposed of by non-MSW means, through industrial or commercial facilities. In any event, thermosets represent a significant opportunity for recovery of resources through various recycle technologies [11, 12].

10.2.1 PURs

PURs have been called the "protean plastic" material because of the wide variety of products obtainable from urethane chemistry. PURs are formed by the reaction of isocyanates with active hydrogen containing compounds such as water, polyols or polyamines. The broad spectrum of products obtainable from this relatively simple chemistry include flexible, energy-management and rigid foams, thermoset and thermoplastic elastomers, automotive reaction injection molding (RIM) exterior body parts, adhesives, coatings, sealants, fibers and films. The breakdown distribution of PURs in the various applications is shown in Table 10.2 [13].

Table 10.2   Polyurethane demand in the U.S.–1990

Application	Demand	
	Million Pounds	(Thousand Tonnes)
Flexible Foams	1732	(787)
Rigid Foams	875	(398)
RIM Elastomers	212	(96)
Cast Elastomers	118	(54)
Other (footwear, adhesives coatings, sealants, misc.)	328	(149)
TOTAL	3265	(1484)

10.2.1.1 Flexible Foams Flexible foams represent the largest volume group of applications of PUR materials. The major categories of flexible foam and the amounts produced are shown in Table 10.3. Furniture cushioning is the largest single application of flexible PUR foam. It has found favor because of its light weight, excellent cushioning properties and economical fabrication. These same characteristics, plus the ability to formulate to a variety of densities and hardnesses, are prime reasons for the success of flexible PUR foam bedding applications such as all-foam mattresses. Carpet underlay may be foam cut to thickness from foam slabs, or foam applied directly to the underside of carpets. Included in the table is the carpet underlay produced by rebonding of flexible foam scrap, an application which consumes most of the scrap produced in molded foam seating operations and the cutting of flexible foam buns for furniture applications. This will be described more fully under Recycling Technologies (Section 10.3).

Automotive seating is generally molded into shape directly. Packaging applications include protection of electronics or delicate instruments, where the foam cushioning may be molded to the exact shape of the product to be protected. Other applications include clothing liners, filters, gaskets, toys, etc.

Table 10.3   Polyurethane flexible foam demand in the U.S.–1990

Market	Demand	
	Million Pounds	(Thousand Tonnes)
Bedding	150	(68)
Furniture	685	(311)
Carpet Underlay	350	(159)
Transportation	400	(181)
Other	147	(67)
TOTAL	1732	(787)

Note: Carpet underlay includes rebond

10.2.1.2 Rigid Foams

The major applications for rigid PUR and the related PIR foams are insulation related and take advantage of the excellent insulation ability or low "K-factor" of these cellular materials. Nearly 900 million pounds (405×10^3 t) of PUR/PIR foams were produced in the United States, apportioned as shown in Table 10.4. Building and construction markets use PUR rigid foam either as boardstock cut from buns or produced continuously or discontinuously as laminates. The laminates can have a variety of facers; either flexible, such as kraft paper, roofing paper or aluminum foil, or rigid, such as aluminum or steel building panels. Refrigeration uses include domestic and commercial refrigerators and freezers, picnic coolers, etc. Packaging applications utilize the pour-in-place capability of urethanes to encapsulate complex shapes in a rigid foam shell for protection. Transportation markets include refrigerated trailers, trucks, tank cars and shipping containers. The "other" category includes marine flotation and pipe insulation among others.

Table 10.4   Polyurethane rigid foam demand in the U.S.–1990

Application	Demand	
	Million Pounds	(Thousand Tonnes)
Building Insulation	480	(218)
Refrigeration	154	(70)
Industrial Insulation	83	(38)
Packaging	66	(30)
Transportation	47	(21)
Other	45	(20)
TOTAL	875	(398)

10.2.1.3 RIM and Cast Elastomers

The bulk of the PUR elastomers produced by the RIM process are used in the automotive industry. Applications include fascia, bumper beams, body panels and modular windows. Non-automotive applications include agricultural equipment, mining equipment, equipment housings, and some sporting goods. About 200 million pounds (90×10^3 t) of PURs were used in these applications in 1990. PUR cast elastomers are used in applications such as industrial tires (fork lifts, earth-moving equipment), skate board wheels, printer rollers, mining applications (pumps, conveyor belts), and mechanical goods (gaskets, belts, seals) in the amount of about 118 million pounds (53×10^3 t).

10.2.2 Phenolics

Phenolic resins are formed in the catalyzed condensation of phenols with formaldehyde. The process was extensively investigated by Leo Baekeland in the early years of this century. Phenolic resins combine low cost with good strength and good retention of mechanical properties over a broad temperature range. These attributes account for the position of phenolic resins as the second leading volume thermoset resin, just behind PURs. Adhesives and binder applications dominate the usage of phenolic resins in the United States. Phenolics are used to bond wood veneers in plywood, as a binder for glass or mineral wool fibers in insulation materials, as a binder for wood particles in particle board, wafer board, and hard board. Other binding applications include friction materials in clutch facings and brake linings, and for abrasive materials in cutting and grinding wheels. Laminate applications include table and computer tops, gears and pulleys, and printed circuit and terminal boards. Phenolic molding compounds are used for a variety of household appliance handles and electrical and automotive parts. Phenolic foam has found some application as a roof insulation to utilize its good insulation values, low flame spread, and low smoke generation under fire situations. Phenolic foams are generally produced as 1–2-inch-thick (25.4–50.8 mm) laminates with fiberglass reinforced foil facers on both sides. Applications are mainly in roofing insulation. The volume breakdown for these applications is detailed in Table 10.5 [14].

Table 10.5   Phenolic resin demand in the U.S.–1990

| Applications | Demand | |
	Million Pounds	(Thousand Tonnes)
Bonding/Adhesive	811	(369)
Plywood	1457	(662)
Laminates	197	(90)
Molding Compounds	168	(76)
Coatings	14	(6)
Other	180	(82)
TOTAL	2827	(1285)

10.2.2.1 Phenolic Molding Resins   Phenolic molding resins may contain as much as 50% filler by weight. Wood or nut flours, glass or graphite fibers among others can be used as fillers or reinforcements. Molded applications include small appliances, housewares, electrical outlet boxes, pulleys, cutlery handles and many others. The market distribution is shown in Table 10.6.

Table 10.6   Phenolic molding resin by application demand–1990

| Market | Demand | |
	Million Pounds	(Thousand Tonnes)
Appliances	27	(12)
Closures	12	(5)
Electrical	62	(28)
Housewares	37	(17)
Industrial	8	(4)
Transportation	17	(8)
Other	5	(2)
TOTAL	168	(76)

10.2.3 Unsaturated Polyesters

Unsaturated polyesters are composed of esters of phthalic/maleic anhydride mixtures with propylene glycol, or vinyl ester resins from bisepoxides and methacrylic acid, dissolved in styrene monomer. The maleic or methacrylic moieties provide unsaturation sites for cross-linking when the styrene is polymerized during the curing process. The result is a cross-linked, hard, glassy polymer. Most of the applications for unsaturated polyester resins are glass-reinforced and highly filled, such that the resin content of the functional part is often less than 30%. Table 10.7 describes the application information for this important, widely used resin system [15].

Table 10.7   Polyester demand in the U.S.–1990*

| Application | Demand | |
	Million Pounds	(Thousand Tonnes)
Reinforced		
Molded, Fil-wound, Pultruded, etc.	720	(327)
Sheet	168	(76)
Other	339	(154)
TOTAL	1227	(557)

* Resin only

Unsaturated polyester applications include reinforced panels, greenhouses and awnings, bathtubs and shower stalls, piping and ducts, and underground storage tanks. Boat hulls for motor and sail boats and automotive body components are important applications, including some of the body panels on the General Motors APV. Other applications include campers, snowmobiles, skis, electronic circuit boards, radomes, etc.

10.2.4 Epoxy Resins

Epoxy resins are based on the reaction products of polyphenols, such as bisphenol A or novolacs, with epichlorohydrin and cured with aromatic amines or anhydrides, among others. Protective coatings constitute the largest application for epoxy resins (Table 10.8) in three principle market applications: transportation, maintenance, and metal containers [16]. Reinforced applications are the next largest segment, and includes laminates for printed circuit boards, composites for radar installations, business equipment and aircraft and automotive applications. Because of their strength and good dimensional stability, epoxy composites are used in tooling for casting and molding applications. Encapsulation of electrical components, flooring and adhesives are other major applications areas.

Table 10.8   Epoxy demand in the U.S.–1990

| Application | Demand | |
	Million Pounds	(Thousand Tonnes)
Bonding/Adhesives	28	(13)
Flooring/Paving	26	(12)
Coatings	195	(87)
Reinforced		
Elect. laminates	55	(25)
Other	31	(14)
Casting/Molding	28	(13)
Other	101	(46)
TOTAL	464	(210)

10.2.5 Melamine and Urea Resins

Melamine and urea resins are formed by the reaction of melamine or urea with formaldehyde. By far the largest application for these materials is as a binder for wood products, plywood, particle board and counter and table top laminates. Paper treating, protective coatings, and crease-proofing treatment for textiles make up the bulk of the remainder. Molding applications include electrical equipment, dinnerware and buttons. The volumes for these applications are shown in Table 10.9 [17].

Table 10.9   Melamine, urea demand in the U.S.–1990

Application	Demand	
	Million Pounds	(Thousand Tonnes)
Bonding Resins	1153	(524)
Protective Coatings	90	(41)
Paper Treatment	51	(21)
Textile Treatment	22	(10)
Molding Compounds	74	(34)
Other	49	(22)
TOTAL	1439	(652)

10.3   Recycling Technologies

Recycling of thermoset resins has been examined fairly broadly as a means of utilizing in-plant scrap initially, and ultimately to recovery of material from postuse or postconsumer sources. The recycling techniques available for thermoset resins are somewhat limited; they include incorporation of regrind in thermoplastic or thermoset systems, recovery of raw materials via hydrolysis or glycolysis, recovery of chemical values via pyrolysis or recovery of heat through incineration. Lately, though, thermal processing techniques have been introduced as an alternative in PUR RIM recycle [18].

10.3.1   Size Reduction

Central to any scheme to reuse thermoset resins as regrind are techniques of size reduction. Grinding systems have been designed to handle the broad spectrum of materials encountered, from the high modulus, highly filled SMC systems to the tough, elastomeric, RIM PURs with low filler levels. Bauer [19] has described the equipment requirements and economics for the grinding of thermoset scrap for incorporation as filler in virgin resin. As described, phenolic scrap is first fed to a series of pulverizers, where the material is reduced to about a 20 mesh size. A second series of hammer mills is used to further reduce the particle size to 200 mesh. The material is then discharged through sorting screens to ensure the proper particle size is achieved and conveyed to a feeder container for blending with virgin resin. According to Modern Plastics Encyclopedia [20] there are at least 15 granulator vendors serving the U.S. market [21]. Plastics World 1990 Directory contains a similar list of vendors.

10.3.2   Material Recycling

10.3.2.1   PUR Flexible Foams   The conversion of flexible PUR foam into cushioning generates considerable waste, 8-12%, depending on the shape of the foam blocks, and the complexity of the cut parts. Molded seat cushion production generates some scrap as well. All of this scrap, plus some imported material, is used in the production of rebonded carpet underlay. About 280 million pounds (126×10^3 t) of domestic flexible foam scrap per year is consumed in this way. And in addition, about 100–120 million pounds (45–54×10^3 t) of scrap foam is imported for this application [22].

As with all scrap recycle processes, the first step in the reuse of flexible foam scrap is separation of the foam from any contaminants, such as wire, fabric, debris, etc. The foam is then chopped into pieces of suitable size, and the foam pieces are coated with binder. Typically the binder is an isocyanate prepolymer, prepared from toluene diisocyanate (TDI) or 4,4 diphenylmethane diisocyanate (MDI) and a polyether polyol. Amounts of binder of 10–20% by weight of foam are used. After catalysis and thorough mixing, the foam binder mixture is placed in a mold and compressed. The foam is maintained in the compressed state during cure with heat and steam. Semicontinuous and continuous process may be used also. By varying the degree of compression, foam densities of 2.5–6.2 lb/ft^3 (40–100 kg/m^3) can be prepared [23, 24].

Another potential technology for the utilization of scrap flexible foam involves recycle of the comminuted scrap back into the foam formulation. In a process described by B. D. Bauman *et al.*, [25], the scrap foam is first ground at cryogenic temperatures to a suitable powder. The powder is then mixed with polyol, typically at levels of 15–20 parts per hundred of polyol. The blended polyol/ground foam slurry can be handled on typical foam processing equipment. Adjustments to catalyst and isocyanate levels may be necessary for optimum performance. The mechanical and physical properties of the foam with regrind are comparable to control materials, Table 10.10. It is claimed that this use of recycle scrap can lower foam costs by 3.5%.

Table 10.10   Physical property comparison–flexible foam regrind

Physical Property	Control		15% Regrind	
Density, lb/ft^3 (kg/m^3)	1.17	(19.0)	1.14	(18.5)
Indent. Deflec. 25%, psi (kPa)	30.50	(210)	30.00	(207)
Indent. Deflec. 56%, psi (kPa)	58.90	(406)	57.90	(399)
Support Factor	1.93		1.93	
Breathability, ft^3/min (m^3/min)	5.00	(0.14)	4.00	(0.11)
Tensile Strength, psi (kPa)	12.30	(85)	13.70	(94)
Tear, lb/linear in (kN/m)	1.70	(0.30)	1.80	(0.32)
Elongation %	114.00		121.00	
90% Compression Set	5.80		6.10	
Resilience %	42.40		42.40	

Source: Bauman et al., 1983

10.3.2.2 PUR RIM   As pointed out in Table 10.2, RIM elastomer demand in 1990 exceeded 200 million pounds (90 × 10^3 t). At an overall scrap rate of 10%, about 20 million pounds (9 × 10^3 t) of RIM plant scrap are generated annually. Several technologies have been proposed recently to deal with this valuable resource; recycling the regrind back into RIM parts [26–31], compression molding of regrind into other parts, and extrusion of scrap into profiles and tubing.

The recycle of RIM scrap back into RIM parts [32] requires a two-stage size-reduction process, first in a granulator, then in an impact disc mill. The latter reduces the initial 0.24–0.35 inch (6–9 mm) particles to a 180 μm powder without the need for cryogenic temperatures. The powder is then mixed with the polyol; 30% by weight in the polyol yields 10% of regrind in the overall polymer. A three-stream process is required, Figure 10.1, because diethyltoluenediamine, a common extender for current PUR RIM fascia, is adsorbed into the regrind. The properties of the RIM elastomer prepared with regrind are identical to the controls, Table 10.11. Most importantly, the surface finish of the painted parts is excellent and comparable to the control materials. Also, cost calculations suggest that use of the regrind saves money, about 5% of the part cost for an 11 lb (5 kg) part.

Compression molding offers another opportunity for recycle of RIM scrap [33]. In this process, the ground scrap, without any added binder, is compression molded into the desired

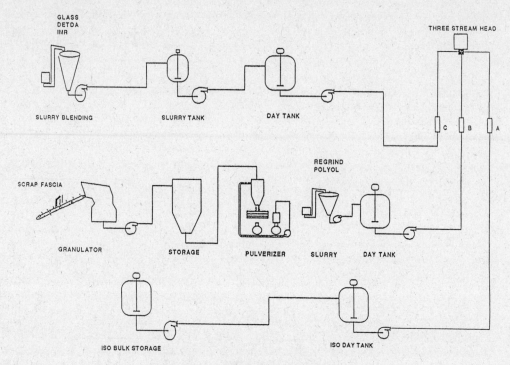

Figure 10.1   Three-stream recycle process

Table 10.11   Properties of PUR RIM fascia containing regrind *

Property	No Regrind	Regrind, Unfilled Unpainted	Regrind, Unfilled Painted	Regrind, Filled Painted
Specific Gravity	1.04	1.04	1.04	1.07
Flexural Modulus				
10^3 psi	50.8	47.7	51.5	54.8
(MPa)	(350)	(329)	(355)	(378)
Tensile Strength				
10^3 psi	3.61	3.62	3.77	3.64
(MPa)	(24.9)	(24.9)	(26.0)	(25.1)
Elongation, %	240	265	245	250
Gardner Impact, -29 °C				
In-lb	>320	>320	>320	>320
(J)	(>36.2)	(>36.2)	(>36.2)	(>36.2)
Tear Strength				
lb/linear in	570	565	580	570
(kN/M)	(100)	(99)	(102)	(100)

* 10% regrind (-80 mesh) added to formulation; regrind had a flexural modulus of 50×10^3 psi (345 MPa)

shape. Under the temperature and shear of the molding conditions, 180–190 °C and 2.9–5.9×10^3 psi (20–40 MPa), the RIM material will flow and the particles knit together. The mechanical properties and surface finish of the resultant part, while not quite equal to the original polymer, Table 10.12, still are sufficient for a number of applications. Suggested uses

Table 10.12   Property comparison of compression molded RIM scrap

Property	Original	Recycled
Density lb/ft^3 (g/cm^3)	69 (1.10)	73 (1.17)
Flexural Modulus 10^3 psi (MPa)	61 (420)	63 (435)
Tensile Strength 10^3 psi (MPa)	2.85 (20)	2.10 (15)
Elongation, %	100	50
Heat Sag, 1 h, 121 °C in (mm)	0.35 (9)	0.39 (10)

include seat shells, air flow deflectors, and spoilers, for example. Part prices are claimed competitive with polyolefin, SMC and engineering thermoplastics.

Some particular types of RIM scrap have been extruded into profiles and shapes at 215–225 °C, in single- and twin-screw extruders [34]. The mechanical properties of the extruded material are quite different from the original RIM material, and swelling measurements indicate a measurable loss of cross-linking. However, the extruded material is still a tough elastomer, Table 10.13, with projected uses in tubing, profiles, floor tiles and body side moldings.

Table 10.13   Properties of virgin and extruded reinforced RIM

Property	Virgin*		Extruded	
Specific Gravity	1.16		1.2	
Hardness, Shore D	60		50	
Flexural Modulus, 10^3 psi (MPa)				
23 °C	65	(448)	22.3	(154)
70 °C	56	(386)	15.2	(105)
-29 °C	125	(861)	35.5	(245)
Tensile Strength, 10^3 psi (MPa)	2.8	(19.3)	2.3	(15.8)
Elongation, %	90		104	
Heat Deflect. Temperature, °C at 0.5 MPa			127	
Coef. Linear Therm. Expan., ppm/ °C -29 °C to 29 °C	45.0		71.0	
Notched Izod, ft-lb/in (J/m) at -29 °C			6.1	(326)
Tear Resistance, lb/in (N/m)	430	(75)	277	(48)
Abrasion Resistance mg/1000 revs.			24.0	

* RIMline E-9031 (ICI Americas, Inc.)

10.3.2.3 Phenolics   Experiments with cured phenolic resins have shown that 5–20% of cured resin can be incorporated as filler without catastrophic deterioration of resin properties. As shown in Table 10.14, some deterioration in overall properties is experienced, with the most serious being the 35% loss in unnotched impact strength, even at the lowest regrind level (5%) and only slightly improved at the finest particle size. Interestingly, notched impact shows some improvement with regrind at some filler levels. The flexural properties are not affected appreciably by the percentage of regrind (Figure 10.2) nor by the particle size (Figure 10.3). Tensile stress, on the other hand, exhibits lower values, even at the smallest loading of regrind, especially with the coarse ground material. Dielectric strength, water absorption and heat deflection temperature are not affected significantly by the amount or the particle size of the regrind within the limits of this study [35, 36].

Table 10.14   Effect of phenolic regrind on properties

	Flexural Strength 10^3 psi (MPa)		Tensile Stress 10^3 psi (MPa)		Notched Impact ft-lb/in (J/m^2)		Unnotched Impact ft-lb/in (J/m^2)		Heat Deflect. Temperature °C
Virgin Powder	12.4	(85.8)	6.7	(46.7)	0.38	(752)	1.58	(3154)	115
Virgin + 5% Coarse	10.7	(74.0)	3.3	(23.1)	0.45	(904)	0.98	(1955)	110
Virgin + 5% Medium	11.8	(81.2)	5.9	(40.6)	0.57	(1135)	1.02	(2039)	107
Virgin + 5% Fine	11.5	(79.1)	5.4	(37.0)	0.48	(967)	1.19	(2376)	109
Virgin + 10% Medium	11.7	(80.5)			0.37	(736)	1.02	(2039)	107
Virgin + 15% Medium	11.4	(78.8)			0.41	(820)	1.01	(2018)	111
Virgin + 20% Medium	11.2	(77.0)			0.37	(749)	1.00	(1998)	109

Source: Kalyon, et al., 1984

RELATIVE
PROPERTY VALUE

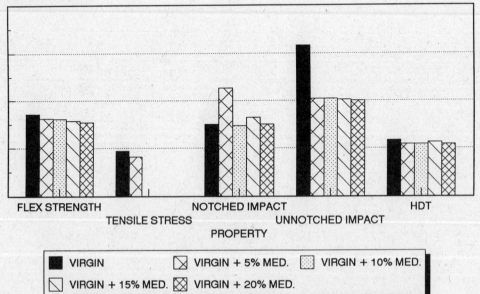

DATA FROM KALYON ET AL.,1984

Figure 10.2   Effect of percent phenolic regrind on properties

PROPERTY VALUE

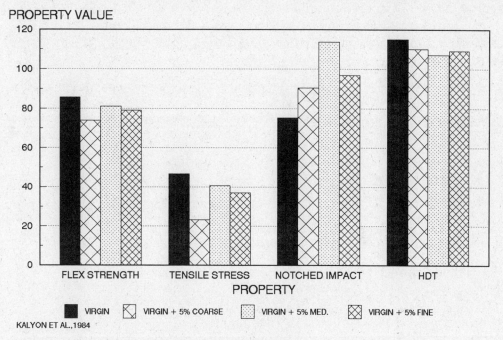

KALYON ET AL.,1984

Figure 10.3   Effect of phenolic regrind particle size on properties

Processing of the uncured resins becomes more difficult with added regrind as filler. The resulting higher shear viscosity will require higher molding pressures. Overall, then, incorporation of regrind material into phenolic resins can be accomplished but with some penalty to ultimate strength values [37, 38]. Flexural, tensile and unnotched impact strengths are affected most severely; dielectric strength, heat deflection temperature (HDT), water absorption and notched impact are less influenced. Therefore, applications of phenolic resins with incorporated regrind need to be selected carefully to accommodate these changes in overall properties.

10.3.2.4 Epoxy Resins

Incorporation of regrind into epoxy resin formulations is attended by similar problems as those encountered with phenolics. In general, the viscosity of the mix is increased, leading to processing problems. Strength and impact values are generally lower also. The results of an investigation of the use of epoxy resin regrind from three different cure systems are shown in Table 10.15. The particle size distribution of the regrind was 70–80% of 200–500 μm size and the remainder less than 200 μm. Composition of the resin mix was 20% regrind and 80% of the corresponding virgin formulation. Regrind was incorporated in two ways: dry blending of the regrind powder in the virgin formulation just prior to cure, and presoaking of the regrind powder in one of the system components for an hour at 90 °C and four days at room temperature prior to cure [39].

From Table 10.15, it is seen that addition of regrind has the expected effect of increasing hardness and decreasing most other properties. Some exceptions include the increase in dart impact for the polyamine cured sample with "soaked" regrind, along with an improved HDT. For the polyamine cured sample, the flex strength is only slightly lowered by the addition of dry regrind and the volume resistivity is increased. The variation in flex strength with cure type and regrind is shown in Figure 10.4. Interestingly, it was observed that the adhesion of the epoxy resin to aluminum was dramatically increased by the addition of presoaked regrind to the anhydride cure formulations.

An interesting recycle effort has been mounted by Peninsula Copper Industries, Inc., which recovers copper from circuit boards [40]. After leaching off the copper from the epoxy/glass

Table 10.15　Effect of regrind on properties of epoxy resins

Material	Hardness Rockwell L	Flexural Modulus 10^3 psi (MPa)	Flexural Strength 10^3 psi (MPa)	Dart Impact ft-lb (J)	Heat Deflec. Temp. °C	Vol. Resistance 10^{15} ohm-cm
Polyamine-Control	120	475 (3272)	16.2 (111)	0.83 (1.13)	103	1.700
20% Regrind-Dry	129	336 (2315)	5.7 (39)	<0.83 (<1.13)	90	1.340
20% Regrind-Soaked	128	325 (2239)	6.7 (46)	1.67 (2.23)	108	0.024
Polyamide-Control	118	383 (2638)	12.2 (84)	2.50 (3.39)	60	0.690
20% Regrind-Dry	121	364 (2507)	11.9 (82)	<0.83 (<1.13)	54	2.300
20% Regrind-Soaked	113	236 (1626)	5.8 (40)	0.83 (1.13)	46	0.870
Anhydride-Control	121	358 (2467)	12.0 (83)	1.67 (2.26)	70	12.100
20% Regrind-Dry	122	368 (2535)	6.8 (47)	0.83 (1.13)	71	7.190
20% Regrind-Soaked	126	244 (1681)	7.3 (50)	<0.83 (<1.13)	65	8.010

Source:　Tran-Buni and Deanin, SPE ANTEC, 1987

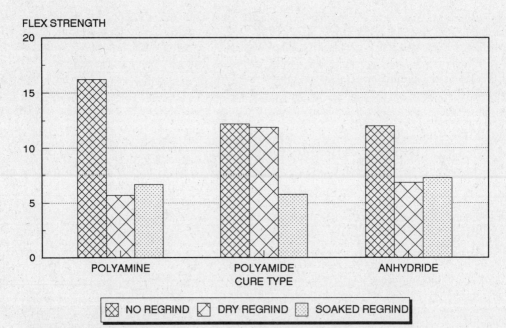

DATA FROM TRAN-BUNI AND DEANIN,1987

Figure 10.4　Effect of regrind on flex strength of epoxy resins with different cure agents

laminates, the remainder is calcined to remove the epoxy binder, which leaves a resin-free glass-fiber cloth. Comparison of the fiber, chopped to 0.4 inch (10.2 mm) lengths, was made with 0.25 inch (6.4 mm) chopped strand in general purpose thermoset polyester resin at a 15% loading. Tensile and flexural strengths were 3626 psi (25.0 MPa) and 11,800 psi (81.3 MPa), respectively, compared with 3600 psi (24.8 MPa) and 12,300 psi (84.7 MPa) for the controls, Table 10.16. Commercialization in the near future is planned. Similar expectations are envisioned for the glass and filler recovered from the pyrolysis of SMC materials (Section 10.3.3).

Table 10.16   Average mechanical properties of polyester glass composites

Property	Ultimate Stress			
	P.C.I.*		P.P.G.	
	10³ psi	(MPa)	10³ psi	(MPa)
Tensile Strength	3.6	(25.0)	3.6	(24.8)
Compressive Strength	13.0	(89.6)	n/a	n/a
Flexural Strength	11.8	(81.3)	12.3	(84.7)
Fiber Length, in (mm)	(avg.) 0.4	(10.2)	0.25	(6.4)

* Peninsula Copper Ind., Inc. Recycled Glass

10.3.2.5 Unsaturated Polyester SMC

The advent of the General Motors APV has rekindled interest in the challenge to recycle SMC scrap materials. Several publications have appeared lately describing the reuse of ground SMC or bulk molding compound (BMC) scrap as filler in SMC and BMC compounds and in thermoplastic formulations [41–43]. News items indicate substantial European activity as well [44, 45]. In addition to regrind techniques, SMC recycle activities also include pyrolysis, with recovery of fuel value and recycle of recovered filler [46, 47].

In one study cured SMC scrap was shredded and granulated in two steps. Two material sizes were explored—a coarse fraction which passed a 3/8-inch (9.5 mm) screen, and a fine fraction passing a 3/16-inch (4.8 mm) screen. Both materials were blended into BMC formulations replacing 10% and 20% by weight of the standard filler. The formulated resins were compression molded in a hot mold with the results shown in Table 10.17. Clearly some loss in tensile stress and modulus are observed with the incorporation of the coarse regrind, with somewhat less reduction observed with the fine regrind, Figure 10.5. Both notched and unnotched Izod impact strengths suffered from the inclusion of regrind, with the coarse material

Table 10.17   Properties of BMC containing recycled SMC

	BMC Control	Std. BMC + Coarse Recycled SMC		Std. BMC + Fine Recycled SMC	
Ground SMC (%)	0	10	20	10	20
Tensile Stress					
10³ psi	4.05	2.33	2.11	2.51	2.33
(MPa)	(27.9)	(16.1)	(14.5)	(17.3)	(16.1)
Tensile Modulus					
10⁶ psi	1.90	1.43	1.33	1.84	1.48
(GPa)	(13.1)	(9.9)	(9.2)	(12.7)	(10.2)
Elongation (%)	0.44	0.68	0.34	0.22	0.14
Flexural Modulus					
10⁶ psi	1.53	1.42	1.37	1.48	1.34
(GPa)	(10.5)	(9.8)	(9.4)	(10.2)	(9.2)
Notched Izod					
ft-lb/in	5.06	5.06	2.96	3.91	4.14
(J/m)	(270)	(270)	(158)	(209)	(221)
Unnotched Izod					
ft-lb/in	6.76	5.20	3.42	5.16	5.63
(J/m)	(361)	(278)	(183)	(276)	(301)

Source: W.D. Graham, R.B. Jutte, SPI Press Mold., May 1990.

VALUE

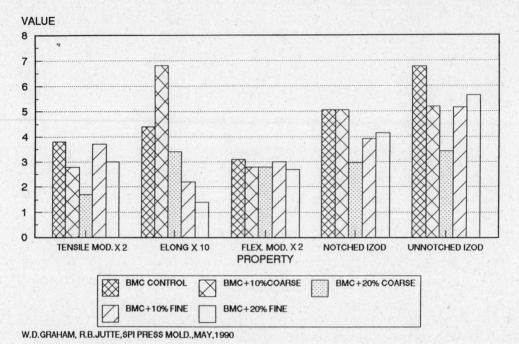

W.D.GRAHAM, R.B.JUTTE,SPI PRESS MOLD.,MAY,1990

Figure 10.5 Properties of BMC containing recycled SMC

being more disadvantageous, at least at the 20% added level [48, 49]. In addition, it was noted that the regrind has high resin demand, and may require system reformulation.

The ground SMC also was added to polyethylene (PE) and polypropylene (PP) at loadings of 15, 30 and 50% by weight. The ground SMC was dry blended with the thermoplastic pellets in a reciprocating screw extruder to produce a variable length log. Some bridging in the feed throat of the extruder was encountered, especially with the fine grind SMC. Samples were stamped from the logs in a cold mold using a fast closure hydraulic press, to yield the results shown in Table 10.18. Some improvement in notched Izod impact strength is noted for the coarse recycled SMC, along with a distinct improvement in HDT, from 63 °C to 94 °C at the highest (50%) loading.

The overall results indicate that the recycle SMC is acting as a filler and not as a reinforcement. By contrast, only 9% chopped glass in PP raises the HDT to 115 °C, and essentially doubles the notched impact strength and tensile modulus [50, 51].

Another approach has been described using SMC scrap ground to -200 mesh (30 μm) size, and used as a replacement for calcium carbonate in SMC formulations [52]. Up to 88 parts of regrind per hundred of resin were used. Viscosity limitations precluded higher levels of usage. Mechanical testing on molded paste and on SMC indicated that the regrind had little effect on mechanical properties especially at lower levels. Some decrease in properties was observed at higher loadings, but none that would severely affect the performance of the material. On the other hand, surface quality was adversely affected at regrind loadings above 30 phr.

One may conclude that regrind SMC can be recycled into both BMC and PP as a filler, with some adjustments in formulation and materials handling required. The changes in mechanical properties will necessitate careful selection of applications appropriate to the recycle-containing formulations.

Another interesting application for scrap from unsaturated polyester spray-up applications involves addition of granulated scrap to a syntactic polyester foam at 10–28% loadings [53]. The mixed blend is then pumped through oversized hoses to a spray gun where the mix is

Table 10.18   Properties of PP containing recycled SMC

	Neat PP	PP + Coarse Recycled SMC			PP + Fine Recycled SMC			PP + 9% Chopped Glass
Ground SMC (%)	0	15	30	50	15	30	50	0
Tensile Stress								
10^3 psi	3.21	3.12	3.05	3.25	2.98	3.04	2.49	5.5
(MPa)	(22.1)	(21.5)	(21.0)	(22.4)	(20.5)	(20.9)	(17.2)	(37.9)
Tensile Modulus								
10^6 psi	0.15	0.25	0.32	0.44	0.21	0.30	0.41	0.29
(GPa)	(1.00)	(1.72)	(2.20)	(3.03)	(1.45)	(2.07)	(2.82)	(2.00)
Elongation (%)	9.72	3.50	2.34	1.61	4.78	2.47	1.25	3.45
Flexural Modulus								
10^6 psi	0.14	0.19	0.24	0.33	0.19	0.24	0.36	0.21
(GPa)	(0.96)	(1.31)	(1.65)	(2.27)	(1.31)	(1.65)	(2.48)	(1.45)
Notched Izod								
ft-lb/in	1.46	1.58	1.67	1.90	1.39	1.54	1.37	2.82
(J/m)	(78.0)	(84.4)	(89.2)	(101.5)	(74.2)	(82.2)	(73.2)	(150.6)
Unnotched Izod								
ft-lb/in	18.14	4.43	2.91	2.78	3.84	3.07	1.95	6.77
(J/m)	(969)	(237)	(155)	(148)	(205)	(164)	(104)	(362)
Heat Deflect. Temp., °C at 0.5 MPa	63.8	67.2	78.3	94.4	66.1	84.4	82.8	115

Source: W.D. Graham, R.B. Jutte, SPI Press Mold., May 1990.

catalyzed. The present application is a core laminate applied between two reinforced polyester skins that back up vacuum-formed acrylic tub and spa shells.

The glass fibers from SMC have been reclaimed by heating the SMC to 350–400 °C, crushing, chopping and treating the residue with hydrochloric acid. The glass fibers are treated with adjuvants and blended with molten polyvinyl chloride (PVC) or PE [54]. Circuit boards have been recycled by stripping the copper and granulating to various particle sizes for use as a filler in thermoplastics [55].

10.3.3 Chemical Recycling: Hydrolysis/Glycolysis/Pyrolysis

Polymers containing carbonyl functions can be cleaved back to monomers by hydrolysis or glycolysis. For polyesters such as polyethylene terephthalate (PET), hydrolysis merely reverses the polycondensation reaction which formed the polymer initially. For PURs, the situation is more complicated, in that one of the polymer building blocks, the diisocyanate monomer, is not obtained but rather its reaction product with water, the diamine, is observed along with the polyol [56-61]. Carbon dioxide is evolved as a hydrolysis byproduct. Of course, the diamine can be converted back to the diisocyanate if desired. These reactions are diagrammed in equations 10.1 and 10.2. The PUR situation is further complicated by the presence of other hydrolyzable functions in the polymer, such as the urea functions in flexible foams, and the isocyanurate functions in some rigid foams. Fortunately, both of these groups can be hydrolyzed back to amine and carbon dioxide, as shown in equations 10.3 and 10.4 [62].

Polyester

$$-R_1\text{-}\overset{\displaystyle O}{\overset{\|}{C}}\text{-}O\text{-}R_2\text{-} + H_2O \longrightarrow -R_1\text{-}\overset{\displaystyle O}{\overset{\|}{C}}\text{-}OH + HO\text{-}R_2\text{-} \qquad (10.1)$$

Polyurethane

$$-R_3-NH-\overset{\overset{\displaystyle O}{\|}}{C}-O-R_4- \; + \; H_2O \longrightarrow -R_3-NH_2 \; + \; HO-R_4- \; + \; CO_2 \quad (10.2)$$

$$-R_3-NH-\overset{\overset{\displaystyle O}{\|}}{C}-NH-R_3- \; + \; H_2O \longrightarrow 2\;-R_3-NH_2 \; + \; CO_2 \quad (10.3)$$

$$+ \; 3H_2O \longrightarrow 3\;-R_3NH_2 \; + \; 3\;CO_2$$

$$(10.4)$$

The appeal of the hydrolysis method for breaking down PURs, especially from sources where both urethane and/or urea or isocyanurate linkages may be present, is the simplicity of converting everything to the di- or polyamine plus the polyol(s). The major disadvantage, of course, is that the diamine(s) and the polyol(s) must be separated before either is reused.

Glycolysis of PURs, on the other hand, converts everything to a mixture of polyhydroxy compounds which can be used directly without further separation, equation 10.5. The situation gets more complicated when urea or isocyanurate functions are involved. In both of these cases, amino functional moieties were produced along with the hydroxycarbamates, as shown in equations 10.6 and 10.7 [63].

$$-R_3-NH-\overset{\overset{\displaystyle O}{\|}}{C}-O-R_4- \; + \; HO-R_5-OH \longrightarrow -R_3-NH-\overset{\overset{\displaystyle O}{\|}}{C}-O-R_5-OH \; + \; HO-R_4- \quad (10.5)$$

$$-R_3-NH-\overset{\overset{\displaystyle O}{\|}}{C}-NH-R_3- \; + \; HO-R_5-OH \longrightarrow -R_3-NH-\overset{\overset{\displaystyle O}{\|}}{C}-O-R_5-OH \; + \; -R_3NH_2 \quad (10.6)$$

$$+ \; HO-R_5-OH \longrightarrow -R_3-NH-\overset{\overset{\displaystyle O}{\|}}{C}-O-R_5-OH \; + \; -R_3NH_2$$

$$(10.7)$$

Even in these cases, however, separation can be avoided by converting the amino functions to hydroxyl with ethylene or propylene oxides, equation 10.8. Note that the ethoxylation (or propoxylation) increases the functionality of the system by converting an amino function to two hydroxyl functions. Both hydrolysis and glycolysis have been explored at some depth and will be discussed more fully below.

$$-R_3NH_2 \; + \; 2 \; \overset{\displaystyle O}{\triangle} \longrightarrow -R_3N \overset{\displaystyle \nearrow^{OH}}{\underset{\displaystyle \searrow_{OH}}{}} \quad (10.8)$$

10.3.3.1 Steam Hydrolysis Flexible PUR foams can be hydrolyzed to diamine, polyol and CO_2 as described in equation 10.2, by the action of high pressure steam. General Motors Research Laboratories has published most of the available information on this process [64, 65]. High pressure steam will hydrolyze flexible foam rapidly at temperatures of 232–316 °C. The diamines can be distilled and extracted from the steam and the reclaimed polyols recovered from the hydrolysis residue. The hydrolysis temperature did have an effect on the polymer quality and yield. The optimum foam degradation temperature for the polyol yield and quality was 288 °C. Recovered polyol could be used in a flexible foam recipe at the 5% level with excellent results, as shown in Table 10.19.

Table 10.19   Comparison of flexible foam physical properties

Property	Control Foam	Foam Made With 5% Recycled Polyol
Density, lb/ft³ (kg/m³)	2.59 (41.6)	2.66 (42.6)
Tensile, psi (kPa)	23.2 (160)	23.9 (165)
Tear, lb/linear in (N/m)	2.31 (405)	2.18 (382)
Elongation, %	160	167

Source: Campbell and Meluch, 1976

A continuous process using a vertical reactor has been designed which yields 60–80% recovery of polyol from foam scrap at 288 °C with residence times of 10–28 minutes [66, 67]. A twin-screw extruder has been described as the reactor to continuously hydrolyze foam scrap. Temperatures and residence times are similar [68]. Ford researchers have explored the hydrolysis of PUR flexible foam seat cushion material, but at somewhat lower temperatures than the General Motors study [69]. A systematic kinetic study showed that the rates of formation of the 2,4- and 2,6-toluene diamines (TDA) and polypropylene oxide followed pseudo first order kinetics. Yields of TDA, isolated from the reaction mixture by vacuum distillation, were 65–80%. The variation in TDA yield with reaction conditions is shown in Table 10.20.

Table 10.20   Yield of pure diamine from reaction of PUR foam

Temperature °C	Time min	Yield %
185	30	48
185	60	61
200	15	64
200	30	86
200	60	77
200	360	93

Source: Mahoney, et al., 1974

A variation of the high pressure steam hydrolysis was explored using diethylene glycol as solvent. Temperatures of 190–220 °C were required for reasonable rates of hydrolysis. Use of a small amount (0.2% by weight of glycol) of lithium hydroxide greatly accelerated the degradation of the foam, and reaction times of a few minutes could be obtained at 170–190 °C. Kinetic analysis of the catalyzed hydrolyses suggested a complex reaction scheme, with two reactions proceeding at different rates, both producing toluene diamine as the product. At 180 °C, for example, the fast reaction is proceeding about five times faster than the slow one. The individual rate constants are shown in Table 10.21. It was surmised that the faster reaction might be the hydrolysis of the urethane linkages and the slower one, the urea linkages [70, 71].

Table 10.21   Rate constants for the lithium hydroxide catalyzed hydroglycolysis of flexible foam

Temperature °C	K'fast min^{-1}	K'slow min^{-1}
190	12.70	3.300
180	4.60	0.940
170	2.77	0.360
170	3.35	0.450
170	2.30	0.426
160	2.00	0.660
160	1.53	0.670
150	0.87	0.025

Source: Gerlock et al., 1984

The presence of the glycol solvent complicates product isolation from the reaction mixture. The procedure adopted was extraction of the polyol with hexadecane. Evaporation of the hexadecane yields a high quality polyol which could replace up to 50% of virgin polyol in a seating formulation. Capital and operating cost were estimated for a plant to consume 4 million pounds (1.8×10^3 t) of waste foam per year in a two shift operation. Approximately 2.4 million pounds (1.1×10^3 t) of polyol could be produced [72].

The hydroglycolysis procedure was applied to flexible foam obtained from an auto shredder operation. The dark-colored foam contained some water, volatile hydrocarbons, oil and finely divided glass, metal and plastic debris. However, the polyol recovered from this contaminated foam could be used as a replacement for 5–20% of the virgin polyol in foam formulation [73, 74].

Cured unsaturated polyester resin has been hydrolyzed under neutral conditions at 225 °C for 2–12 hours. Filtration of the solids remaining after hydrolysis yielded isophthalic acid (60% of theoretical), styrene-fumaric acid copolymer, and unhydrolyzed starting material. The copolymer could be recovered by acetone extraction in high yield. Propylene glycol recovery was quite low, under 50% [75].

10.3.3.2 Glycolysis   As described above, hydrolysis of PURs and polyureas produces both a polyol and a diamine, which require separation prior to reuse. Glycolysis of PURs, on the other hand, can yield a mixture of polyols (eq. 10.5), which could be reused directly. A series of papers and patents by Upjohn have shown the utility of this approach [76–80].

Scrap from a variety of sources, rigid foam, flexible foam, RIM, and microcellular elastomers have been digested with glycols to yield reusable polyols. As noted in eq. 10.6 and 10.7, urea and isocyanurate linkages can lead to polyamine formation, which can be converted to polyol with alkylene oxide (eq. 10.8). The process consists of digesting the ground material with an equal weight of a 90/10 mixture of di(alkylene) glycol/diethanolamine at 190–210 °C for several hours. After cooling and treatment with propylene oxide, the polyol mixture is filtered to remove insoluble materials and is ready for use.

The polyols produced by this method have fairly low equivalent weights (95 ± 5) and are most suitable for rigid foams. The reclaimed polyols could be substituted for up to 40% of the virgin polyol in typical rigid foam formulations, as shown in Table 10.22 [81]. To reuse these polyols in flexible foam formulations would require considerable chain extension with alkylene oxide.

Table 10.22   Rigid foam formulations and properties with recycle polyol

Components	A		B		C	
Polymeric iso., pts. by wt	140		140		140	
Aromatic Tetrol, pts. by wt	100		60		60	
Recycle Polyol, pts. by wt			40		40	
Properties	**A**		**B**		**C**	
Density, lb/ft^3 (kg/m^3)	2.03	(32.5)	2.09	(33.5)	2.09	(33.5)
Compressive Strength, psi (kPa)	17.3	(119)	21.5	(148)	20.9	(144)
Oxygen Index, %	23.0		24.5		24.3	

Notes: B = recycled polyol from aromatic tetrol foam; C = recycled polyol from sorbitol-based foam
Source: Ulrich, et al., 1978.

RIM scrap could be glycolyzed to a polyol in similar fashion. Reuse of the polyol in RIM formulations was limited to about 5%. Above this level, physical properties deteriorated, as illustrated in Table 10.23. However, polyols derived form RIM scrap could be used at levels of 25% in rigid foam formulations without detrimental effects,; Table 10.24 [82].

Table 10.23   Property comparison of RIM systems containing recycle polyol

Property	Control		4% Recycle Polyol		RIM Containing: 8% Recycle Polyol		12% Recycle Polyol	
Flexural Modulus, 10^3 psi (MPa)								
-29 °C	64.10	(442)	65.90	(454)	75.04	(517)	98.70	(680)
24 °C	15.10	(104)	15.00	(103)	15.40	(106)	10.60	(73)
70 °C	4.54	(31)	4.23	(29)	3.02	(21)	2.47	(17)
Flexural Strength, 10^3 psi (MPa)								
-29 °C	3.84	(26.5)	3.91	(26.9)	4.46	(30.7)	5.24	(36.1)
24 °C	1.11	(7.6)	1.05	(7.2)	1.00	(6.9)	0.80	(5.5)
70 °C	0.38	(2.6)	0.36	(2.5)	0.28	(1.9)	0.21	(1.4)
Tensile Modulus, 10^3 psi (MPa)								
50%	1.66	(11.4)	1.13	(7.8)	1.39	(9.6)	1.34	(9.2)
100%	1.61	(11.0)	1.31	(9.0)	1.51	(10.4)	1.56	(10.7)
200%	1.83	(12.6)	1.74	(12.0)				
Tensile Strength,10^3 psi(MPa)								
	2.00	(13.7)	1.71	(11.7)	1.61	(11.1)	1.90	(13.0)
Elongation, %	170		210		130		180	
Heat Sag, in (mm) at 121 °C	0.40	(10.2)	0.50	(12.7)	0.74	(18.8)	0.48	(12.2)

Source: Ulrich, et al., 1979

Simioni has explored the glycolysis reaction further, using potassium acetate as catalyst [83, 84]. Rigid foams were prepared with the reclaimed polyol, with reasonable properties obtained using 100% of the reclaimed material, Table 10.25. The good thermal conductivity indicates good cell structure and high closed cell content of the foams. Interestingly, the use of the recovered polyol at intermediate levels seems to yield improved compressive strengths, Figure 10.6.

Table 10.24   Property comparison of rigid foams containing recycle polyol from RIM scrap

Foam Property	Control	Rigid Foam Containing:		
		10% Recycle Polyol	25% Recycle Polyol	50% Recycle Polyol
Density, lb/ft^3 (kg/m^3)	1.96 (31.8)	2.24 (36.3)	2.16 (35.0)	2.05 (33.2)
Compressive Strength, psi (kPa)	20.3 (140)	23.6 (163)	22.6 (156)	15.8 (109)
Oxygen Index, %	20.6	20.6	20.6	21.1
Dimension Stability				
Humid Aging, %	6.9	5.4	7.4	17.2
Dry Aging, %	2.9	1.5	2.0	3.7

Notes: Humid aging , 14 days at 70 ˚C, 100% RH; Dry aging, 14 days at 93 ˚C
Source: Ulrich, et al., 1979

Table 10.25   Properties of foams obtained from polyol blends containing glycolysis polyol

Property	Control Polyol Sorbitol-based	30% Recycle Polyol	50% Recycle Polyol	70% Recycle Polyol	100% Recycle Polyol
Density					
lb/ft^3	1.75	1.81	1.81	1.81	1.81
(kg/m^3)	(28)	(29)	(29)	(29)	(29)
Compressive Strength					
psi	24.2	24.8	28.4	29.0	19.9
(kPa)	(167)	(171)	(196)	(200)	(137)
Therm. Cond.					
Btu-in/ft^2-h-˚F	0.137	0.134	0.136	0.130	0.126
(w/m˚K)	(.0198)	(.0194)	(.0196)	(.0187)	(.0182)
Oxygen Index, %	21.5	21.7	22.0	22.3	22.7

Source: Simioni, et al., 1983

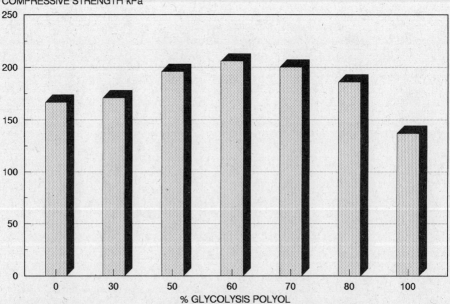

Figure 10.6   Compressive strength improvement in rigid foams with glycolysis polyol

Microcellular elastomers can be glycolyzed with dipropylene glycol and tetrabutyl titanate as catalyst [85, 86]. The resultant, near-linear polyols yield PUR foams of low dimensional stability. These stability characteristics can be greatly improved by using higher functionality isocyanates and isocyanurate formulations to increase cross-link density. The effect of isocyanurate content on volume change under humid age conditions is shown in Figure 10.7.

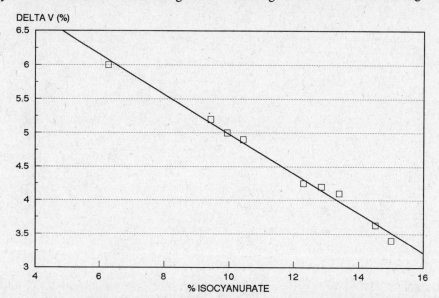

Figure 10.7   Effect of trimer content on the volume change in foams obtained from glycolysis polyols

Although a considerable amount of research and development effort has been expended on the hydrolysis/glycolysis of PUR foams [87–94], including production facilities at the million pound-per-year (450 t) level, the perception remains that the processes are too costly relative to virgin polyol production [95]. However, recently The Ford Motor Company, Europe, has described a recovery process which yielded acceptable polyol at claimed costs 30% less than new polyol [96, 97]. Studies are continuing at the Technical University, Aalen, where the process was developed. Prof. G. Bauer has described a pilot facility to recover polyols from automotive seating by glycolysis [98, 99]. The process can utilize the polyester fabric as well. At a ratio of flexible foam to polyester of 1.5:1 and a ratio of scrap to added glycol of 2:1, a polyol with the properties shown in Table 10.26 was obtained. The pilot facility has an output capacity of 198 lb/h (90 kg/h) of polyol. A schematic of the process is shown in Figure 10.8.

Table 10.26   Properties of glycolysis polyol

Property	
Viscosity, cP (MPa·s)	5500
Hydroxyl Number (mg KOH/g)	380
Acid Number (mg KOH/g)	1.5
Water Content (%)	0.1
Aromatic Amine Content (%)	0.5

10.3.3.3 Pyrolysis   In addition to chemical treatment, thermal energy can be used to break down polymers to recover chemical values. In general, only vinyl addition polymers yield appreciable amounts of monomer via "unzipping" or depolymerization mechanisms [100,

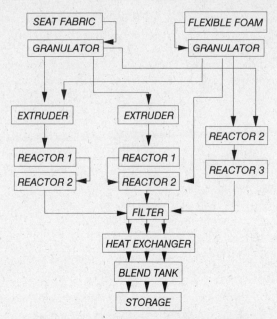

Figure 10.8   Flow diagram of alcoholysis pilot plant

101]. Thus, polymethyl methacrylate (PMMA) can be pyrolyzed to methyl methacrylate (MMA) in 90% yield. Polystyrene yields some styrene, along with other products [102]. The pyrolysis of polyolefins produces a mixture of saturated and unsaturated hydrocarbons similar to naphtha cracking products [103–106]. The presence of chlorine in the polymers presents complications. PVC, for example, eliminates hydrogen chloride gas which presents corrosion problems. In addition, hydrogen chloride can catalyze polymerization and condensation reactions among other pyrolytic products [107]. Bases such as calcium hydroxide can moderate these complications. Braslaw and coworkers [108] have examined the plastic mixture from automotive waste under pyrolytic conditions. As shown in Table 10.27 the product composition was about 45% char (25% expected based on non-volatile content), 35% liquid, and 20% gas, at a final temperature of 550–600 °C. Attempted scale-up experiments gave disappointing results, with a much higher gas to liquid weight ratio. In the authors' opinion, the high nitrogen and sulfur content of the liquids would make them less desirable as a fuel.

Table 10.27   Pyrolysis of auto shredder residue: bench-scale experiments

Final Temperature °C	Char % of Feed	Liquid % of Feed	Gas % of Feed
462	47.0	36.8	16.2
563	45.7	33.9	20.4
657	46.3	35.2	18.5

Source: Braslaw, et al., 1983

The pyrolysis of PUR materials has been examined in some detail. Pyrolysis conditions have ranged from 250–1200 °C and have included inert as well as oxidative atmospheres [109–113]. Pyrolysis of PURs based on polypropylene glycol and TDI in an inert atmosphere at 200–250 °C occurs by random scission of PUR bonds to isocyanate and hydroxyl [114, 115]. At higher temperature, scission of the polyether chains occurs to yield a variety of oxygenated products. Under similar conditions, flexible foams lose most of their nitrogen at about 300 °C, concurrent with the loss of about one-third of their mass. For rigid foams, the higher the temperature (200–500 °C), the greater the nitrogen and weight loss. At 200-300 °C, rigid PUR foams produce isocyanate and polyol in about equal proportions [116]. Other studies have indicated polyureas from TDI-based flexible foams, and polycarbodiimides from MDI based rigid foams [117]. Above 600 °C, both the polyureas and polycarbodiimides decompose further to nitriles, olefins and aromatic compounds.

Considerable effort has been devoted to large scale pyrolysis trials of SMC scrap over the last two years by the SMC Auto Alliance and General Motors Corp. [118]. The equipment and processes were originally designed for scrap tires and have presented some operational problems which may have been avoided with proper design. Nonetheless, the trials did show that SMC pyrolysis could generate a fuel gas in quantities sufficient to sustain the pyrolysis. Also, a heavy oil pyrolysate of potential heating value was obtained. The solid byproducts, carbon, calcium carbonate and glass fibers exit the reactor and are cooled and isolated for recovery and reuse. A laboratory evaluation demonstrated that up to 20% of the ground solid byproduct could be substituted for calcium carbonate in SMC formulations with no loss of properties or surface quality. The results are encouraging enough that the SMC industry is exploring the permit requirements for siting a properly designed pyrolysis unit in Indiana along the Interstate 59 corridor [119].

10.3.4 Energy Recovery

A few studies of incineration of thermoset polymers have been conducted. It appears that the technology exists to safety incinerate these materials economically, with or without heat recovery, in an environmentally safe manner. In particular, incineration of PUR foams [120], mixed plastics containing PUR and PVC [121], and auto shredder residue (ASR) [122], have been combusted under various conditions with emissions well within environmentally acceptable limits.

For example, the combustion of rigid PUR foam waste has been examined. Experiments showed that full material decomposition is achieved at temperatures of 700 °C and that mass reduction exceeds 85%. Stack gas emissions showed 80 ppm NOx, 0.25 ppm HCl, and a trace of trichlorofluoromethane. There was no trace of CO, free isocyanate, HCN, phenol, formaldehyde or phosgene [123].

Mixed plastic waste from an auto shredder operation has been subjected to the Voest-Alpine High Temperature Gasification process [124]. In this process, the plastic waste is gasified on a coke bed previously heated at 1600 °C, producing hydrogen and carbon monoxide. These very hot gases are cooled and the heat used to generate steam. Noncombustible material forms a liquid slag which exits the reactor into a water bath where it freezes into glassy granules. The granules, which have very low leachables, can be used as a building material. The fuel gas from the process can be burned to generate still more energy, for an overall recovered energy efficiency of 80–85%. The emissions from the combustion of this gas are quite low and meet very strict air quality standards, Table 10.28.

Energy recovery from PUR RIM scrap, which has an energy content equivalent to coal, about 12–14,000 BTU/lb (38–32 MJ/kg), has been reported recently, using several different combustion technologies; fluidized bed, rotary kiln and mass burn combustors. Although each technique had different feed preparation requirements, clean combustion could be achieved in each. Significantly, the stack gas emissions were well within permitted limits and below what would be expected from fuels such as coal or oil. Figure 10.9 compares the total emissions from various fuels for the production of the same quantity of steam [125].

10.4 Applications

Most of the applications for the material recycle of thermoset resins fall into two broad categories; ground thermoset resin plus adhesive binder; and ground thermoset as filler in a

Table 10.28 High temperature gasification comparison of flue gas emissions

	MWCP[a] 1981	MWCP[b] 1988	TA-LUFT[bc] 1986	HTV Gas[d]
Particulates[e]	50	10	30	<1.0
Cl 450	5	50	1.7	
SO_2	230	25	100	<60
NOx	220	100.0	500	60
Heavy Metals				
Class 1: Cd, Hg, Ti	2	0.1	0.2	0.01
Class 2: As, Co, Ni, Se, Ti	4	0.5	1.0	0.01
Class 3: Sb, Pb, Cr, Cu, Mn, Pt, Pd, Rh, V, Sn	N.D.	1.0	5.0	<1.00
2,3,7,8-TCDD Equivalent	$>>0.1 \times 10^{-6}$	0.1×10^{-6}	N.D.	$<0.1 \times 10^{-6}$

Notes: [a] MWCP = Municipal Waste Combustion Plant–1981; [b] 1986; [c] TA-Luft = German Gov't Air Quality Standard–1986; [d] HTV Gas = Emissions from combustion of pyrolysis gas; [e] Figures in mg/Nm^3, related to 11% oxygen content

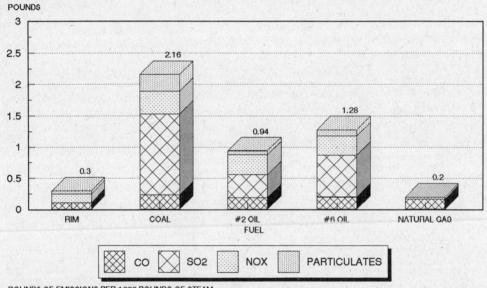

POUNDS OF EMISSIONS PER 1000 POUNDS OF STEAM
RIM RESULTS FROM FLUIDIZED BED COMBUSTOR
DATA FROM NORTH AMERICAN COMBUSTION HANDBOOK

Figure 10.9 Stack gas emissions from steam production with various fuels

thermoplastic or uncured thermoset matrix. A notable exception is the thermoforming of RIM PUR materials described in Section 10.3.2.2. In the first category, a large amount of the thermoset material is bound together with a small amount of the adhesive. The properties of the composite are characteristic of the original thermoset. An example of this system is PUR flexible foam rebond used as carpet underlay. In the second example, a lesser amount of the thermoset is bound as filler in a relatively large amount of a thermoplastic. In this case the resultant composite exhibits the properties of the matrix material modified by the thermoset filler. An example of a thermoplastic system is the SMC-filled PP referred to in Section 10.3.2.5. Various thermoset examples are also described in this section with reground PUR, phenolic, epoxy and SMC scrap as fillers.

Other applications of thermoset with thermoplastic binders include carpet backing with ethylene-vinyl acetate (EVA) or ethylene-methyl acrylate copolymers and ground foam scrap [126] and carpet backing from ground PUR bumper scrap with EVA binder [127].

The ground plastics scrap from a decommissioned automobile has been bound together with an isocyanate bonding agent [128]. The plastic materials from a 1977 Pontiac Trans Am were ground to a minus 20 mesh particle size, and formed into a composite panel with aspen wafer board skins using a polymeric MDI-based isocyanate as the binder. The composite possessed a much higher flexural modulus than a wafer board made with aspen wafers alone, Table 10.29.

Table 10.29   Comparison of properties of composite panel wafer board*

Property	Composite		Aspen Waferboard	
Hardness: Shore D	60		60	
Density, lb/ft^3 (g/cm^3)	43.7	(0.70)	43.0	(0.69)
Flexural Modulus,10^3 psi (MPa)	384.00	(2646)	130.00	(896)
Mod. of Rupture, 10^3 psi (MPa)	3.36	(23.2)	6.02	(41.5)
Water Immersion, 24 h at 25 °C				
% Change, volume	21		20	
% Change, weight	50		92	
Taber Abrasion, 1000 cycles/1000 g				
H22 Wheel, Loss, 10^{-3} oz (mg)	8.8	(250)	9.9	(280)

* Composite panel made with ground auto plastic and waferboard
Source: Carroll, McClellan, U.S. Patent 4 382 108, 1983.

Another attack on the utilization of automotive plastic waste involves a separation process to recover less heterogeneous materials. The flexible PUR foam has been separated from ASR. Solvent extraction and detergent washing served to remove oil, grease, and dirt contaminants, to produce an odor free foam perhaps suitable for rebond or other recycle techniques such as hydrolysis [129].

Carrier has described recently a technology to recycle postconsumer waste plastic utilizing a 50% or higher thermoplastic material as carrier, to produce a new, reusable, high-quality material [130]. The heart of the process is a plastification roller which carefully crushes, kneads and melts together the mixed plastics at about 160 °C to minimize thermal material damage. The resultant homogeneous mix is extruded and compression molded in automatic presses. Mechanical properties of a 50% scrap/50% PE molded flat sheet are compared in Table 10.30 with low-density polyethylene (LDPE). The material produced with this technology is similar to LDPE in physical properties and may be suitable in some applications replacing wood, metal

Table 10.30   Comparison of properties of industrial waste plastic/PE (50/50) and LDPE

Physical Property* (units)	Recycled Plastic (50% PE)		LDPE	
Density, lb/ft^3 (kg/m^3)	62.4	(1000)	57.4	(920)
Elast. Mod., 10^3 psi (MPa)	159.5	(1100)	72.5	(500)
Tensile Strength, 10^3 psi (MPa)	2.61	(18)	3.33	(23)
Notch Sens., ft-lb/in^2 (kJ/m^2)	1.43	(3)		
Max. Service Temperature (°C)	85		90	
Min. Service Temperature (°C)	-50		-50	
Water Absorption	0.3		0.1	

* Testing done on compression molded samples
Source: Carrier, ANTEC, 1989

concrete or virgin plastics. Although the materials discussed are principally thermoplastics, thermosets are not specifically excluded from the process.

Thermoset granules can be used as the blasting medium to selectively remove paint from various substrates, particularly from SMC or other plastic auto parts. Plastic blasting media cause less substrate damage and can save SMC parts otherwise destined for scrap [131].

The cutting of flexible PUR foam slabstock into cushioning generates considerable waste, 8–12%, depending on the shape of the foam blocks, and the complexity of the cut parts. Molded seat cushioning production generates some scrap as well. All of this scrap, plus some imported material, is used in the production of rebonded carpet underlay as described in Section 10.3.2.1. About 280 million pounds (126×10^3 t) of scrap flexible foam per year is consumed in this way. In addition, about 100–120 million pounds ($45–54 \times 10^3$ t) of scrap foam is imported for this application.

Flexible foam rebond for carpet underlay represents a most successful recycle application, as described above. S. Miyama has also discussed how foams may be separated from scrap containing PVC skins for rebonding into shapes for various automotive applications [132]. In similar fashion, ground PUR bumper scrap is mixed with a PUR adhesive prepolymer for use as a repair material for PUR auto parts [133]. Foam-asphalt composites have been described as useful noise absorbing and heat insulating materials in automobiles [134].

An interesting emerging application for rebonded PUR foam is its use as a substrate for hydroponic gardening. When compared to rock wool substrates, the less hydrophilic PUR has much lower moisture content and higher air content. Thus the PUR foam has less tendency to overwetting, and is suitable for high growing temperatures. Also, it can be steam sterilized at 100 °C for reuse, up to four times. This growing application threatens to exceed the supply of flame retardant free scrap foam [135].

PUR foam scrap has been described as an excellent material for the clean up of oil spills. Chopped PIR foam is stuffed into porous bags, which are linked together to form a boom. The oil is adsorbed on the foam and can be recovered by squeezing the boom between rollers [136, 137].

Although applications for recycled thermosets are limited today, the variety of rapidly emerging recycle technologies will surely change this situation. Helping to advance progress in the recycle of thermoset materials is the heightened activity of the appropriate trade organizations. The role of the SMC Auto Alliance has been described in the pyrolysis section. The Polyurethane Division of SPI has formed recently a Polyurethane Recycle and Recovery Council (PURRC) to accelerate technology developments in PUR recycle, particularly of postconsumer scrap [138]. And the Council for Solid Waste Solutions of SPI has formed a Durables Committee to explore recycle opportunities from durable goods plastic waste [139]. See Appendix 3 for further discussions of these organizations.

10.5 Global Activities

Much of the activity discussed previously mirrors activities in Europe and Japan. BMW has proclaimed its intent to produce a fully recyclable car, including plastic materials, [140, 141]. Toyota is recycling some RIM scrap into mud flaps [142]. DSM and BASF have discussed SMC regrind recycle [143, 144]. And, of significance to all plastics recycling, Germany's big three— Bayer, BASF, and Hoechst—have announced their intention to form a research joint venture for recycling plastic [145]. Hopefully, these worldwide efforts will help to establish all plastics as recyclable materials. We need to emphasize also the real energy savings of plastics over competing materials of construction [146, 147], and the enormous energy savings achievable through the use of lightweight plastics and composites in transportation applications [148]. It would be folly, indeed, if we lose these real benefits because of a lack of public understanding of the recyclability of plastic materials.

References

[1]   Curlee, T. R. *The Economic Feasibility of Recycling*. New York: Praeger, 1986.
[2]   Leidner, J. *Plastics Waste: Recovery of Economic Value*. New York: Marcel Dekker, Inc., 1981.
[3]   Schreiber, H. P. *SPE ANTEC, 1977*, p. 526.
[4]   Tran-Buni, C. V., Deanin, R. D. *ACS Symposium Ser. 367*, 237 (1988).
[5]   Kinstle, J. F. in *The Impacts of Material Substitution on the Recyclability of Automobiles*. Blelloch, R. A., Ed. New York: ASME, 1984.
[6]   Meyer, L. G. *British Plastics and Rubber 1985* (Oct.): p. 32.
[7]   Seymour, R. B. *Ind. and Eng. Chem., Ann. Rev. 2* (1974): p. 424.
[8]   Kreuziger, K., Lubish, H-J. *Plaste Kautsch, 27*(1) (1980): p. 47 .
[9]   Dueweke, P. W. "Plastics Fare Well in Energy Use." *Plastics News* (9 July 1990): p. 8.
[10] Franklin Associates, LTD. "Characterization of Plastic Products in Municipal Solid Waste." Prepared for the Council for Solid Waste Solutions, Feb. 1990.
[11] Manno, P. in *Reaction Polymers*. Gum, W. F., Ed. Munich, Vienna, New York: Hanser Publishers (in press).
[12] *Modern Plastics 68* (1) (Jan. 1991): p. 113.
[13] Reference 12, p. 117.
[14] Reference 12, p. 116.
[15] Reference 12, p. 116.
[16] Reference 12, p. 114.
[17] Reference 12, p. 119.
[18] Meister, B. and Schaper, H. *Kunststoffe, 80* (1990): p. 1260.
[19] Bauer, S. H. *Plastics Engineering, 33*(3) (1977): p. 44; *SPE ANTEC, 1976*, p. 650.
[20] Bartas, B. *Modern Plastics Encyclopedia 1989*, p. 336.
[21] *Modern Plastics 67* (5) (May 1990): p. 51.
[22] Hull & Co. "End Use Market Survey of the Polyurethane Industry in the United States and Canada." Prepared for the Society of the Plastics Industry, Inc., Polyurethane Division, 29 May 1990.
[23] *Polyurethane Handbook*. Oertel, G., Ed. Munich, Vienna, New York: Hanser Publishers, 1985, p. 176.
[24] Woods, G. *The ICI PURs Book*. New York: J. Wiley, 1987, p. 203.
[25] Baumann, B. D., Burdick, P. E., Bye, M. L., Galla, E. A. *SPI International Tech. Conf. 1983*, p. 139.
[26] Morgan, R. E., Weaver, J. D. *SAE International*. Paper Number 910580. 25–28 Feb. 1991.
[27] Vanderhider, J. A. International Polyurethane Forum, Nagoya, Japan, 1990.
[28] Cornell, M. C. *Polyurethanes 1990, 33rd SPI Annual Technical/Marketing Conference*, Oct. 1990, p. 440.
[29] Reference 18.
[30] Taylor, R. P., Eiben, R., Rasshofer, W., Liman, U. *SAE International*. Paper Number 910581. 25–28 Feb. 1991.
[31] Singh, S. N., Piccolino, E., Bergenholz, J. B., Smith, R. C. *SAE International*. Paper number 910582. 25–28 Feb. 1991.
[32] Reference 26.
[33] Reference 30.
[34] Reference 31.
[35] Kalyon, D. M., Hallouch, M., Fares, N. *SPE ANTEC 1984*, p. 640.
[36] Kalyon, D.M., Fares, N. *Plast. Rubber Process Appl., 5*, (1985): p. 369,.
[37] Deanin, R. D., Ashar, B. V. *Org. Coatings & Plastics, 41* (1979): p. 495.

[38] Kruppa, R. A. *Plastics Tech., 23*, (5) (1977): p. 63.
[39] Tran-Buni, C. V. Deanin, R. D. *SPE ANTEC, 1987*, p. 446.
[40] Hanson, D. G. *SPI Composites Institute, 46th Annual Conference*, 18–21 Feb. 1991.
[41] Graham, W. D., Jutte, R. B. *SPI Composites Institute, Press Molders Conference*, 2 May 1990.
[42] Jutte, R. B., Graham, W. D. *SPI Composites Institute, 46th Annual Conference*, 18–21 Feb. 1991.
[43] Butler, K. *SPI Composites Institute, 46th Annual Conference*, 18–21 Feb. 1991.
[44] Short, H. "DSM to Expand Auto Recycling Efforts." *Plastics News* (28 May 1990): p. 1, 5.
[45] Short, H. "BASF Offers Recycling Progress Report." *Plastics News* (9 July 1990): p. 13.
[46] Norris, D. *SPI Composites Institute, 45th Annual Conference*, 12–15 Feb. 1990.
[47] Vernyi, B. "Thermoset Recycling May Be Possible." *Plastics News* (26 Feb. 1990): pp. 1, 7.
[48] Reference 41.
[49] Reference 42.
[50] Reference 41.
[51] Reference 42.
[52] Reference 43.
[53] *Modern Plastics 67* (2) (Feb. 1990): p. 124.
[54] Morel, E., Richert, G., Marting, C. *Chem. Abst. 96* 163775C (1982). Commission of European Communities. Report No. ERU 6833FR.
[55] Rosett, L. K., Rosett, K. *Plast. Compd. 6* (7) (1983): p. 47.
[56] Chapman, T. M. *J. Polym. Sci., A. Polym. Chem., 27* (1989): p. 1993.
[57] Grigat, E. *Kunststoffe, 68* (1978): p. 281.
[58] Ulrich, H., Odinak, A., Tucker, B., Sayigh, A. A. R. S. *Polym. Eng. Sci., 18* (1978): p. 844.
[59] Sheratte, M. B. *SPI Tech. Conf.* (1977): p. 59.
[60] Kinstle, J. F., Forshey, L. D., Valle, R., Campbell, R. R. *Polymer Preprints 24*(2) (1983): p. 446.
[61] Ulrich, H., *Advances in Urethane Sci. Tech., 5* (1978): p. 49.
[62] Reference 58.
[63] Reference 58.
[64] Campbell, G. A., Meluch, W. C. *Environ. Sci. Tech., 10*(2) (1976): p. 182.
[65] Meluch, W. C., Campbell, G.A. U.S. Patent 3 978 128, 1976.
[66] Campbell, G. A., Meluch, W. C. *J. Applied Polym. Sci., 21* (1977): p. 581.
[67] Salloum, R. J. *SPE ANTEC, 1981*, p. 491.
[68] Grigat, E., Hetzel, H. U.S. Patent 4 051 212, 1977.
[69] Mahoney, L. R., Weiner, S. A., Ferris, F. C. *Environmental Sci. & Tech., 8* (2) (1974): p. 135.
[70] Gerlock, J., Braslaw, J., Zimbo, M. *Ind. Eng. Chem. Process Des. Dev. 23*, 545, (1984).
[71] Japan Patent 54-068883, 1979.
[72] Braslaw, J., Gerlock, J. L. *Ind. Eng. Chem. Process Des. Dev., 23* (1984): p. 552.
[73] Reference 69.
[74] Reference 70.
[75] Reference 60.
[76] Reference 58.
[77] Ulrich, H., Tucker, B., Odinak, A. Gamache, A. R. *Elastomers & Plastics, 11* (1979): p. 208.
[78] Frulla, F. F., Odinak, A., Sayigh, A. A. R. U.S. Patent 3 709 440.
[79] Frulla, F. F., Odinak, A., Sayigh, A. A. R. U.S. Patent 3 738 946, 1973.

[80]   Tucker, B., Ulrich, H. U.S. Patent 3 983 087, 1976.
[81]   Reference 58.
[82]   Reference 77.
[83]   Simioni, F., Bisello, S., Cambini, M. *Macplas, 8*(47) (1983): p. 52.
[84]   Simioni, F., Bisello, S., Tavan, M. *Cellular Polym., 2*(4) (1983): p. 281.
[85]   Simioni, F., Modesti, M., Navazio, G. *Macplas, 12*(88) (1987): p. 157.
[86]   Simioni F., Modesti, M., Brambilla, C.A. *Cellular Polym., 8* (1989): p. 387.
[87]   Prajsnar, B. *Recycling* (1979): p. 11213.
[88]   Ten Broeck, T. R., Peabody, D. W. U.S. Patent 2 937 151, 1960.
[89]   Bridgestone Tire Co. French Patent 1 484 107, 1967.
[90]   Schutz, W. U.S. Patent 4 339 358, 1982.
[91]   Wolf, H.O. U.S. Patent 3 225 094, 1965.
[92]   Pizzini, L. C., Patton, J. T. U.S. Patent 3 441 616, 1969.
[93]   Yukuta, T., Ishiwaka, T., Usui, K., Akoh, M. Great Britain Patent 2 062 660, 1980.
[94]   Iwasaki, K., Kawakami, H., Hirose, T. Japan Patent 77-95849, 1977.
[95]   *Chem. Market Rep. 233*(1) (1988): p. 5.
[96]   *Urethanes Technology* (Apr./May 1989): p. 19.
[97]   APME Newsletter, No. 29 (1988).
[98]   Bauer, G. Recycle 1990, Davos, Switzerland, 29–31 May 1990.
[99]   Bauer, G. Germany Patent 2 738 572; Europe Patent 948.
[100] Reference 5.
[101] Tesoro, G. *Polymer News, 12* (1987): p. 265.
[102] Roy, M., Rollin, A. L., Schreiber, H. P. *Polym. Eng. Sci., 18* (1978): p. 721.
[103] Bhatia, J. *Polymer Preprints 24,* (1983): p. 436.
[104] Bhatia, J., Rosi, R. A. *Chem. Eng.,* (10) (1982): p. 58.
[105] Kaminsky, W., Sinn, H. *Kunststoffe, 68* (1978): p. 284.
[106] Hammerling, L., Kaminsky, W., Reichanrdt, R. *Recycle Int., 4* (1984): p. 603.
[107] Reference 106.
[108] Braslaw, J., Gealer, R. L., Wingfield, R. C. *Polymer Preprints, 24* (1983): p. 434.
[109] Madorsky, S. L., Straws, S. *J. Polym. Sci. 36,* (1959): p. 183.
[110] Ingham, J. D., Rapp, N. S. *J. Polymer Sci., A2,* 689 (1964): p. 4941.
[111] Tilley, J. N., Nadeau, H. G., Reymore, H. E., Waszeciak, P. H., Sayigh, A. A. R. *J. Cell. Plast. 4,* (2) (1968): p. 56.
[112] Levin, B. C. NBSIR 85-3267, 1986.
[113] Maya, P., Levin, B. C. *Fire Mater. 11,* (1) (1987) and references cited therein.
[114] Woolley, W. D., Fardell, P. J. *Fire Res. 1,* (1977): 11.
[115] Woolley, W. D. *Br. Polym. J.4,* (1972): p. 27.
[116] Wooley, W. D., Fardell, P. J., Buckland, I. G. Fire Research Note No. 1039 (Aug. 1975).
[117] Chambers, J., Jiricny, J., Reese, C. B. *Fire and Mater, 5* (1981): p. 133.
[118] Reference 46.
[119] SMC Auto Alliance. *SPI Composites Institute, 46th Annual Conference,* 18–21 Feb. 1991.
[120] Hilyard, N. C., Kinder, A. I., Axelby, G. L. *Cellular Polymers, 4* (1985): p. 367.
[121] Binder, J. J. Proc. Nat. Waste Process, Conf. 11, 1984, p. 458.
[122] Freimann, P. Voest-Alpine Report on "High Temperature Gasification of Shredder Residue," 1989.
[123] Reference 120.
[124] Reference 122.
[125] Myers, J. I., Farrissey, W. J. *SAE International.* Paper Number 910583. 25–28 Feb. 1991.
[126] Alexander, R. L. U.S. Patent 3 406 127, 1968.

[127] Mitsubishi Motors, Japan Patent 58-073320, 1983.

[128] Carroll, W., McClellan, T. R. U.S. Patent 4 382 108, 1983.

[129] Jody, B. J., Daniels, E. J., Bonsignore, P. V. *SAE International*. Paper Number 910854. 25–28 Feb. 1991.

[130] Carrier, K. *SPE ANTEC, 1989*, (1989): p. 9807.

[131] Ackermann, K. "Plastics Granules 'Take It All Off' Auto Parts." *Plastics News* (23 July 1990): p. 5.

[132] Miyama, S. *Conservation and Recycling, 10* (1987): p. 265.

[133] Toyota Motor Co. Japan Patent 58-098374, 1983.

[134] Kumasaka, S., Ishiro, K., Kamanaka, T., Horikoshi, S. Japan Patent 61-272250.

[135] Benoit, F., *Kunststoffe, 79*(4) (1989): p. 348.

[136] Golding, G. R., Boaggio, R. J. U.S. Patent 4 366 067, 1982.

[137] Faudree, III, T. L. U.S. Patent 4 230 566, 1980.

[138] Noble, H. L. *SPI 33rd Annual Polyurethane Technical/Marketing Conference*, 2 Oct. 1990; *Plastics World* (Nov. 1990): p. 12.

[139] The Council for Solid Waste Solutions, Washington, D.C. 20005.

[140] Bregar, W. "BMW To Scrap Z1, But Still Will Pursue Recyclable Cars" *Plastics News* (3 Dec. 1990): p. 10.

[141] *Mechanical Engineering* (Nov. 1990): p. 66.

[142] Morgan, R. E., Weaver, J. D. *SAE International*. Paper Number 910580. 25–28 Feb. 1991.

[143] Vernyi, B. "Premix Regrinds SMC, BMC for Uses as Fillers." *Plastics News* (28 May 1990): p. 5.

[144] Reference 45.

[145] Short, H. "Chemical Giants Plan Joint R & D." *Plastics News* (21 May 1990): pp. 1, 24.

[146] Gum, W. *SPI 33rd Annual Polyurethane Technical/Marketing Conference*, 2 Oct. 1990.

[147] Fussler, C. *SAE International*. 25–28 Feb 1991.

[148] Farrissey, W. J. *SAE International*. Paper Number 910579. 25–28 Feb. 1991.

Abbreviations

ABS	Acrylonitrile-butadiene-styrene
ART	Advanced Recycling Technology
ASR	Automatic shredder residue
ASTM	The American Society for Testing and Materials
BHET	Bis-hydroxyethyl terephthalate
BMC	Bulk-molding compound
BPF	British Plastics Federation
BPI	British Polyethylene Industries
C-PP	Copolymer-polypropylene
CAS	Chemical Abstracts Service
CFC	Chlorofluorocarbon
CFR	Code of Federal Regulations
CPRR	Center for Plastics Recycling Research
CPVC	Chlorinated polyvinyl chloride
CSWS	The Council for Solid Waste Solutions
DEG	Diethylene glycol
DMT	Dimethyl terephthalate
DOT	Department of Transportation
DTDA	4,4′dithiodianiline
EG	Ethylene glycol
EPA	Environmental Protection Agency
EPDM	Ethylene-polypropylene-diene monomer
EPS	Expandable (expanded) polystyrene
ESCR	Environmental stress crack resistance
EVA	Ethylene-vinyl acetate
EVC	European Vinyls Corporation
EVOH	Ethylene-vinyl alcohol
FDA	Food and Drug Administration
GE	General Electric
GECOM	Groupe d'Etude pour le Conditionnement Moderne
H-PP	Homo-polypropylene
HCFC	Hydro-chlorofluorocarbon
HCl	Hydrochloric (acid)
HDPE	High-density polyethylene
HDT	Heat deflection temperature
HIPS	High-impact polystyrene
I-PP	Impact-modified polypropylene
ICS	International Container System
IPA	Isophthalic acid
IV	Intrinsic viscosity
LDPE	Low-density polyethylene
MDA	Methylene dianiline
MDI	4,4′diphenylmethane diisocyanate
MDPE	Medium-density polyethylene
MFR	Melt flow rate
MI	Melt index
MMA	Methyl methacrylate

M_n	Number average molecular weight
MRF	Material recovery facility
MSW	Municipal solid waste
M_w	Weight average molecular weight
n	Degree of polymerization
NITSE	Nonintermeshing twin-screw extruder
NJCT	New Jersey Curbside Tailings
non-MSW	Nonmunicipal solid waste
NMP	n-methyl-2-pyrrolidinone
NPRC	National Polystyrene Recycling Company
NRT	National Recovery Technologies
NYSERDA	New York State Energy Research and Development Authority
OPP	Oriented polypropylene
OPS	Oriented polystyrene
P&G	Procter & Gamble
PA	Phthalic acid
PBT	Polybutylene terephthalate
PC	Polycarbonate
PE	Polyethylene
PET	Polyethylene terephthalate
PG	Propylene glycol
PIA	Plastics Institute of America, Inc.
PIR	Polyisocyanurate
PMMA	Polymethyl methacrylate
PO	Polyolefin
PP	Polypropylene
PPO™	Polyphenylene oxide
PRF	Plastics Recycling Foundation
PS	Polystyrene
PUR	Polyurethane
PVC	Polyvinyl chloride
PVDC	Polyvinylidene chloride
R2B2	Recoverable Resources Boro Bronx/Bronx 2000
RCRA	Resource Conservation and Recovery Act
RIM	Reaction injection molding
RNJCT	Recycled New Jersey Curbside Tailings
RPI	Rensselaer Polytechnic Institute
RPS	Recycled polystyrene
SAE	The Society of Automotive Engineers, Inc.
SAN	Styrene-acrylonitrile
SBR	Styrene-butadiene rubber
SEM	Scanning electron microscope
SMC	Sheet-molding compound
SN	Stadtereinigung Nord
SPI	Society of the Plastics Industry, Inc.
TDA	2,4 and 2,6 toluene diamines
TDI	Toluene diisocyanate
TPA	Terephthalic acid
VAc	Vinyl acetate
VC	Vinyl chloride
VCM	Vinyl-chloride monomer
VDC	Vinylidene chloride

VHDPE	Virgin high-density polyethylene
VI	Vinyl Institute
VPE	Virgin polyethylene
VPS	Virgin polystyrene
VRI	Vermont Republic Industries
WEI	Wellman Engineers, Inc.
WKR	Wormser Kunstoff Recycling GmBH
XRF	X-ray fluorescence

Unit Abbreviations

dl/g	deciliters per gram
DM	German Deutsche Mark(s)
F50	fifty percent failures
FF	French franc
ft-lb/in	foot-pounds per inch
ft-lb/in^2	foot-pounds per inch squared
g/cm^3	grams per centimeter cubed
GPa	gigapascal(s)
h	hour
hp	horsepower
in-lb	inch pound
J/m	joules per meter
J/m^2	joules per meter squared
kg/h	kilograms per hour
kg/HH-yr (mo)	kilograms per household year (month)
kg/m^3	kilograms per meter cubed
km	kilometer
kPa	kilopascal(s)
kW	kilowatt
lb/ft^3	pounds per foot cubed
lb/h	pounds per hour
lb/HH-yr (mo)	pounds per household year (month)
lb/yd^3	pounds per yard cubed
£	British pound
mg/Nm^3	milligrams per newton meters cubed
mm	millimeter
MPa	megapascal(s)
N/m	newtons per meter
nm	nanometer(s)
Pa	pascal(s)
phr	parts per hundred (resin)
ppm	parts per million
psi	pounds per square inch
t	tonne(s), metric
μm	micron
W	watt
W/mK	watts per meter Kelvin
yr	year

Appendices

Appendix 1: Standard Guide for the Development of Standards Relating to the Proper Use of Recycled Plastics
ASTM Designation: D 5033-90 [1] [1]

Summary of Definitions

commingled plastic, n—a mixture of plastics, the components of which may have widely differing properties.

industrial plastic scrap, n—material originating from a variety of in-plant operations that may consist of a single material or a blend of materials.

off-spec or off-grade virgin plastics, n—resin that does not meet its manufacturer's specification.

performance standard, n—a document that defines levels of performance and provides both evaluation techniques and end-use criteria.

plastic container, n—a receptacle used to hold material for shipment, transport, or storage and composed of thermoplastic or thermoset plastic materials.

plastic recycling—a process by which plastic materials that would otherwise become solid waste are collected, separated, or processed and returned to use.

post-consumer materials, n—those products generated by a business or consumer that have served their intended end uses, and that have been separated or diverted from solid waste for the purpose of collection, recycling, and disposition.

purge (plastic), n—material resulting from the passing of polymer through a molding machine or extruder to clean the machine, or when changing from one polymer to another, or one color or grade of polymer to another.

recovered material, n—materials and byproducts that have been recovered or diverted from solid waste, but not including those materials and byproducts generated from, and commonly reused within, an original manufacturing process.

reconstituted plastic, n—a material made by chemical or thermal breakdown of plastic waste into components followed by their conversion into a final composition by chemical action.

recycled plastic, n—those plastics composed of post-consumer material or recovered material only, or both, that may or may not have been subjected to additional processing steps of the types used to make products such as recycled-regrind, or reprocessed or reconstituted plastics.

regrind (plastic), n—a product or scrap such as sprues and runners that have been reclaimed by shredding and granulating for use in-house.

reprocessed (plastic), n—regrind or recycled-regrind material that has been processed for reuse by extruding and forming into pellets or by other appropriate treatment.

reuse—the use of a product more than once in its original form.

reworked plastic, n—a plastic from a processor's own production that has been reground, pelletized, or solvated after having been previously processed by molding, extrusion, etc.

---

[1] Excerpted and reprinted, with permission, from the Annual Book of ASTM Standards, copyrighted ASTM.

source reduction, n—a system including design, manufacturing, acquisition, and reuse of materials (including product and packaging), that reduces the quantity of waste produced.

virgin plastic, n—plastic material in the form of pellets, granules, powder, floc, or liquid that has not been subjected to use or processing other than that required for its initial manufacture.

Appendix 2:   Coding and Labeling Plastics for Recycling

(A)   SPI's Voluntary Plastic Container Coding System [2] [1]

The code is a three-sided triangular arrow with a number in the center and letters underneath. The three-sided arrow was selected to isolate and distinguish the number code from other markings. The number inside and the letters indicate the resin from which the container is made; containers with labels or base cups of a different material may, if appropriate, be coded by their primary, basic material:

1 = PETE	(polyethylene terephthalate)
2 = HDPE	(high-density polyethylene)
3 = V	(vinyl)
4 = LDPE	(low-density polyethylene)
5 = PP	(polypropylene)
6 = PS	(polystyrene)
7 = Other	

The code is intended for molding into or imprinting on or as near to the bottom of the bottle or container as is technically feasible. This is to achieve a consistent location and to avoid interference with other functional requirements.

It is recommended that the symbol be molded into or imprinted on all bottles and jars which have a capacity of 16 ounces or more, and other containers such as tubs or trays with a capacity of 8 ounces or more. These sizes have been chosen because recyclers indicated that the effort to sort smaller containers (which often could not accommodate a readable code) is disproportionate to their value. These smaller containers can still be recycled through mixed plastics recycling processes.

(B)   Standard Practice for Generic Marking of Plastic Products
ASTM Designation: D 1972-91 [3] [2]

The purpose of this standard is to provide a system for uniform marking of products which have been fabricated (excluding additives) from polymeric materials.

---

[1] Excerpted and reprinted, with permission, from The Society of the Plastics Industry, Inc.

[2] Excerpted and reprinted, with permission, from the Annual Book of ASTM Standards, copyrighted ASTM.

The acronyms (abbreviations) used are to provide for generic identifications of the polymer(s).

This marking system is to provide assistance in identification of products for making subsequent decisions as to handling, disposal, or waste recovery.

The system is based on standard acronyms relating to plastics published by the International Organization for Standardization (ISO 1043) and those shown in Classification D 4000 and Terminology D 1600.

Plastic products may be marked at some point on the exterior surface with the appropriate symbols below a triangular outline, for example:

ABS

Copolymers or a polymer blend or alloy may be labeled with the appropriate symbols for the component polymers separated by a slash (/) below a triangular outline, for example:

PVC/PA

The marking shall be by injection molding, embossing, melt imprinting, or other legible marking in the surface of the polymer.

Some of the most frequently used plastics and the acronyms to be used with the marking as given in ASTM D 1972-91 are presented here. For a copy of the standard, contact: American Society for Testing and Materials, 1916 Race St., Philadelphia, PA 19103.

Plastic "Family" Name	Acronyms
Acrylonitrile-butadiene-styrene	ABS
Acrylonitrile-styrene-acrylate	ASA
Chlorinated polyethylene	CPE
Chlorinated poly(vinyl chloride)	CPVC
Epoxide; Epoxy	EP
Ethylene-ethyl acrylate	EEA
Ethylene-vinyl acetate	EVA
Ethylene-vinyl alcohol	EVAL
Polyamide	PA
Polybutene-1	PB
Polycarbonate plastics	PC
Phenol-formaldehyde	PF
Polybutylene terephthalate	PBT
Polyethylene terephthalate	PET*
Polyester, thermoset (unsaturated)	UP
Polyethylene (low density)	LDPE
Polyethylene (high density)	HDPE
Polyethylene (linear low density)	LLDPE
Poly(methyl methacrylate)	PMMA
Poly(4-methylpentene-1)	PMP
Polyoxymethylene; polyacetal	POM
Polypropylene	PP
Polystyrene	PS
Polystyrene-Impact Modified	IPS

Polyurethane	PUR (PU)
Poly(vinyl acetate)	PVAC
Poly(vinyl alcohol)	PVAL
Poly(vinyl chloride)	PVC*
Poly(vinylidene chloride)	PVDC
Styrene-acrylonitrile	SAN
Styrene-butadiene	SB
Styrene-maleic anhydride plastics	S/MA
Styrene-α-methylstyrene	SMS
Thermoplastic Elastomers	
Polyolefinic	TPO
Polyurethane, thermoplastic	TPUR
Styrene block copolymer	TES
Vinyl chloride-vinyl acetate	VCVAC
Vinyl chloride-vinylidene chloride	VCVDC

*Alternate acronyms (that is, PETE for PET or V for PVC)

Commercial Blends	Acronyms
Acrylonitrile-butadiene-styrene + poly(vinyl chloride)	ABS/PVC
Acrylonitrile-butadiene-styrene + styrene maleic anhydride	ABS/SMA
Acrylonitrile-styrene-acrylate + poly(methyl methacrylate)	ASA/PMMA
Poly(butylene terephthalate) + poly(ethylene terephthalate)	PBT/PET
Poly(butylene terephthalate) + poly(phenylene ether)	PBT/PPE
Poly(butylene terephthalate) + rubber	PT/RBR
Poly(vinyl chloride) + chlorinated polyethylene	PVC/CPE
Poly(vinyl chloride + nitrile-butadiene rubber	PVC/NBR
Poly(vinyl chloride) + poly(methyl methacrylate)	PVC/PMMA
Styrene-maleic anhydride + high impact polystyrene	SMA/IPS
Polyurethane + polyisocyanurate	PUR/PIR

(C)   Marking of Plastic Parts—Materials Practice SAE J1344 Revised March 1988 [4] [3]

This SAE Recommended Practice provides a system for marking plastic parts to designate the general type of material from which the part was fabricated.

The purpose of this recommended practice is to provide information to facilitate selection of optimum materials and procedures for repairing and repainting plastic parts, as well as recycling.

The system is based on standard symbols for terms relating to plastics published by the International Organization for Standardization (ISO 1043) and those shown in ASTM D 4000 [D 1600]. The standard symbol for molding markings into parts is:

with the height of letters at 3.0 mm. In view of the wide variety of parts used in different assembly situations, this recommended practice does not prescribe size, location, and/or method of marking. It is intended that the designer specify such details on the engineering drawing.

The acronyms given for the plastic family names and commercial blends are the same as that is ASTM D-1972-91.

---

[3] Excerpted and reprinted, with permission, copyright 1988 Society of Automotive Engineers, Inc.

Appendix 3:   Organizations Involved in Plastics Recycling

1.   Council on Plastics and Packaging in the Environment (COPPE) [5]

COPPE was founded in 1986 by a small group of plastic packaging companies. It has grown to its present 60 member organization representing plastics and packaging suppliers, manufacturers, marketers, users, recyclers, and related trade associations. The purpose of the group is to provide information to policy makers, the media, and consumers on plastics and packages in the solid-waste environment. This is done through press kits, media briefings, news releases, monthly newsletters, quarterly reports, and informational fact sheets. The organization sponsors technical and scientific information exchange forums on solid-waste issues for industry, government, and other interested parties. COPPE promotes source reduction, recycling, composting, incineration, and landfilling as a balanced integrated approach to solid-waste management.

2.   National Association for Plastic Container Recovery (NAPCOR) [6]

NAPCOR is a not-for-profit trade association whose member companies include PET producers and bottle and container fabricators. The NAPCOR's purpose is to aid in the economical recovery of PET plastic containers through collection, reclamation, and development of end-use markets for post-consumer containers. Initial emphasis is on the PET beverage bottle but ultimately the recycling of all plastic containers will be included. To fulfill its purpose NAPCOR assists communities starting new recycling PET programs or adding PET to an already established program. Promotional support is provided through educational brochures and media coverage. Technical information assistance is given for processing, reclamation, and remanufacturing. NAPCOR also helps communities locate buyers for collected PET bottles. Since the program started, NAPCOR has assisted 30 states in their recycling efforts.

3.   Plastics Institute of America, Inc. (PIA) [7, 8]

The PIA, a not-for-profit organization comprised of members from industry, academia, and government, has as its primary goal the advancement of education and research in polymers and plastics. The PIA is one of the oldest organizations devoted exclusively to plastics education and research. As such, the organization has become both a strong voice and a pivotal factor in helping to define the long-term needs and direction of the plastics industry and has been a pioneer in its efforts in plastics recycling. During the last decade, under a grant from the U.S. Department of Energy, the PIA conducted a major research effort into the use of automotive shredder residue (ASR) as an energy source and for reprocessing into noncritical applications. The results of the work are published in a two-volume report [9], [10] which provides an excellent basis for future work on ASR. The PIA's annual RecyclingPlas Conference on Business Opportunities in Plastics Recycling is widely considered the leading forum on this subject. The Conference, now in its seventh year, brings together international authorities to provide technical and business insight in the latest developments in the world on recycling. The PIA is the only organization currently offering a course on plastics recycling technology. It is an intensive two-day advanced technical course offered three to four times a year. The course content includes an up-to-date overview of collection and separation techniques, secondary and tertiary reclamation processes, and recycled-product applications.

4.   Plastics Recycling Foundation (PRF) [11, 12]

The PRF was formed in 1985 as an independent not-for-profit organization. Members of the PRF include companies manufacturing resins, plastics products, and packaging goods; soft-drink companies, machine suppliers, recyclers, and The Society of the Plastics Industry, Inc.

The PRF has a membership of 57 companies. There are three membership levels: directorate, associate, and supporting. The PRF's mission is to sponsor research to improve the technology and economics of plastics recycling, to demonstrate pilot programs in recycling technology, and to provide information on recycling technology and product markets. Funding is provided by the member companies, state and federal agencies, and academic institutions.

In 1985 the PRF established the Center for Plastics Recycling Research (CPRR) at Rutgers University, Piscataway, New Jersey. The CPRR serves as the focal point of research projects funded by the PRF and is itself responsible for many of the major projects. Research projects at 16 different universities have been funded during the past six years. The CPRR operates demonstration pilot programs in collection, sortation, and reclamation. The reclamation pilot plant is a fully equipped pilot facility for separating, cleaning, and reclaiming PET beverage bottles. To date, the CPRR has signed 19 licenses for resin reclamation technology. A current project at the CPRR having important future significance is the development of an automatic system for the separation of clear PVC, clear and green PET, natural HDPE.

5.   Polystyrene Packaging Council, Inc. (PSRC) [13]

The PSRC represents polystyrene foam resin suppliers and manufactures of polystyrene foam products. Member companies also include molding and extrusion equipment manufacturers and trade associations. The mission of the PSRC is to foster the effective use and recycling of polystyrene and to provide accurate, reliable information on solid-waste disposal practices. The PSRC sponsors a number of polystyrene recycling projects throughout the country and provides information and technical assistance to public officials, consumers, and local businesses on effective solid-waste management. It has also assisted in the formation of the National Polystyrene Recycling Company.

6.   The Council for Solid Waste Solutions (CSWS) [14]

The CSWS was established in 1988 by companies in the plastics industry concerned with the solid-waste management issue and the role of plastics in this issue. The primary objective of the CSWS is to develop programs and provide guidance and assistance on the disposal and recycling of plastics. It does this through comprehensive and well-planned technical, governmental affairs, and communications programs. The council's technical program includes supporting research to develop new and improved technology for plastics recycling, waste-to-energy incineration, and enhanced degradability. In 1990 a "Blueprint for Plastics Recycling" [15] was developed which affords a comprehensive action plan to provide guidance to all segments of the growing plastics recycling industry. The plan covers the collection, handling, reclamation and end-use markets/applications aspects of the industry. In its governmental affairs programs, the CSWS helps legislators at the local, state, and federal levels on solid-waste management legislation, sponsors educational symposia to share new information and technology with legislators, and assists municipalities, community groups, and organizations in developing waste-management programs. The council's communication programs are designed to inform the public about plastics in the solid-waste stream, to provide factual information to the news media and others on plastics, and to foster public acceptance of reasonable solid-waste disposal solutions.

The CSWS has three membership levels: executive board, associate, and supporting. Currently, there are 12 executive board member companies. The CSWS is a Special Purpose Group of The Society of the Plastics Industry, Inc. discussed later.

7.   The Society of Plastics Engineers, Inc. (SPE) [16]

The objective of the SPE is to promote scientific and engineering knowledge relating to plastics by conducting technical meetings, publishing technical reports and papers, encouraging exchange of technical information, and cooperating with institutions to establish and maintain high quality technical education. Membership in the SPE is available to anyone with qualifying work and/or educational experience in plastics, sponsorship by a member, and supplying the necessary information on an approval application. The SPE is organized on a section and division basis. Each member becomes associated with a local section which conducts monthly meetings providing members with technical information and other activities. A member also chooses to become affiliated with a division which is composed of other members having an interest in a specific area of plastics technology. A new division recently formed in the SPE is the Plastics Recycling Division [17]. This division is concerned with all aspects of the recycling, reclamation, resource recovery, and disposal of plastic materials. Its mission is to foster the advancement of process and product technology in the development of recycled plastics as a valuable raw material resource. The division strives to promote and provide factual information on plastics recycling to its members and the general public. In fulfilling its mission, the division has in the past and continues to work with local sections to foster interest in plastics recycling on a community level. The division also conducts technical sessions during the society's annual technical conference, co-sponsors one or two regional technical conferences each year, and supports recycling fairs in various parts of the country. A list of speakers, a technical paper index, an annual best technical paper/presentation award, and other educational, governmental, industrial, and international recycling activities are all a part of the division's functions.

8.   The Society of the Plastics Industry, Inc. (SPI) [18]

The SPI is a trade association composed of member companies from all segments of the plastics industry—material suppliers, fabricators, machinery manufacturers, compounders. The SPI functions centrally but also through a network of operating units which consists of 27 divisions, regions/sections, and special purpose groups. Established in 1937 the SPI has grown to more than 2200 company memberships and is now considered the "voice of the plastics industry." The SPI provides central services to the members in information, communications, technical, regulatory, and governmental affairs. But member companies also pursue specific programs of special interest through the operating units of the SPI. In this way units have formed separate entities, special task groups, or redefined objectives to become actively involved in specific issues of interest, e.g. the solid-waste management issue. As mentioned above, the CSWS was formed as a special purpose group specifically to provide solutions to this problem.

Most recently the Polyurethanes Division of the society formed the Polyurethanes Recycle and Recovery Council (PURRC) [19]. Its objective is to develop the technology to commercially recover and/or recycle 25% of all polyurethanes produced annually by 1995. The estimated amount to be recycled would be from 250 to 350 million pounds (113 to 158×10^3 t) in 1995. The PURRC is to work toward that goal by (1) establishing a collection system, (2) identifying the sources and amount of polyurethanes to be recycled and recovered, (3) evaluating current technologies, (4) developing commercial reclamation systems, and (5) developing uses for the recovered products. The PURRC is also to promote and communicate the benefits and viability of recycling polyurethanes.

The SMC Automotive Alliance (SMCAA) [20] is an activity of the Composites Institute, another division of the SPI. Members of the SMCAA are suppliers and molders of sheet-molding composite (SMC) products for the automotive industry. Although the primary goal of the SMCAA is to promote new automotive applications for SMC, recycling has become a top priority of the group. Pyrolysis and regrinding for filler application are being investigated by the SMCAA to reclaim and reuse SMC automotive products. The initial phases of this work have been reported at a recent SAE conference [21] and is discussed in Chapter 10.

The Vinyl Institute (VI) [22] was founded in 1982 as a division of the SPI by manufacturers and suppliers of PVC, and additives/modifiers supplier companies. The VI has a very active program addressing various issues of PVC in solid waste. It is supporting and promoting the recycling of PVC bottles, sponsoring research on commingled plastics, providing consumer education on PVC recycling and incineration, and participating in industry-wide programs on the issue. The VI has recently formed VIGOR—the Vinyl Institute Group on Recycling.

Other divisions of the SPI with programs in plastics recycling include the Plastic Bottle Institute, Rigid Plastic Container Division, and the Vinyl Siding Institute.

Further information on these organizations may be obtained from the following:

1. Council on Plastics and Packaging in the Environment
   1001 Connecticut Avenue, N.W.
   Suite 401
   Washington, DC 20036
   (202) 331-0099

2. National Association for Plastic Container Recovery
   4828 Parkway Plaza Boulevard
   Suite 260
   Charlotte, NC 28217
   (704) 357-3250

3. Plastics Institute of America, Inc.
   277 Fairfield Road
   Suite 100
   Fairfield, NJ 07004
   (201) 808-5950

4. Plastics Recycling Foundation
   1275 K Street N.W.
   Suite 400
   Washington, DC 20005
   (202) 371-5337

5. Polystyrene Packaging Council, Inc.
   1025 Connecticut Avenue N.W.
   Suite 513
   Washington, DC 20036
   (202) 822-6424

6. The Council for Solid Waste Solutions
   1275 K Street N.W.
   Suite 400
   Washington, DC 20005
   (202) 371-5319

7. The Society of Plastics Engineers, Inc.
   14 Fairfield Drive
   Brookfield, CT 06804
   (203) 775-0471

8. The Society of the Plastics Industry, Inc.
   1275 K Street NW
   Suite 400
   Washington, DC 20005
   (202) 371-1022

9. Polyurethanes Recycle and Recovery Council
    355 Lexington Avenue
    New York, NY 10017
    (212) 351-5422

10. The SMC Automotive Alliance
    355 Lexington Avenue
    New York, NY 10017
    (212) 351-5410

11. The Vinyl Institute
    Wayne Plaza Interchange 11
    155 Route 46 West
    Wayne, NJ 07470
    (201) 890-9299

Appendix 4:   University Programs

An overview of some of the most recent published and current research being carried out at a number of universities and colleges is present here.

Cornell University

Researchers at Cornell are investigating solution processing as a means to obtain virgin-grade polymer from recycled products. The process is an alternative to depolymerization as a method to obtain polymers free of deleterious impurities and degraded polymer. In a series of papers, Rodriguez and Vane [23–25] have reported on studies of the dissolution of PET in n-methyl-2-pyrrolidinone (NMP), the behavior characteristics of the solution, and polymer recovery by slow or quench crystallization and by non-solvent precipitation. A process combining a solution-cleaning system with a wet or dry system for cleaning PET bottles has been proposed as a method to produce high purity recycled polymer that could be used directly for virgin-polymer applications. The process takes over after the PET has gone through a "normal" shred-wash-dry operation. Subjecting the "clean" PET to an NMP solvent wash at 130 °C followed by a solvent dissolution at 160 °C removes any remaining adhesives, PS, PVC, aluminum, polyolefins. Depending on the pretreatment, PET of 99.99% wt % purity is easily obtained.

Michigan State University

Work on plastics recycling at the University is done at the School of Packaging and has focused on wood fiber/recycled polyolefin composites. Two studies on composites with aspen wood fibers were reported recently. One study [26] involved the processing-property relationships of treated and untreated aspen fibers blended with recycled HDPE using a co-rotating intermeshing twin-screw extruder. Mechanical properties were found to be sensitive to screw configuration and compounding temperature; untreated and acetylated fibers exhibited higher strength characteristics than untreated fibers; fiber dispersion appeared as the most important factor throughout the study. In the other study [27] mechanical properties were determined on composites of untreated aspen fibers/reground multilayer PP ketchup bottles and also on composites with virgin PP. The reclaimed PP composites generally showed better mechanical properties than the virgin composites; properties increased with fiber content and orientation up to 30% fiber content. The reclaimed composite also exhibited greater dimensional stability than the virgin PP composite under ambient and extreme environmental conditions. The retention of strength under wet conditions is indicative of a contribution by the other components in the reclaimed PP composite material.

Northwestern University

Petrich [28] is studying the pyrolysis of SBR, neoprene rubber, shredded tires, RIM polyure-thane scrap, ABS scrap, and carpet-mill scrap. A scheme for recovery of aluminum from aluminized ABS has been devised. The major emphasis of this work is a post acid treatment of the pyrolysis chars to produce low-ash carbon to form valuable carbon end products as in pigments and activated adsorbents. A recent paper [29] describes the process and discusses the pyrolysis of polyurethane RIM scrap and truck tires.

Polytechnic University

Research on a unique approach to the tertiary recycling of thermosets has been published by Tesoro. A series of papers [30–32] published in 1990 describes the formation of reversible cross-links in epoxy resins and the mechanical and thermal properties of the resins. Substi-tuting disulphide-containing compounds, specifically 4,4 dithiodianiline (DTDA) for meth-ylene dianiline (MDA) as a cross-linking agent in an epoxy resin showed no significant differences in properties although the ratio of resin to curing agent was slightly different. Reduction of the disulphide bonds in the DTDA-containing resin gave completely soluble polymer within a limit of cross-link density. The soluble polymer is recross-linked by oxidation or by reaction with the polyfunctional thiol compounds. The properties of the final product can be altered by modifying the re-cured resin with different cross-linking agents during the reaction.

In the latest publication [33] on this research, properties of polyimide copolymers contain-ing DTDA as a cross-linking agent were compared with commercial two component polyimide copolymer systems. Reduction of the disulphide-containing copolymer gave soluble products and these products were shown to be re-cured by reaction of the thiol groups with polyfunctional epoxides.

State University of New York at Binghamton

Researchers [34] are using statistical-design concepts to study the effect of compatibilizer type and concentration and waste plastic composition on the mechanical properties of the extruded mixtures. PP, PS, and Kraton™ compatibilizers have been examined with virgin and reclaimed PET and HDPE. Preliminary results of compatibilizer additions to several compositions show strength and elongations greater than predicted by a simple mixture rule. Also small amounts of compatibilizers increased the fatigue resistance of mixed virgin/recycled materials.

The Pennsylvania State University, The Behrend College

An active program on plastics recycling exists at the Plastics Technical Center of this college. Some initial results of current projects were reported recently. One paper [35] discussed the effects of multiple re-extrusions (regrinds) of PS, HDPE, and PP on the mechanical and rheological properties of these materials. Another [36] presented initial results showing that mixtures of 10 and 25% postconsumer recycled HDPE with virgin HDPE satisfied the molding and mechanical property requirements for fabrication of a recycling collection container. A third paper [37] related some unsuccessful attempts to use recycled newspaper as a filler for plastic. A novel product concept evolved, however, i.e. encapsulating recycled paper between plastic film for potential packaging or insulation applications.

University of Florida

Recycling research at the University is being carried out by two groups in two very different and defined areas. Beatty, of the Department of Materials Science and Engineering, has investigated the mechanical properties of posts made of commingled recycled plastic and has recently reported on this [38] and some accelerated environmental tests [39]. This work is a part of a collaborative effort with the Florida Department of Transportation. The Department is seeking new applications for recycled plastics within the state as mandated by a Florida law. Evaluation of flexural and impact properties together with detailed studies of the outer, or skin, and core sections of posts showed that the best structure on a weight, material, and property basis was composed of a high-density outer section and an inner core of lower density material. Beam theory analysis indicates the flexural properties are influenced by the outer section while the inner core is an energy absorber. Accelerated warpage tests at temperature of 120–130 °C were conducted on four different manufacturers fence posts and a wooden post. With one exception, the plastic posts showed no sign of stress relaxation. Plastic posts also absorbed very little boiling water and salt water compared to wooden posts.

In the Department of Chemistry and Center for Macromolecular Science and Engineering, Wagener is investigating the depolymerization of diene polymers to low-molecular weight oligomers by a metathesis reaction [40]. In the presence of excess ethylene, the depolymerization yields monomeric diene products. For example, polybutadiene was depolymerized to oligomers and in the presence of ethylene to 1,5 hexadiene. The process is adaptable to other unsaturated polymers and could provide a means of obtaining useful chemicals from this type of waste polymers.

University of Lowell

This university has had an ongoing program related to plastics recycling for many years through Deanin and his pioneering work in this field. He has studied many different polymers and has published on HDPE [41], phenolic scrap [42], solid and foam polystyrene scrap [43], vulcanized rubber scrap [44], epoxy resins [45], and mixed plastics fraction from junked autos [46–48]. More recent work from the University discusses a study on the development of a foam composite HDPE, waste paper board stock by a foam extrusion laminating process. Virgin HDPE was used in this initial trial [49]. Another report details a process for the reclamation of computer housings made of Noryl R™, a modified PPO™ [50] into pelletized product. According to the report, ASTM mechanical property tests conducted on the molded pellets showed the product has properties in the range of virgin PPO mechanical properties. An investigation on the extrusion of reground LDPE sheet scrap having a bulk density of 15.8 lb./ cu. ft. using a single-screw extruder was reported by Malloy [51] of Lowell. Improvements in the dimensional stability and quality of the extrudate was quantified when a feedback process control was incorporated and a gear pump was added in series respectively. Another study [52] showed the effects of incorporating fibers and fillers in plastic lumber on the mechanical properties of the lumber. Long glass fiber reinforced PP increased the flexural strength at room temperature (and below) while compression and flexural moduli were about equivalent to that of wood lumber. Increases in these properties were noted also with mica and to a less, but significant degree, with calcium carbonate. In a separate study [53], EVA was blended as a compatibilizer at levels of 1 and 2% with mixtures of recycled PET/PE from beverage bottles. Properties of the mixture were improved at a low level of EVA. This result gives an indication that EVA as an adhesive or a bottle cap liner can function as a compatibilizer in a mixed PET/ PE waste stream.

University of Texas at Austin

Research on the utilization of recycled PET in unsaturated polyesters for polymer concrete applications has been reported by researchers from the University [54]. Polymer concrete formulation consisting of 11 different experimental polyester resins containing 15 to 40% recycled PET were prepared. The polymer concrete formulations were optimized for precast or thin-layer overcast applications. For precast uses, rigid unsaturated polyester-type resins with high modulus, low elongation at break and good dimensional stability were used in the polymer concrete formulations; for overcast applications, flexible unsaturated polyesters with low modulus, high elongation and good bond strength were employed. The final formulated polymer concrete had a wide range of properties comparable to concretes formulated with unsaturated polyesters contain virgin materials.

The use of recycled PET in polymer concrete presents another important potential application for this material. The PET need not be purified to the same extent as for other applications. Colored PET can be used and the final product does not reenter the solid-waste stream for many years. The authors of this work conclude that the fast curing time and excellent mechanical durability properties allow polymer concrete products to be cost efficient materials.

References

[1] *Standard Guide—The Development of Standards Relating to the Proper Use of Recycled Plastics*. Designation D 5033-90. American Society for Testing and Materials, Philadelphia, PA.

[2] *SPI's Voluntary Plastic Container Coding System*. Brochure. The Society of the Plastics Industry, Inc., Washington, DC, 1988.

[3] *Standard Practice for Generic Marking of Plastic Products*. Designation D 1972-91. American Society for Testing and Materials, Philadelphia.

[4] *Marking of Plastic Parts*. SAE J 1344. The Society of Automotive Engineers, Inc, Warrendale, PA, revised March 1988 .

[5] Informational Brochures. Council on Plastics and Packaging in the Environment, Washington, DC.

[6] Membership Information Kit. National Association for Plastic Container Recovery, Charlotte, NC.

[7] *This Is PIA*. Plastics Institute of America, Inc., Fairfield, NJ.

[8] *Annual Report 1990*. Plastics Institute of America, Inc. Fairfield, NJ.

[9] *Secondary Reclamation of Plastics Waste*. PIA Research Report. "Phase I, Development of Techniques for Preparation and Formulation." Technomic Publishing, Lancaster, PA, 1987.

[10] *Secondary Reclamation of Plastics Waste*. PIA Research Report. "Phase II, Evaluation of Industrial Processes: Financial Analysis and Potential Markets." Technomic Publishing, Lancaster, PA, 1987.

[11] *Annual Report 1990*. Plastics Recycling Foundation, Washington, DC.

[12] Membership Information Packet. Plastics Recycling Foundation, Washington, DC.

[13] Informational Brochure. Polystyrene Packaging Council, Washington, DC.

[14] Various Brochures. The Council for Solid Waste Solutions, Washington, DC.

[15] *The Blueprint for Plastics Recycling*. The Council for Solid Waste Solutions, Washington, DC.

[16] Membership Information. The Society of Plastics Engineers, Inc., Brookfield, CT.

[17] Policy Manual. Plastics Recycling Division, The Society of Plastics Engineers, Brookfield, CT.

[18] Member Services Guide 1991-1992. The Society of the Plastics Industry, Inc., Washington, DC.

[19] Fact Sheet. Polyurethanes Recycle and Recovery Council, New York, NY.

[20] Top Priority for SMC, Ward's Auto World.

[21] Cucuras, C. N., Flax, A. M., Graham, D., Harth, G. H. International Congress and Exposition, Detroit, MI, Mar. 1991.

[22] Various Brochures. The Vinyl Institute, Wayne, NJ.

[23] Vane, L. M. and Rodriguez, F. The Society of Plastics Engineers, Inc., Regional Technical Conference, Nov. 1990, pp 100–109.

[24] Vane, L.M. and Rodriguez, F. American Chemical Society. *Polymer Preprints* Vol. 32, No. 2(June 1991): pp 138–141.

[25] Rodriguez, F., Vane, L. M. and Clark, P. Presented at American Institute of Chemical Engineers, Chicago, Nov. 1990.

[26] Simpson, R., Selke, S. E. American Chemical Society. *Polymer Preprints* Vol. 32, No. 2 (June 1991): pp 148–149.

[27] Yam, R. L., Gogoi, B. K., Lai C. C., Selke, S. E. *Polymer Engineering and Science* Vol. 30, No. 11 (Mid-June 1990): pp 693–699.

[28] Petrich, M. A. Private Communication. 25 Feb. 1991.

[29] Petrich, M. A. Presented at the Environmental Technology Expo, Chicago, Apr. 1991.

[30] Tesoro, G. C., Sastri, V. R. *J. Appl. Poly. Sci.* Vol. 39 (1990): pp. 1425–1437.

[31] Sastri, V. R., Tesoro, G. C. *J. Appl. Poly. Sci.* Vol. 39(1990): pp. 1439–1457.

[32] Engelberg, P. I., Tesoro, G. C. American Chemical Society. *Polymer Engineering and Science* Vol. 30, No. 5 (Mid-March 1990): pp 303–307.

[33] Wu, Y., Tesoro, G. C. American Chemical Society. *Polymer Preprints* Vol. 32, No. 2 (June 1991): pp 140, 141.

[34] Clum, J. A., Emerson, C. R., Lofthouse, R. Private Communication. Feb. 1991.

[35] Dzeskiewicz, L. The Society of Plastics Engineers, Annual Technical Conference Proceedings, Vol. 37, 1991, pp. 2558–2560.

[36] Bowes, G. S. The Society of Plastics Engineers, Annual Technical Conference Proceedings, Vol. 37, 1991, pp. 2556, 2557.

[37] Hanes, C. The Society of Plastics Engineers, Annual Technical Conference Proceedings, Vol. 37, 1991, p. 2561.

[38] Cao, L., Byun, S. G., Baugh, D. W., Beatty, C. L., Ramer, R. M. American Chemical Society. *Polymer Preprints* Vol. 32, No. 2 (June 1991): p. 129, 130.

[39] Liedermooy, I., Fitzgerald, S., Beatty, C. L., Ramer, R. M. American Chemical Society. *Polymer Preprints* Vol. 32, No. 2 (June 1991): pp. 131, 132.

[40] Wagener, K. B., Puts, R. D. American Chemical Society. *Polymer Preprints* Vol. 32, No. 1 (Apr. 1991): pp. 379, 380.

[41] Deanin, R. D., Doshi, S. J. The Society of Plastic Engineers, Annual Technical Conference Proceedings, 1978, p. 772.

[42] Deanin, R. D., Ashar, B. V. American Chemical Society. *Org. Coat. Plast. Chem.*, 41, 495 (1979).

[43] Deanin, R. D., Amran, A. American Chemical Society. *Polymer Preprints* Vol. 24, No. 2 (1983): p. 430.

[44] Deanin, R. D., Hashemiolya, S. M. American Chemical Society. *Polymer Materials Science and Engineering,* 57 (1987): p. 212.

[45] Tran-Bruni, C. V., Deanin, R. D. American Chemical Society. *Polymer Materials Science and Engineering,* 56 (1987): p. 446.

[46] Deanin, R. D., Nadkarni, C. S. American Chemical Society. *Polymer Materials Science and Engineering,* 50, (1984): p. 408.

[47] Deanin, R. D., Yniguez, A. R. American Chemical Society. *Polymer Materials Science and Engineering,* 50 (1984): p. 413.

[48] Deanin, R. D., et. al. American Chemical Society. *Polymer Materials Science and Engineering,* 53 (1985): p. 826.

[49] McBride, D. J., Schott, N. R., Giacobbe, J. M. Presented at the American Institute of Chemical Engineers, Mar. 1990.

[50] Schott, N. R. Presented at the American Institute of Chemical Engineers, Nov. 1990.

[51] Malloy, R. A., Chen, S. J., Orroth, S. A. The Society of Plastics Engineers, Annual Technical Conference, 1990, pp. 1430–1433.

[52] Sales, M., Johnson, M., Malloy, R., Chen, S. The Society of Plastics Engineers, Annual Technical Conference, 1986, pp. 901–908.

[53] Malloy, R. A., Lai, F., Condo, M., Boudreau, K. Presented at the American Institute of Chemical Engineers, Nov. 1990.

[54] Rabiez, K. S., Fowler, D. W., Paul, D. R. American Chemical Society. *Polymer Preprints* Vol. 32, No. 2 (June 1991): pp. 142, 143.

Index

abbreviations 263–265
ABS 4, 5, 9, 58, 153, 155–158, 163, 164
acetaldehyde 58
acrylic(s) 4, 5, 11, 171–185, 247
(see also depolymerization, methyl
    methacrylate, PMMA)
  applications 174
  .cast sheet 172, 175, 176
  cell cast 173
  cracking (see depolymerization)
  extruded sheet 172, 173
  markets 174, 175
  recycling 175, 176, 183–185
  regrinds 176, 183, 185
  scrap 171, 176, 177, 179, 181
adhesive(s) 51, 53, 54, 56–58
  removal 53, 54, 56, 57
Advanced Recycling Technology 93, 196,
    225
AKW Apparate and Verfahren GmbH 99
Allied Chemical Corp. 13, 164
Allied Signal 63
alloys and compounds 63
aluminum 53–55, 59, 67, 140, 198
  removal 53, 54, 67
American Commodities, Inc. 83
American Plastics Recycling Group 95
American Society for Testing and
    Materials 5–7, 199, 267, 268
  definitions of recycling terms 267
  marking of plastic products 268
Amoco Chemical Co. 127, 211
amorphous (PET) 48
appendices 267–280
AP Shrinker 119
Arco Chemical Co. 127
ASOMA 142
AT&T 9, 164
Australia 69, 70
Automated Recycling Corp. 57, 58
Automotive shredder residue 4, 14, 155,
    156, 176, 184, 185, 254–257
  composition 156
  incineration 255, 256

pyrolysis 254
automotive 4, 14, 155, 157, 166, 183,
    235, 256, 257
  disassembling 157
  plastic waste 4, 144, 155, 157, 257
  recycling 166, 256, 257

BASF Corp 100, 124, 166, 258
Battelle Laboratories 102
battery cases 92, 98, 99
(see also PP)
  legislation 92
Bauer, G. 253, 261
Bauer, S. H. 238, 239
Bauman, B. D. 239, 259
Bayer AG 12, 100, 166, 258
B. F. Goodrich Co. 141, 143, 145, 223
Beatty, C. L. 277, 279
Beckman, E. S. 224, 228
bisphenol-A 165
blasting media 183, 257
bottle bill 4, 13, 19
bottle(s) 21, 34, 49, 50, 60, 65, 66–68, 70,
    133, 139, 141, 143, 190
(see also PET soft drink, containers, PVC,
    HDPE)
  closures 49
  five-gallon water 158
  "merchant" 70
  PC 165
  "step-on-it" program 21, 23
Braslaw, J. 254, 2
British Polythene Industries 98
British Plastics Federation 98, 99
bulk molding compound 245, 246
Buss-Condux Inc. 119

Camden County, NJ
(see case studies)
Canada 42, 43, 62, 69, 76, 77, 100, 101,
    224
  commingled plastics 224

PET recycling   69
polyolefin recycling   100
Canadian Recycling Directory   101
Carpco Inc.   53, 54
Carrier, K.   257, 261
case studies   36–41
  Camden County, NJ   37, 41
  San Jose, CA   36, 37
Center for Plastics Recycling Research
  (CPRR)   36, 37, 79, 81, 82, 142,
  157, 160, 162, 192, 196, 200, 201,
  208, 210, 211
  process licensees   82
  reprocessing economics   218, 219
  resin reclamation system   79–82
Cep Adour Industry   141
Chardonol Div. of Cook Composites and
  Polymers   64
Chevron Chemical Co.   127
Clean Tech reclaiming process   84
closed-loop   65, 88
coextrusion blow molding   97
Coca-Cola Company   66
coding   6, 7, 268–270
collection   19–42, 52, 102, 125, 126, 153,
  191
  containers   20, 21
  curbside   4, 19–27, 32, 33, 36–42, 52,
    101, 102, 125, 141, 153, 192
  frequency of   22
  special   41, 43
  types of programs   40
  vehicles   24–29
commingled plastics   154, 189–225
(see also separation, NJCT)
  composition   197
  compression molding   195
  continuous extrusion   194, 195
  CPRR/WEI refine technology   211–218
  economics   196
  elemental analysis   197
  feedstock   192
  global activities   224, 225
  intrusion process   193, 194
  manufacturing processes   193–195
  new processes   210–218
  "Reverzer" process   195
  structure-property relations   196–200
  waste   191, 195
compatibilizing agents   157, 160, 210, 218
compositional quenching   219
compression molding   195

container(s)   20, 22, 24, 30, 41, 50, 65, 77–
  86, 88, 101, 189–191, 196, 210
  beverage   22, 24
  chemical   41, 42
  collection   20
  commingled   30
  polyolefin reclaiming plants   83
  "step-on-it" program   21, 23
Continental Can   65
Cookson Industrial Materials Ltd   99
Cornell University   275
Council for Solid Waste Solutions
  (CSWS)   4, 7, 167, 258, 272, 274
Council on Plastics and Packaging in the
  Environment (COPPE)   271, 274
CPRR/WEI refined commingled
  technology   211–218
  blends of polystyrene and PET   212
  economics   218, 219
  properties of blends   214, 216
  resin recovery process   211
cryogenic grinding   54
crystal polystyrene   111, 112
crystallinity   48, 75
Custom Pak   97

Day Products Inc   53, 58, 82
Deanin, R. D.   277, 279
decomposition (see depolymerization)
definitions   5, 6, 76, 153, 267
densification (see EPS)
depolymerization   9–2, 47, 51, 58–60, 64,
  66, 163–165, 171, 172, 176–182, 184,
  244, 247–253, 255
(see also hydrolysis, glycolysis, PET,
  PMMA, pyrolysis)
  continuous   180
  cost of   176
  dry distillation   177
  extrustion methods   181, 182
  fluidized bed technique   180, 181
  molten metal, metal salts   178, 179
  retort method   180
  superheated steam   177, 249
diethylene glycol   60, 64, 249
dimethyl terephthalate   47, 48, 59, 163
Dow Chemical Co.   56, 89, 127, 136,
  165
DuPont   12, 14, 58, 59, 63, 83, 88, 160,
  177, 183
Durables   4, 5, 126, 153, 158, 166

Eaglebrook   83
Eastman Kodak Co.   12, 50, 58
Eastman Chemical Products Inc.   12, 59,
        61, 66
economics   32 , 35, 66–68, 82, 125, 126,
        166, 176, 218, 219, 221–223, 239
(see also individual polymers, MRF, refined
        NJCT, recycling)
Embrace Systems Technologies, Inc.   61,
        98
engineering thermoplastics   5, 153–166
(see also ABS, modified PPO™, Nylon,
        PBT, PC, PET, polyacetals)
    definition   153
    depolymerization   164
    disposal   154, 155
    global-recycling activities   166
    recycling   159–165
    recycling economics   166
    reprocessing   158
    reprocessing companies   159
    reprocessing machinery   161 , 162
    sources   155–158
Environmental Protection Agency
        (EPA)   88, 142, 145
environmental stress-crack resistance of
        recycled HDPE   94, 95, 97
epoxy resin(s)   233, 237, 243, 244
    applications   237
    recycled regrind   243, 244
    volume   233, 237
EPS (expandable polystyrene)   24, 111,
        113, 115, 117–122, 124, 126, 127,
        200 (see also PS)
    collection   24
    densification   117–121, 126, 127
    markets   115
    preparation   113
    recycled products   124–126
Erema-RM   121
ET/1   193, 194, 196, 199, 200, 209, 225
ethylene glycol   12, 47, 59
Europe   68, 76, 77, 99, 127, 139, 141, 144,
        155, 166, 224, 258
    commingled plastics   224
    plastics recycling   155
    polyolefin recycling   99
    PS recycling   127
    PVC recycling   139, 141, 144
    thermoset recycling   258
EVA   51, 53, 256
Exide Corp.   92

expandable polystyrene (see EPS)
extruder   12, 115, 117, 123, 181, 182, 211–
        213
    depolymerizing   181, 182
    non-intermeshing twin screw   211–213
    single-screw   117, 123, 181
    twin-screw   117, 123, 182
Extrusion   115, 117
    continuous   193, 194
    depolymerization   181, 182
Exxon Chemical   210

Farbwerke Hoechst   12
Far East   70
Fiberfil   61
film   41, 50, 51, 58, 89, 113, 153
(see individual polymers)
    coating   89
    oriented polystyrene   113
    recycling of   8, 145, 146
Fina Oil & Chemical Co.   127
flotation (see separation)
fluff (see automotive waste)
fluidized bed   180 , 181, 255
foam   42, 63, 64, 90, 91, 114, 115, 117,
        121, 122, 234, 235, 238, 239, 246,
        249, 257
(see individual polymers, EPS)
    cups   117, 123
    extrusion   114, 115
    shapes & loose fill   117
    sheet   63, 113, 115, 117, 121, 122
    syntactic polyester   246
Food and Drug Administration   123
Ford Motor Company   12, 184, 249, 253
Frankel, H.   88, 104
Franklin Associates Ltd   136, 234
Fuji Recycle Industry K.K.   101

GECOM   140
Gelimat System   121
Gemark   58
General Electric   10, 63, 157, 158, 165
General Motors   12, 237, 245, 249, 255
General Polymers   158
geotextiles   61
Germany   11, 43, 62, 69, 98 , 100, 127,
        141, 180, 195, 225, 258
Gibbs, M.   93–95, 106
Glenro-Densifier   119

glycolysis   12, 58, 59, 60, 64, 66, 163, 247, 248, 250–253
Goodyear   47, 59, 79, 163
Graham Recycling reclaiming process   85
GREPP   140

Hamburg University   11
Hammer, F.   13, 193, 194, 226
HDPE (high-density polyethylene)   4, 6, 12–14, 22, 41, 50, 53, 54, 66, 67, 75, 78, 83, 88, 93–95, 97–101, 136, 139–141, 145, 160, 171, 185, 190–192, 194, 197, 198, 200, 207, 210, 224
(see PE, polyolefins)
    agricultural film   41
    base cup   48, 49, 50, 54, 57, 67, 97
    bottles   76, 78, 83, 88, 93–95, 97, 98, 101
    chemical containers   42
    coextrusion blow molding   97
    curbside recovery rates   38
    recycled products   96, 97, 99
Heat transfer media   177–179
High-density polyethylene (see HDPE)
Herbold GmbH   100, 142
Hoechst, AG   100, 166, 258
Hoechst Celanese   59, 66, 160, 163
Huntsman Chemical Co.   127
hydrocyclone   51, 53–55, 68, 81, 99, 101
hydrolysis   12, 47, 51, 59, 163, 247–250

Image Carpets   61
impact modifiers   160, 219
impact polystyrene   111–113
IMPET™   160
incineration   6, 8, 102, 137, 138, 154, 166, 255, 256
Instamelt System   119
Institute of Scrap Recycling Industries   83
intrinsic viscosity (IV)   48–51, 53, 58, 61, 62 (see also PET)
    definition   48
Insulblank™   61
intrusion process   193, 194

Japan   11, 70, 76, 77, 101, 127, 145, 154, 166, 224, 258
    commingled plastic   224
    plastics recycling   154

polyolefin recycling   101
PS recycling   127
PVC recycling   145
thermoset recycling   258
Japanese Plastic Waste Management Institute   101
John Brown   83
Johnson Controls   53, 54, 65, 68, 92

Klobbie, E.   193, 225
Klobbie   193, 224

labeling   7
labels   49, 51, 54, 57
landfilling   102, 145, 154
    in Japan   155
landfills   1, 8, 154
Lankhorst Recycling Technology   193, 224
LDPE (low-density polyethylene)   9, 11, 75, 93, 185, 194, 197, 210, 224, 257
lead   145, 178, 179
Leidner, J.   5, 6, 8, 15
low-density polyethylene (see LDPE)

M. A. Industries   63, 85, 92
    reclaiming process   85
M. K. Recycling Inc.   121
Malloy, R. A.   277, 280
Mandish Research International   124
Martin Color-Fi   62
material recovery facility (MRF)   4, 29–36, 41, 43, 66, 125, 143, 192
    economics   32–35
    functions   29
    in Germany   43
    workings of   30, 31
melamine and urea resins   233, 238
    application   238
    volume   233, 238
melt index   76
melt processing   102
melt filtration devices   82
melt-flow rate   76, 117
methanolysis   59, 66, 163
methyl methacrylate   171–173, 176–180, 182–184
(see also acrylics, PMMA)
    polymerization   172, 173

Michigan State University 275
"Micronizing" and Micronyl SA   140, 141
Milgrom, J.   5, 6, 15
Mitsubishi Petrochemical Co.   195
Miyama, S.   258, 262
Mobay (Miles Inc.)   165
Mobil Chemical   13, 89, 121, 127, 200
Mobil thermoplastics densifier   121
modified polyphenylene oxide
    (PPO™)   156, 165
morphology   48, 207, 209
MRC Polymers Inc.   63, 158, 160
MRF (see material recovery facility)
municipal solid waste   1, 4, 101, 113, 116,
    136, 137, 144, 145, 153–155, 189–
    191, 210, 234

National Association for Plastics Container
    Recovery (NAPCOR)   8, 65, 271,
    274
National Polystyrene Recycling
    Company   111, 119, 123, 127
National Recovery Technologies   142
Nauman, E. B. 102, 106
Neutron activation 197
New Jersey Curbside Tailings
    (NJCT)   196–200
(see also NJCT/RPS blends)
    composition   197
    compression properties   199
    specific gravity   199
    voids 198
New York State Department of
    Environmental Conservation   7
New York State Energy Research &
    Development Authority   137
NJCT/Recycle polystyrene (RPS)
    blends   200–208
    crystallinity in   203
    microstructure   202–205
    microvoids in   205
    molecular packing in   205
    morphology   207
    scanning electron microscopy
        studies   206–209
    structure properties 200–201, 208–210
Northeast Recycling Council   7
Northwestern University   276
Noryl™   165
Nylon-6/6   69, 155, 158, 160
Nylon-6   160, 164

nylon(s)   4, 13, 153, 154, 156, 158, 160,
    164
    hydrolysis   13
    fiber   158

Organizations   271–275
(see specific name)
Owens-Illinois Corp.   13, 79
Oxychem   143

Packaging   101, 103, 113, 116, 125, 126,
    128, 139, 141, 143, 144, 146, 153,
    189, 190, 221, 234, 235
Partek Corp. reclaiming process   86
Partnership for Plastics Progress   272
PBT (polybutylene terephthalate)   153,
    155, 156, 163
PC (polycarbonate)   4, 42, 153, 155, 156,
    158
PE (polyethylene)   4, 9, 11, 13, 75, 88,
    136, 154, 180, 198, 208, 211, 224,
    246, 247
(see also HDPE, polyolefin(s))
    classification   75
    onsumption   76
    drum reconditioning   88
    film   89, 98–100
    melt index   76
    polymerization techniques   75, 76
    pyrolysis   102
    reprocessing effects   89–95
    thermal decomposition   101
Pelo Plastique Inc   69
Peninsula Copper Ind.   243, 244
Pennsylvania State University–Behrend
    College   276
Pepsi Cola Co.   66
PET (polyethylene terephthalate)   4, 12,
    13, 19, 22, 29, 36, 47–71, 78, 88, 100,
    139, 141, 153, 155, 156, 158, 160,
    163, 171, 180, 190–192, 194, 197,
    200, 210, 224, 247
(see recycling, recycled)
    adhesive removal   53, 54
    amorphous 48
    chemistry   47, 48
    crystallinity   48
    curbside recovery rate   38
    depolymerization   47, 59
    fiberfil   61

film 50, 58
foam   63
hydrolysis, methanolysis,
    glycolysis   58–60
global recycling activities   68–71
market   49, 50
preparation of   47, 48
recycling economics   66–68
recycling rates   50, 51
reprocessing methods   51–58
separation   35, 36, 221, 222, 224
soft drink bottle   48–51, 70, 78, 79,
    134, 190, 191, 210
Petrich, M. A.   276, 279
Phenolic(s)   4, 5, 233, 235, 236, 242,
    243
applications   236
molding resins   236
recycled regrinds   242, 243
volume   233, 236
Phillips 66   Co. 83
Plastic Recovery Systems   92, 145
Plastic(s)   3–10, 20–30, 38, 42, 88, 102,
    113, 153–166, 180, 189, 224
(see also collection, separation)
blasting media   183, 257
coding and labeling   6, 7, 268–270
commingled   157, 189–224
commodities   4–6, 154
curbside recovery rates   38
depolymerization   180
disposal   155, 234
drum recycling   88
durables   5, 153
engineering   5, 153–166
foam   42
identification techniques   30
in automotive   155, 157
in household products   192
in MSW   11
lumber   157, 194, 196, 198, 209
packaging   4, 9, 101, 113, 189, 190,
    210, 221
production   8, 9
recovery rates   3
scrap   6, 8
waste   6, 196
Plastics Institute of America, Inc.   8, 14,
    271, 274
Plastics Recycling Alliance reclaiming
    process   86
Plastics Recycling Corp.   193, 194

Plastics Recycling Foundation   7, 79, 271,
    274
Plastics Recycling Ltd.   224
Plastics Waste Management   99
PMMA (polymethyl methacrylate)   5, 11,
    171–173, 176–185, 254
(see acrylics, methyl methacrylate,
    pyrolysis)
blasting medium   183
depolymerization   171, 176–182, 184,
    254
recycling   171, 176
polyacetals   153, 154
polybutylene terephthalate (see PBT)
polycarbonate (see PC)
polyester (see unsaturated polyester)
polyethylene (see PE)
Polymer Resource Group   53, 87, 100
reclaiming process   87
Polymerix   13
Polymerland Service Centers   158
polymethyl methacrylate (see PMMA)
polyolefin(s)   75–103, 145, 153, 192, 224
(see also recycling, recycled, PE, PP)
coated materials recycling   89
film recycling   89
foam recovery   90, 91
global recycling activities   98–101
packaging   101
postconsumer containers,
    recycling   78–88
pyrolysis   254
reclaiming plants   83
reclaiming processes   83–88
reclamation system   82
reprocessing cost   82
reprocessing effects   77–96
polyols   12, 58, 60, 64, 65, 239, 248–
    253
(see PET, PUR)
polypropylene (see PP)
Polysar Inc.   127
polystyrene (see PS)
Polystyrene Packaging Council   7, 8, 272,
    274
Polytechnic University   276
polyurethane (see PUR)
Polyurethane Recycle and Recovery
    Council   8, 258, 273, 275
polyvinyl chloride (see PVC)
polyvinylidene chloride (see PVDC)
positron lifetime studies   202–205

PP (polypropylene)   4, 9, 10, 54, 76, 77,
    89, 92, 96, 99, 100, 136, 139, 166,
    171, 180, 185, 194, 197, 210, 211,
    224, 246, 247
    battery cases recycling   92, 99, 200
    consumption   77
    melt-flow rate   76
    multilayer   89
    oriented film   89, 100
    reclaiming (see polyolefins)
    reprocessing effects   96
    strapping   100
    types   76
primary recycling   6, 154
Procter & Gamble   65, 99
PS (polystyrene)   4, 9, 11, 13, 111–128,
    136, 153, 154, 171, 185, 197, 202,
    207, 208, 210, 224
(see also EPS, foam, recycling, recycled)
    capacity   112
    densification   117–121
    global recycling activities   127, 128
    markets   113–116
    oriented film   113, 114
    packaging   113–115, 125, 126
    processes   111, 112
    producers   112
    products   112, 113
    pyrolysis   253
    recyclable products   116
    recycled products   121–127
    recycling economics   125, 126
    reprocessing   116–121
    sources and volumes   113–16
PUR (polyurethane)   4, 5, 12, 59, 60, 64,
    155, 233–235, 238–241, 247–250,
    253–255
(see also reaction injection molding (RIM))
    applications   234, 235
    flexible foam   12, 64, 234, 235, 238, 239
    glycolysis   250–253
    hydrolysis   247–250
    incineration   255
    pyrolysis   254, 255
    recycled   256–258
    recycling   238–241
    rigid foam   235
    volume   233, 234
Pure Tech Int.   55, 56
PVC (polyvinyl chloride) 4, 6, 9, 11, 22, 36,
    41, 59, 62, 88, 91, 100, 133–147, 180,
    197, 198, 224, 247, 255

(see also recycling, recycled)
    bottles   133, 134, 139–141, 143
    copolymers   136
    disposal streams   136
    film   145
    flexibles   133–135, 144, 145
    global recycling activities   139–141
    incineration   137, 138, 255
    markets   133–136
    plastisols   133, 135, 136, 146
    production   133–136
    pyrolysis   253
    reprocessing   139, 140, 142, 145, 146
    rigids   133, 134, 139–144
    separation   36, 52, 139–142
    wire and cable jacketing   91, 92, 145
PVDC (polyvinylidene chloride)   50, 56,
    58, 66, 136
pyrolysis   9–12, 165, 180, 184, 244, 253,
    255
Puffiber™   61

Quantum Chemical   83, 87, 210
    reclaiming process   87
quaternary recycling   6
(see also incineration)

Ramaplan   225
Rastra Building Systems   124
reaction injection molding (RIM)   234,
    235, 238–241, 250–252, 255
    glycolysis 251, 252
    incineration 255
    recycled regrinds· 239–241
R2B2   141, 142
recycled
    acrylics   176, 183, 185
    HDPE   96–99
    PE   91, 97–99
    PET characteristics   58
    PET markets and applications   60–66
    PP   89, 92, 98, 99
    PS   121–125, 200–208, 212–218
    PVC   139, 140, 143
    thermosets   255–258
recycling   4, 6–14, 18–29, 32–35, 39, 41,
    42, 50–70, 77–101, 116–128, 138–
    147, 154, 159–166, 175, 176, 191–
    196, 210–224, 238–255, 271–278
    acrylics   175, 176

agricultural film　41
automotive　18, 166
chemical containers　41, 42
closed-loop　65
commingled plastics　19, 191–196,
　　210–224
curbside　19–29
economics　32–35, 66–68, 82, 125, 126
　　166, 176, 218, 219, 221–223
education and promotion　21–23
engineering resins　159–166
history　8–13
industrial scrap　77
organizations involved　7, 8, 271–275
PET　50–70
plastics　4, 14
polyolefins　77–101
PP batteries　98
primary　6, 154
programs, characteristics of　39
PS　116–128
PUR　238–241, 247–255
PVC　139–147
quaternary　6
residential　19, 20
secondary　5, 6, 154
tertiary　5, 6
thermosets　238–255
university programs　8, 275–278
Recycloplast　195
Refakt　142
refined NJCT　211–224
blend formulations　212
blend morphology　216, 217
blend properties　216
economics　218, 219
properties　214, 215
Reko　4, 62, 65, 68, 100
Rensselaer Polytechnic Institute　167, 210,
　　219
Repete™　59, 66, 163
Reprise International Plastics
　　Recycling　141
Repro Machine Industries　108
Retech　69
Retort　180
"Reverzer" process　193, 195
Rhode Island　7
RIM (see reaction rejection molding)
Rodriguez, F.　275, 279
Rohm & Haas　160
Roy, M.　11, 16

Ryborz, D. H.　93, 99
Rynite™　60

San Jose, CA (see case studies)
Santana Plastic Products　97
scanning electron microscopy　206–209,
　　215–218
Schott, H.　96, 105
scrap plastics　6, 8
secondary recycling　5, 6, 154
selective dissolution/devolatilization　102,
　　167, 219–223
economics　221–223
process　220
with tetrahydrofuran　221
with xylene　221, 222
Sentinel EPS Recycler　119
separation　9, 10, 29, 30, 36, 41, 51–56,
　　68, 81, 88, 91, 92, 99, 101, 102, 125,
　　139–141, 145, 153, 157, 160, 167,
　　185, 191, 219–224
automatic　140
flotation　9, 10, 51, 53, 56, 92, 145
hydrocyclone　51, 53–55, 68, 81, 99, 101
magnetic　99
manual　29, 30
mechanical　88, 141
selective dissolution/devolatiliza-
　　tion　102, 167, 219–223
solution/washing　55, 56
special　41
using near and supercritical fluids　167,
　　224
Sheet molding compound (see SMC)
sheet　62, 113, 117, 143, 173, 174, 176, 183
acrylic　173, 174, 176, 183
calendered　143
extruded　113
Shenoy, A.V.　93, 96, 105
Signode　62
Sikora, R.　93, 105
Simioni, F.　251, 260
SMC　155, 238, 245–247, 255, 258, 273,
　　275
(see also unsaturated polyesters)
Auto Alliance　255, 258, 273, 275
pyrolysis　244, 255
recycled regrinds　245–247
Society of Automotive Engineers　7, 157,
　　270
marking of plastic parts　270

Society of Plastics Engineers   8, 272, 274
Society of the Plastics Industry, Inc.   6, 7,
  88, 157, 191, 268, 273, 274
  plastic container coding system   268
"Solid stating" or solid-state polymeriza-
  tion   48
Solo Cup   63
Solvay   140
Sorema   100
sortation   139, 140, 142, 153, 191
  automatic   140
sorting   84, 85, 88, 99, 125, 185
  mechanized   88
St. Jude Polymer   13, 53
State University of NY–Binghamton   276
Stadtereinigung Nord   225
Stokbord   224
Storopak Co.   127
strapping   46, 62, 100
Superwood   13, 99
Superwood International Ltd   193, 224
Swedlow, Inc.   11

tailings   191, 192
Technical University, Aalen   253
terephthalic acid   12, 47, 48, 59, 163
terminology   5, 6
tertiary recycling   5, 6
(see also hydrolysis, glycolysis,
  methanolysis, pyrolysis)
Tesoro, G. C.   276, 279
thermoplastics   4–6, 154, 233, 257
(see also engineering themoplastics)
  commodity   4–6, 154
  definition   5
thermoset(s)   5, 233–258
(see also expoxy resins, phenolics, PUR,
  unsaturated polyesters)
  blasting medium   257
  definition   5
  global recycling activities   258
  ground   255–256
  recycled applications   255–258
  recycling technologies   238–255
  size reduction   238
  sources, volumes   233
Toronita densifier   121
Tri-Max   195
"Umwelt service"   127
Union Carbide   9, 83, 89
Union Chemical Laboratories   210

United Kingdom   98, 141
United Resource Recovery   82
U.S. Bureau of Mines   9, 10
University of Florida   276, 277
University of Lowell   277
University of Pittsburgh   167, 224
University of Texas at Austin   278
University of Virginia   197
nnsaturated polyesters   12, 64, 153, 233,
  234, 236, 237, 244, 245–247
(see also sheet molding compound (SMC))
  applications   237
  hydrolysis   247
  recycling   245–247
  volume   233, 237
Upjohn Co.   12, 250
USS Chemicals   10

vehicles   4–29
  body types   24–27
  on-board compaction   28
Vermont Republic Industries   141
vinyl (see PVC)
Vinyl Institute   8, 141, 142, 273, 275
Voest-Alpine high-temperature gasification
  process   255

Wagener, K. B.   277, 279
Walex System   120
Washington and Lee University   196
Weichenreider, E.   195
Welding Engineers, Inc.   211–213
Wellman   13, 53, 61, 63, 69, 158, 164
Western Environmental Plastics   54
"white goods"   153, 158
wire and cable jacketing   91, 92, 145, 224
  "rising current" separator   91
Wormser Kunststoff Recycling
  GmbH   225
Worms Technology   225

X-ray film   12, 51, 58, 66
X-ray fluorescence spectroscopy
  (XRF)   142
X-ray irradiation   88, 140